TEST AND DESIGN-FOR-TESTABILITY IN MIXED-SIGNAL INTEGRATED CIRCUITS

Test and Design-for-Testability in Mixed-Signal Integrated Circuits

Edited by

José L. Huertas
IMSE-CNM, Spain

KLUWER ACADEMIC PUBLISHERS
BOSTON / DORDRECHT / LONDON

A C.I.P. Catalogue record for this book is available from the Library of Congress.

ISBN 1-4020-7724-6 (HB)
ISBN 0-387-23521-3 (e-book)

Published by Kluwer Academic Publishers,
P.O. Box 17, 3300 AA Dordrecht, The Netherlands.

Sold and distributed in North, Central and South America
by Kluwer Academic Publishers,
101 Philip Drive, Norwell, MA 02061, U.S.A.

In all other countries, sold and distributed
by Kluwer Academic Publishers,
P.O. Box 322, 3300 AH Dordrecht, The Netherlands.

Printed on acid-free paper

Printed in the Netherlands.

Table of Contents

Foreword

This book is the third in a series of three dedicated to advanced topics related to Mixed-Signal IC design methodologies. It is one of the results achieved by the Mixed-Signal Design Cluster, an initiative launched in 1998 as part of the TARDIS project, funded by the European Commission within the ESPRIT-IV Framework. This initiative aims to promote the development of new design and test methodologies for Mixed-Signal ICs, and to accelerate their adoption by industrial users.

As Microelectronics evolves, Mixed-Signal techniques are gaining a significant importance due to the widespread of applications where an analog front-end is needed to drive a complex digital-processing subsystem. In this sense, Analog and Mixed-Signal circuits are recognized as a bottleneck for the market acceptance of Systems-On-Chip, because of the inherent difficulties involved in the design and test of these circuits. Specially, problems arising from the use of a common substrate for analog and digital components are a main limiting factor.

The Mixed-Signal Cluster has been formed by a group of 11 Research and Development projects, plus a specific action to promote the dissemination of design methodologies, techniques, and supporting tools developed within the Cluster projects. The whole action, ending in July 2002, has been assigned an overall budget of more than 8 million EURO.

The novelty of the TARDIS initiative is that in addition to the standard R&D work, the participating projects have a compromise to publicize the new methodological results obtained in the course of their work. A Cluster Coordinator, Instituto de Microelectrónica de Sevilla, in Sevilla (Spain) has the role to coordinate and promote actions to carry out effectively the dissemination work and foster cooperation between the participating projects. All public results from the dissemination action are available from the Cluster Web site (http://www.imse.cnm.es/esd-msd).

Mixed-Signal design is a critical part for many IC designs. The advantages brought by System-on-Chip will only be fully successful if techniques are developed that allow coexistence of high-performance analog functions sharing a common substrate with large blocks of digital functions. Interfaces between the analog and the digital world, materialized in data converters will always be present in any mixed-signal design, and he verification of those embedded analog functions, may be in many cases the factor limiting the production-test throughput. New technologies, like Silicon-on-Insulator (SOI), offer interesting possibilities for the design of mixed-signal ICs, but require the mastering of new design techniques. The work of projects in the Cluster has been focused on four main areas (Substrate Noise Coupling, Advanced Data Converters, Testability and Special Technologies).

This book addresses the specific problem of testing and enhancing testability in Mixed-Signal circuits and incorporates results achieved by some Cluster projects with activity in that area, complemented by contributions from external experts that have occasionally participated in activities organized by the Cluster and/or have contributed through other research projects funded by the EU.

We hope that the reader will find this book useful, and we would like to thank all partners of the MSD Cluster for contributing to the success of the initiative. Special thanks are given to all the authors and to the editors for their effort to make this book a reality.

José Luis Huertas, Juan Ramos-Martos; Sevilla, September 2003

Projects in the Mixed-Signal Design Cluster

ABACUS: Active Bus Adaptor and Controller for Remote Units
The objective of this project is the development of an integrated circuit for space applications, that implements the analog/digital interface between the spacecraft On-Board Data Handling (OBDH) bus, and the Remote Terminal Units (RTUs). The design will use 0.8□m SOI technology.

BANDIT: Embedding Analog-to-Digital Converters on Digital Telecom ASICs
The goal of BANDIT is to develop a general design methodology for embedding high-speed analog/digital converters (ADCs) on large digital telecom ASICs, with special attention to the problems caused by mixed-signal integration.

HIPADS: High-Performance Deep Sub-micron CMOS Analog-to-Digital Converters using Low-Noise Logic
The aim of this project is to develop different A/D Converters in deep sub-micron digital CMOS process, using a new Current Steering Logic (CSL) family approach that has the property of inducing a very low substrate noise . The converters are intended to become integrated components of larger systems, and should be considered presently as products under specifications covering end-user applications.

MADBRIC: Mixed Analog-Digital Broadband IC for Internet Power-Line Data Synchronous Link
The project main objective is the development of prototype building blocks of a chipset for high-speed communications through the power lines, that will improve achievable data rates using state of the art mixed-signal integrated circuits and DSP techniques.

MIXMODEST: Mixed Mode in Deep Submicron Technology
The technical target of the MIXMODEST project is to develop design techniques that permit the implementation of mixed-signals systems in the most advanced 0.35μm and 0.25μm deep sub-micron digital CMOS technology.

OPTIMISTIC: Optimisation Methodologies in Mixed-Signal Testing of ICs
The OPTIMISTIC project, concerned with Optimisation Methodologies in Mixed-Signal Testing of ICs, aims at the development and introduction of advanced test generation in mixed-signal IC design. Building upon existing advanced tools for control and test systems, a new approach is to be developed

that will allow the mixed-signal chip designer to take large responsibility in the generation of test as part of the design activity.

RAPID: Retargetability for Reusability of Application-Driven Quadrature D/A Interface Block Design
This project is concerned with the development of an advanced methodology for the design of a mixed-signal application-driven quadrature D/A interface sub-system, aiming at its reusability by a retargeting procedure with minimal changes to their structural sub-blocks.

SUBSAFE: Substrate Current Safe Smart Power IC Design Methodology
The overall technical objective of this project is to develop a design methodology that employs device and circuit simulation to assure IC digital functionality under current injection in the substrate produced by forward bias conditions in N-wells (i.e. during switching of power stages driving inductive loads). The design methodology will change from the current largely empirical approach to Computer-Aided Design guided critical parameter evaluation, validated by a relatively small number of measurements.

SYSCONV: Systematic Top-Down Design and System Modeling of Oversampling Converters
This project develops a system-level model for oversampling delta-sigma converters suitable for use in mixed-signal system simulations and verifications. It addresses the development of a model of the entire converter as a block on its own, that can then be used in efficient mixed-signal system simulations where the converter is only a block in the overall system

TERMIS: High-Temperature / High-Voltage Mixed Signal SOI ASICs for Aerospace Applications
The project addresses the development of a fully integrated high-voltage driver IC for two different electromagnetic micro-motors which are dedicated for satellite applications. Each circuit, in die form, will be packaged in the corresponding micro-motor. The systems must operate at 200°C under a 30V power supply and must survive space irradiation.

VDP: Video Decoder Platform
This project develops a prototype video decoder platform. The result will be an IC that captures video signals and decodes the information for use in, for instance digital TV, set top boxes, and PC video capture. It will exploit innovative architectures trading signal to noise ratio versus accuracy, decoding both analog and digital video sources.

Contributors

Florence Azaïs received M.Eng., and Ph.D. degrees in electrical engineering from the University of Montpellier, France in 1993 and 1996, respectively. She is currently working in the microelectronics department of the Laboratory of Computer Science, Robotics and Microelectronics of Montpellier (LIRMM) as a researcher of the National Council of Scientific Research (CNRS). Since 1993, she has been interested in the general domain of integrated circuit testing, and in particular in fault modeling and mixed-signal circuit testing. She has authored or co-authored more than 70 international papers.

Martin John Burbidge worked towards a HND in mechanical and electrical engineering at Nene technical college, Northampton, from 1985 to 1987. From 1993 to 1996 he studied technology, mathematical modelling, and embedded micro controller design with the Open University. He received a first class honours degree in Electronic and Communications Engineering from Lancaster University in 1999. During and after studying for his degree Kongsberg Simrad employed him as a designer of embedded hardware and software systems used for sonar, and control based applications. Since March 2000 he has been employed as a Research associate at Lancaster University and is investigating test techniques for Phase locked loops. His general interests include high frequency board level and ASIC design, design for test, and built in self-test relating to mixed signal circuits

Gloria Huertas was born in Sevilla, Spain, in 1974. She received the B. S. degree in electronic physics in 1997 and the M. S. degree in microelectronic in 1999 both from the University of Sevilla, Spain. She is currently pursuing the Ph.D. degree from the Instituto de Microelectrónica de Sevilla (IMSE, CNM). She is presently a part-time Assistant Professor at the University of Sevilla, Spain, and working towards her PhD degree. Her research focuses on designing electronic mixed-signal circuits and systems including techniques for testability.

José L. Huertas received the Licenciado en Ciencias degree and the Doctor en Físicas degree in 1969 and 1973, respectively, both from the Universidad de Sevilla, Spain. Since 1971, he has been with the Electronics Department, Universidad de Sevilla, Spain, where he is a full Professor since 1981. In 1989, Dr. Huertas joined the Instituto de Microelectrónica de Sevilla, Centro Nacional de Microelectrónica, CSIC, Spain, for which he is the director since then. Dr. Huertas has served as the general chair of the 1993 European Solid-State Circuits Conference, the 1996 Cellular Neural Networks Workshop, and the 2003 Mixed-signal Test Workshop as well as

the program co-chair of the 2002 SBCCI. He has co-authored several books and published more than 300 papers in international journals and conferences. His current research interests are focused on the design and test of analog and mixed-signal integrated circuits based in standard CMOS technologies.
Dr. Huertas was an Associate Editor for IEEE Transactions on Circuits and Systems, and is a member of the Editorial Board of the Journal of Analog Integrated Circuits and Signal Processing and of the Journal of the Franklin Institute. He received the 1995 IEEE Guillemin-Cauer award, the 1995 Kelvin prize from IEE, the 1998 Torres-Quevedo prize (Spanish National Award on Technological Research) and the Garcia-Cabrerizo Medal for technological research in 2003. Dr. Huertas is an IEEE Fellow from 1994.

Andreas Lechner holds the position of Research Fellow in Mixed Signal Test Engineering at Lancaster University and is the project manager for Lancasters contribution to the Framework V European Project Testability of Analogue Macrocells Embedded in System-on-Chip. Andreas received his Dipl.-Ing. degree in microelectronics from the University of Hannover in 1997 and is a member of the IEEE. Andreas is one of the key contributers of a new portfolio of continuing and proffessional development courses in mixed signal and analogue testing with ISLI, Scotland. His main interests are in understanding and modelling of defect and degradation mechanisms in deep submicron technologies and test strategies for analogue to digital converters.

Gildas Léger was born in St Brieuc, France, in 1976. He received the *Ingénieur en Physique* degree in 1999 from the *Institut National des Sciences Appliquées* (INSA) of Rennes, France. He is currently working as a research engineer at the *Instituto de Microelectrónica de Sevilla* (IMSE, CNM) and pursuing the Ph.D. degree from the University of Sevilla, Spain. His research focuses on design for testability for mixed-signal circuits and system, specially in the domain of analog to digital conversion.

Veikko Juhani Loukusa was born in Posio 26th September 1957. He is married to Heli (nie Kakko), and they have three children (Juho, Kerttu ja Iida Maria). He earned his Master of Science (Eng.) degree in 1983 and Lic.Eng. degree in 1999 from Oulu University. From 1984 to 1987 Mr Loukusa was working as test engineer in Nokia Telecommunications and in 1987 holds also a position of researcher in the State Technical Research Center of Finland. He has been with the Oulu Institute of Technology from 1987 to 1996, from 1996 to 2000 was principal lecturer in measurement technology at Oulu Polytechnic. Since 2001 Mr Loukusa is a Senior specialist in Nokia Mobile Phones

Pascal Nouet received the PhD in Microelectronics from the University of Montpellier, France in 1991. Since 1992, he is a researcher at LIRMM

(Laboratory for Computer Sciences, Automation Control and Microelectronics) and a professor (Associate Professor until 2003 and Full Professor since) at the Institute for Engineering Sciences of the University of Montpellier (ISIM).
He previously worked on Electron-Beam Testing of ICs, on electronic irradiation effects on CMOS circuits and on characterization and modeling of CMOS devices and interconnects.
His current research interest concerns: design and test of mixed signal circuits and MEMS, reliability and failure analysis of integrated systems. He's mentoring or has mentoring a total of 11 PhD thesis and has been author or co-author of numerous papers. Among them more than sixty appeared in journals or in major international conferences.

Eduardo J. Peralías Macías was born in Seville, Spain in 1964. He received the *Licenciado en Físicas* degree in 1992 and the *Doctor en Ciencias Físicas* degree in 1999, both from the University of Seville, Spain. From 1999 to 2001, he was with the *Departamento de Electrónica y Electromagnetismo* at the University of Seville as Assistant Professor. Since 2001, he has been with the *Instituto de Microelectrónica de Sevilla* (IMSE-CNM, Sevilla, Spain), where he is currently a Tenured Scientist. His research interests are in the areas of Mixed Design with emphasis on Analog-to-Digital converters, Test and Design for Testability of Analog and Mixed-Signal Circuits, and Statistical Behavioral Modeling.

Dr Andrew Richardson is currently a Reader in Microsystems Engineering at Lancaster University, Director of the Centre for Microsystems Engineering and Scientific Director of Dolphin Integration, Grenoble. He graduated with an honours degree in Electronic Engineering in 1983 from Manchester University and a PhD from Lancaster University in 1991. In the late 80's he worked on electrical measurements for early failure rate screening of digital VLSI circuits and made significant contributions in the field of IDDQ testing. Over the past 10-years he has put considerable effort into the problems of testing analogue circuits and microtechnology based devices and holds a number of collaborative projects with European industry & academia. More recently he has been focusing on Design for Manufacture technology for heterogeneous micro and nano technology based systems that involves integrating design-for-test and design for reliability strategies into the core functions and enhancing modelling technology to support this process. He is the coordinator of a 4 year European program "Fr. VI Network of Excellence in "Design for Micro & Nano Manufacture" that addresses this area that aims to integrate the activities of 24 of the key institutes active across Europe.

Adoracion Rueda (M'87) obtained the Ph.D. degree in 1982 from the University of Sevilla, Spain. Since 1976, she has been with the *Departamento de Electrónica y Electromagnetismo*, *Universidad de Sevilla*, Seville, Spain,

where she is now a Full Professor in Electronics. In 1989, she became researcher at the Department of Analog Design of the National Microelectronics Center (CNM), now Institute of Microelectronics at Seville (IMSE).
She has participated in several research projects financed by the Spanish Government, and by different programs of the European Community. She has published many technical papers in international journals and major conferences or books. In 1992 she won the Best Paper Award of the 10th IEEE VLSI Test Symposium. Her research interests currently focus on the topics: Design and Test of Analog and Mixed-Signal Circuits, Behavioral Modeling of Mixed-Signal Circuits, and development of CAD tools. She is a member of the IEEE.

Birger Schneider is a founder of microLEX Systems A/S and has been managing the Company since its inception in 1992. Previously, Mr. Schneider worked 18 years at the Danish research centre for applied electronics, ElektronikCentralen, for the first 13 years as responsible for microelectronic testing, and in the latter 5 years as part of the upper management, serving a position a director of technology. During 1968 and 1974, he received his education from Odense Teknikum and Technical University of Denmark. Birger Schneider has participated in several R&D projects in the area of test of microelectronics, e.g. under European cooperative programmes such as the ESPRIT programme. He has published many papers on testing of microelectronics and electronics in general. Over the years, he has held several positions in national and international engineering societies, including vice chair of the "Electrotechnical Science", Danish Academy of Technical Sciences, chairman of the European wide computer aided design for VLSI workshops (CAVE), chairman of the IEEE Computer Society "European Test Technology Technical Committee", member of the EU "Microelectronics Advisory Board", etc. Mr. Schneider is also a founder and director of microLEX Consult ApS

Diego Vázquez-García de la Vega was born in El Coronil, Seville, Spain in 1966. He received the Licenciado en Física degree in 1989 and the Doctor en Ciencias Físicas degree in 1995, both from the University of Seville, Spain. Currently, he is a professor for the Department of Electronic and Electromagnetics at the University of Seville and a researcher at the Microelectronic Institute of Seville-National Microelectronic Center (IMSE-CNM). His research interests include Design, Fault Tolerance, Test and Design for Test of Analog and Mixed-Signal Circuits.

Chapter 0

INTRODUCTION

José L. Huertas
IMSE-CNM

Abstract:

Key words:

1. TESTING ANALOG AND MIXED-SIGNAL INTEGRATED CIRCUITS

The integration level of systems is continuously increasing. In many application contexts the target is to reduce the system just to one chip -a so-called System-on-Chip (SoC)-, where different technologies are mixed up (analog, digital, sensors, RF...). Designing such systems is undoubtedly a major problem in today microelectronics, because of the inherent complexity of dealing with so heterogeneous technologies. However, another task becoming the real bottleneck of present and future IC projects is how to test those complex and heterogeneous systems in an efficient manner, in terms of test time, cost and proficiency.

We are still far from a global solution to this problem -if any finally exist-. For the last twelve years a lot of attention has been paid to the problem of testing mixed-signal integrated circuits. International research projects have been carried out focusing on different problems, specially on how to test and on how to perform design-for-test for the main analog and mixed-signal integrated blocks. This is a first and necessary step to cope with the problem of testing complex SoCs.

J.L. Huertas (ed.), Test and Design-for-Testability in Mixed-Signal Integrated Circuits, 1–6.

Unfortunately, most of the emerging concepts and techniques have been developed in an academic frame, not being widely spread in an industry environment. A better knowledge of this effort by industrials may probably lead to the adaptation, modification or refinement of some of the techniques under development.

There are several active topics in the research on mixed-signal testing. One of these is the development of methodologies for the efficient generation of the stimuli to be used for a given test. This is quite a difficult task, since the goal is to devise a minimum number of test signals to be applied during a small time and in a sequence that minimizes the overall test time. This goal is still far to be reached, but techniques are emerging based on the use of DSP and software tools.

Standardization of a test support architecture is a practical point also addressed since the last few years. After the effort carried out for digital systems, a similar task has been accomplished for mixed-signal counterparts. Although the present standard is far for being of general use, its implementation for medium accuracy and medium speed mixed-signal systems is worth considering, in order to understand both its capabilities and limitations.

Another issue is related to the connection of design and test. The development of a design-and-test flow is under the scope of many researchers. Unfortunately, there are significant problems to solve, one of the most important being the huge simulation times required for linking a design with the test experiments to check its specifications. Resorting to high-level models that can be inserted in behavioral simulators should alleviate this drawback. However, the use of behavioral models has to be made in a manner that does not compromise accuracy, ensuring that all the significant second-order effects are also modeled at high level.

For analog and mixed-signal circuits there are no general test methodologies. Instead, techniques have been developed for every specific class of circuits. Among these, data converters and phase-locked loops are the most representative since they are widely used, many different techniques are known for testing them, and there is still a high demand for faster and more accurate methods improving those existing nowadays. These methods and techniques are mostly procedures for the functional verification of the circuits; all the main specifications of a given circuit are measured under the operational conditions of the intended application.

Although functional or specification-driven test is (and will continue to be) the dominant test style for mixed-signal circuits, the experience on how digital systems evolved is motivating a strong interest in developing structural or fault-driven test techniques. The main reason is a potential reduction of the test application time and an improvement in test quality and yield. Among the emerging techniques and methodologies, those that do not require a complex on-chip support are considered very promising. A significant example of them is the so-called Oscillation-based Test (OBT) strategy, its more appealing advantage being that no external stimuli are needed and that interpreting the test outcomes is usually quite straightforward.

Other two concepts borrowed from digital are Design-for-Testability and Built-In-Self-Test. Although their extension to analog and mixed-signal is difficult and may be questionable, there is a strong interest in exploring what can be done in this area. Methods are coming up with a different degree of practical applicability. Probably, the field where more techniques and strategies have been reported is the field of analog filters, both continuous- and discrete-time. Most of these techniques have an ad-hoc flavor, but they can lead to interesting implementations enhancing the testability of integrated filters.

The purpose of this book is to provide a glimpse on recent research results in the area of testing mixed-signal integrated circuits, specifically in the topics mentioned above. Much of the work reported herein has been performed within cooperative European Research Projects, in which the authors of the different chapters have actively collaborated. It is a representative snapshot of the current state-of-the-art in this emergent field. This volume closes a series of three books, and is formed by 8 contributions that will be introduced next.

2. BOOK OVERVIEW

The book is organized in four parts, each one formed by two chapters. The first two chapters are dealing with **general test support techniques**, intended for software and hardware, respectively.

Chapter 1 presents the main concepts and the basic principles for Digital Signal Processing-based measurements and their use in a test context. DSP-based test is gaining a large acceptance because of its versatility. A deep knowledge of DSP techniques is required since both stimuli generation and test interpretation will strongly depend on the way spectral techniques are understood by the test engineer. Fourier methods are reviewed in this Chapter and related to coherent and non-coherent testing. Windowing and spectral leakage are extensively considered and examples are given to illustrate practical aspects.

Chapter 2 addresses the hardware standardization of test support for mixed-signal ICs. Since the early nineties, there is an effort to develop a Standard Test Access an a Boundary Scan Architecture for both digital and analog circuitry. The result is the IEEE 1149.4, described in detail in this Chapter starting by its basic architecture at high, low and control levels. Then, the basic instruction set is reviewed including mandatory and optional instructions. Attention is paid to both the Bus Interface Circuit and the Analog Boundary Module, detailing its mission and the different configurations to achieve. Finally, test methodologies are discussed when the standard is used.

The next two chapters are related to **specific test techniques** for two kind of mixed-signal modules of special interest: the Analog-to-Digital Converter (ADC) and the Phase Locked Loop (PLL).

ADC is probably the most frequently used mixed-signal module, its ubiquity relying in its use as front-end of a majority of DSP-based systems. *Chapter 3* reviews the static and dynamic performance parameters of any ADC. The aim of this Chapter is to serve as a summary of the available test techniques for ADCs. The basic test setup is presented as well as methodologies for testing most of these parameters. Finally, an introduction to BIST structures and methods close this chapter.

Similarly, PLLs are broadly used in large-scale systems on chip, no matter they are digital or mixed-signal. *Chapter 4* starts by discussing the PLL operation and its typical test specifications. Because of its complexity and due to its use in many different application contexts, PLLs may include IP cores from several sources. This makes essential an efficient verification of the PLL during its design, debug and production. Different test techniques are reviewed both for IC characterization and for production testing. Among the former, lock and capture range, transient response, jitter transfer function and block characterization measurements are described. Regarding production testing, DfT techniques are mentioned, although the emphasis is placed on jitter measurements. Finally, specific BIST techniques are discussed.

Chapters 5 and 6 form the third part of the book and focus on **behavioral models and testing**. These two chapters aim to address the state-of-the-art of behavioral modeling as a tool for testing and for design-for-test analog and mixed-signal blocks.

Chapter 5 describes a novel approach for top-down Design-to-Test methodology for mixed-signal test generation. This approach relies on high-level simulation of a Device-Under-Test (DUT) to come out with the stimuli that will allow the efficient testing of the actual hardware when fabricated. It allows an anticipation of the test needs reducing the overall test debugging time. A behavioral model is typically a description using differential equations. Using behavioral models, a significant speedup of simulations can be obtained, including design verification and allowing software prototype exchange. The implementation of such a model is detailed in this Chapter, considering as an example the case of oversampled modulators. A generic model is presented as well as a way to extract a behavioral test model. Finally, a practical case study illustrates the methodology.

Chapter 6 presents a behavioral model for representing the static behavior of ADCs. Closed-form formulae are presented for describing the ADC. Since the Chapter is focused on multistage ADCs, the structure of these converters is modeled through the use of new analytical expressions introduced in the Chapter. Later on, error correction and calibration techniques are discussed and incorporated into the models, providing a general frame for both design and test of multistage converters no matter the techniques used for improving the converter performance. These features allow the development of DFT and BIST strategies which account for any calibration and correction technique without the risk of fault masking, a typical feature when DFT or BIST is used in multistage converters.

Finally, the last two chapters present **techniques for Built-In-Self-Test** (BIST) of analog and mixed-signal circuits.

Chapter 7 deals with integrated filters and carries out a review of different techniques developed in the last years for improving the testability of these circuits. Methods and strategies for Design-for-Testability (DFT) and for BIST are evolving trying to bring to analog and mixed-signal circuits the capabilities now available for digital. In this Chapter the basic rules and guidelines for DFT are considered and then several approaches for off-line test are summarized: analog registers, analog test buses, stage reconfiguration, use of swopamps. Next, attention is paid to DFT methods for on-line test likewise replication (total or partial), cheksums, and fully differential circuits. BIST is also addressed paying attention to both spectrum- and oscillation-based techniques. Finally, an example intends to clarify how different techniques can be merged to enhance the testability of a filter.

Chapter 8 presents an overview of the so-called Oscillation-based Test (OBT) techniques. In essence, this is a structural test technique evolving as an intent of advancing in the use of defect-driven techniques for analog and mixed-signal ICs. The operational principles are detailed as well as the evolution of the basic concept towards a practical implementation. A key issue is the application to complex systems, since this technique is much easier to apply to small structures. Decomposition can be very efficient when a supporting framework is provided, as discussed in this Chapter. A complete example corresponding to an integrated system is detailed to illustrate an ad-hoc methodology extensible to many other cases. The OBT technique is shown to be practical in circuits and systems where an adequate filtering action is provided. However, circuits where such an action does not exist can incorporate OBT through a modified procedure, which is discussed in the Chapter by considering the case of $\Sigma-\Delta$ modulators.

The editor aims to motivate an interest in the reader for the many problems still open in the test of mixed-signal integrated circuits, as well as to encourage her or him to develop new ideas, to adopt and/or adapt the existing ones, to check the practical feasibility of existing methods in industrial designs, and to qualify the real usefulness of the reported strategies and methodologies.

There is no doubt that testing will be more and more related to the common design practice as time evolves. This means that the work performed by researchers, who are presently trying to advance the state-of-the-art in this field, is paving the way for the leverage of microelectronics, and will serve as the basis of the future of integrated system design and test flows, probably incorporating heterogeneous technologies (nanodevices, .mechanical, magnetical, optical, RF, etc) in a common frame.

The editor also would like to acknowledge with thanks the effort done by all chapter authors and the quality of their contributions to this book.

Chapter 1

MIXED-SIGNAL TEST

Birger Schneider
Microlex

Abstract: DSP-based test offers important savings in test cost. In addition, it enables a flexible set-up, allowing to emulate complex instruments and to combina software tools and hardware instrumentation in an efficient manner. This chapter reviews the basic concepts for DSP-based test by considering both coherent and non-coherent test approaches.

Key words: Mixed-signal test, DSP-based test, Coherent testing, Non-coherent testing

1. INTRODUCTION

Modern mixed-signal testing builds upon digital signal processing. The digital signal processing in turn employs sampling of signals, and the sampling can be classified in one of two different principles:

- Coherent sampling
- Non-coherent sampling

Of the two, non-coherent sampling is the more basic and commonly used. However, coherent testing, when feasible offers higher advantages in mixed-signal testing, and it is a challenge for designers and test engineers to turn the mixed-signal test into a coherent test.

A major benefit of a DSP based test approach is that the device response is memorised. Hence it can be analysed for many different parameters, i.e. a reat number of parameters may be extracted from a single measurements without repeating the measurement. Since modern computers are very fast and produce even complicated calculations in a fraction of a ms, this offers

J.L. Huertas (ed.), Test and Design-for-Testability in Mixed-Signal Integrated Circuits, 7–44.

significant throughput savings in mixed-signal testing – not least in production testing.

Measuring the response of a device under test (DUT) may for just a single measurement take several ms, actually often hundreds of ms depending of the type of measurement. Once a measurement is captured, the signal processing can be done in a negligible amount of time in comparison to performing the measurement itself, thus saving valuable time. For example, in a video analyser test application [Ref. 1], a PAL video signal (625 lines, 50 Hz) is captured and analysed in less than 200 ms applying DSP techniques. Actually 4 line arecaptured and in the analysis about 52 parameters are extracted, characterising the composite video signal.

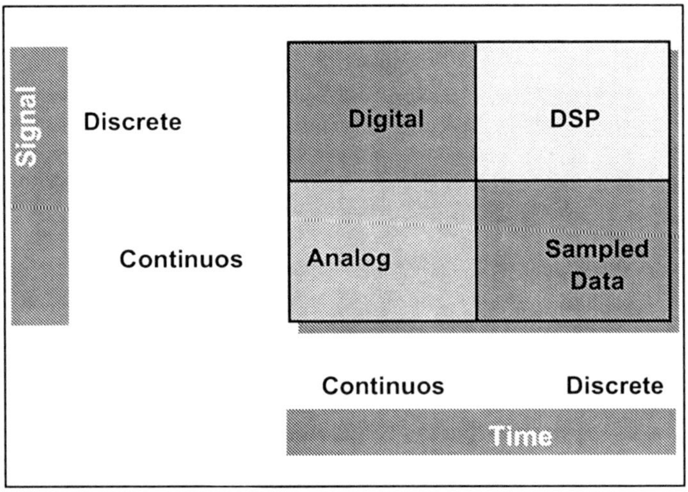

Figure 1-1. A signal in a mixed-signal can be discrete or continuous in time and signal amplitude

One advantage of DSP based testing is that it enables a flexible, low cost set-up to test e.g. ADCs and DACs since it allows to emulate complex and dedicated instruments in software in addition to the hardware test instrumentation itself. For example, a digitiser and associated DSP processing capabilities, either on-board or in an attached computer system, allows the test instrumentation to serve as a multimeter, a digitising oscilloscope, or a spectrum analyser. Similarly, an arbitrary waveform generator (AWG) coupled with DSP processing capabilities can be made to operate as generators for sine, square, triangle, ramp and pulse waveforms, complex TV signals, PDM signals for baseband mobile phones, etc.

A DSP based system needs to face a requirement of having test signals and spectra that are simultaneously periodic and discrete. This may be a challenge, but it can easily be overcomed by making a mixed-signal system into a *coherent* one.

2. BASIC TESTING PRINCIPLES FOR DSP MEASUREMENTS

A DSP based tests typically builds upon a conversion of a signal from the continuous time domain to a frequency domain representation, or spectrum. The signal in the time domain (continuous signal and time) is sampled and thus turned into a signal, discrete in time continuous in voltage. The signal is then converted to the frequency domain through use of a Fast Fourier Transform (FFT) or a Discrete Fourier Transform (DFT), which converts the signal to a voltage-frequency signal.

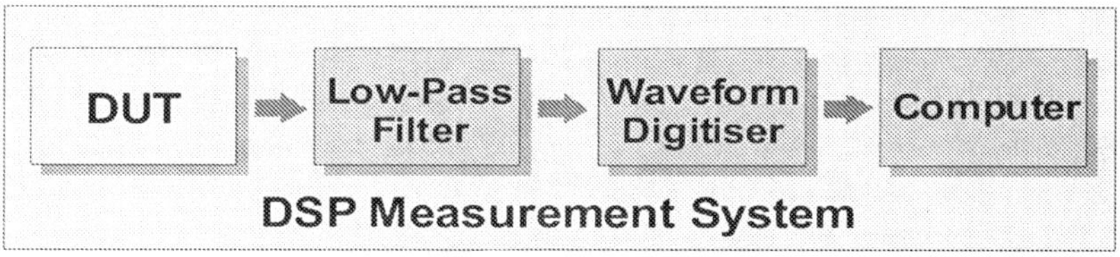

Figure 1-2. A basic measurement principle for a DSP based measurement of an analog signal from a device under test (DUT)

A test set-up for a DSP based measurement system follows the basic principles shown in [Ref. 1]. Here the analog output is filtered through a low-pass filter, e.g. an anti-aliasing filter as part of the digitiser, if available. Sometimes the filter may also band-pass filter to ensure that the DC level is filtered as well. The function of the filter is partly to avoid folding of frequencies in higher bands, partly to reduce the contribution of noise from a broadband spectre. The output of the filter is passed to the waveform digitiser itself. The waveform digitiser is a sampling A/D converter, where the input is true continuous in time and signal but the output is only continuous in signal and discrete in time due to sampling operation taking place.

A computing device receives the output from the waveform digitiser and this is where computation of an FFT or DFT transformation takes place. The computing unit may either be a computing unit as part of the instrument itself, e.g. a DSP processor, or it can be a central processor in a test system, e.g. a PC. In the 1980s and 1990s, the trend for more advanced digitiser instruments was to have an on-board DSP processor. In modern instrumentation based on for example fast interfaces like PCI-bus, PXI-bus and VXI-bus, and given the high computational speed of modern computers, the trend is that the DSP processing itself is done in the PC. The computation speed of PCs, e.g. using fast DMA transfer, tends to match that of DSP processors, and programming is a lot simpler. So the tradition of DSP processors is typically only carried on in stand-alone instruments like GPIB or USB controlled units.

In some test applications, an additional element may be included in the DSP Measurement chain of Figure 1-2. To obtain higher measurement resolution that what is offered by the digitiser itself, a programmable notch filter may be added in the chain [Ref. 2]. Typically a programmable, low noise and low-Q notch filter is used to suppress the fundamental signal, thus allowing a higher resolution for measurements of noise and harmonics (the basic operation is to remove say 8-10 bits of the fundamental signal). An advantage is that the notch filter is a rather low cost component compared to employing a much higher resolution A/D element in a digitiser at a given frequency. Further the notch filter can be located close to a DUT and it can be software controlled. However, it requires a little more expertise to use.

It should be noted that the output of the DUT is not restricted to analog signals. In the case of a digitised output of an analog signal, for example in testing of D/A converter, the DUT output can be captured by a Digital Vector Memory and then transferred to the DSP processing unit. The DSP processing is in principle indifferent since in both situations the signal processing is performed on an analog sampled signal presented as a digitised signal.

The digitising operation of a waveform digitiser is one of transforming the original continuous analog signal to a set of discrete sampled values. It is done through the use of a sample window, or windowing function, and as shall be shown later on this operation introduces a source of error. Once a sample set exists, as shown in Figure 1-3, the FFT transformation is exact. However, the operation of obtaining the sample values may introduce errors, not least because of the use of the windowing function, and this is one area where the user should assess the results and consider, whether the obtained results seem to make sense.

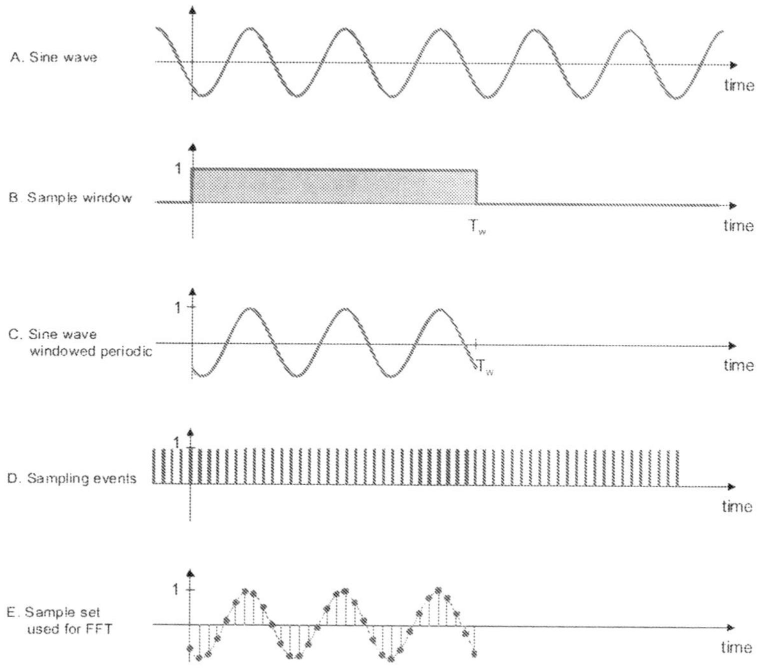

Figure 1-3. The principle of a Waveform Digitiser. A continuous waveform(A) is windowed for a given length of time(B) through the uses of a sample window. The two functions are multiplied, resulting in a waveform for a windowed period (C). This signal is sampled (D) during the windowed period, resulting in a sample set (E). For simplicity we assume equal-spaced sampling.

3. THE FOURIER SERIES AND FOURIER INTEGRAL

The mathematics of the Fourier Transforms assumes that the signal is continuous from $-\infty$ to $+\infty$. The FFT algorithm further assumes that the sampled portion of the waveform represents a segment of an infinitely repeated signal. One can think of the mathematics in the FFT conversion in a way, where the endpoint of the sampled segment wraps around from the last sample to the first sample. If the signal(s) consists of a complete number of cycles, then the signal(s) will wrap smoothly, and there will be no discontinuities across the wrapping points. In this situation, the signal is referred to as a *coherent signal*, typically a multitone, since it will also contain e.g. harmonics of each test tone. In a coherent measurement, each test tone falls into a unique Fourier spectral bin. If there is a discontinuity in the endpoints of the sampled signal, then the signal is *non-coherent*.

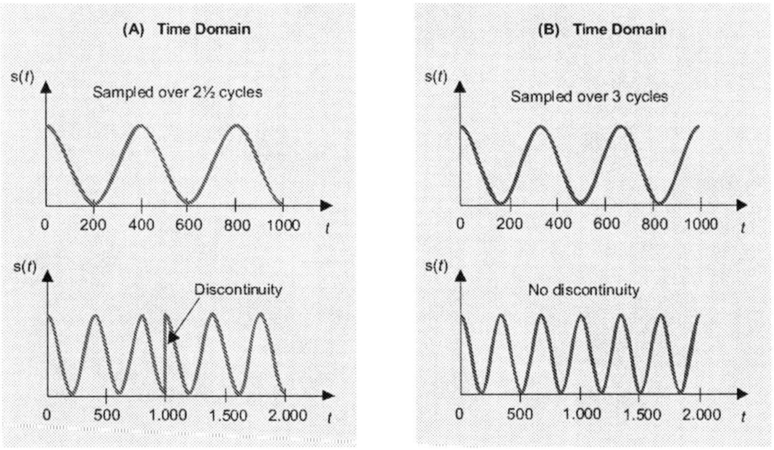

Figure 1-4. The effects of a FFT wrapping. In figure (A) 2½ period are sampled. This leads to a discontinuity in the wrapping point for the sampled segment, shown in lower of the two curves. In figure (B), a complete 3 cycle segment is used. As a result, no discontinuity will be seen in the end-points, see lower curve.

As can be seen from Figure 1-4, a situation of discontinuities emerge, if a waveform exposed to FFT transformation does not consist of an integer number of complete periods of the signal. An analysis performed on the Figure 1-4. A will result in an error in the frequency spectrum results, since it introduces *Spectral Leakage*, a phenomena that we will deal with later on. Hence, the best results in an FFT transformation can be obtained, if the periodicity assumed by the FFT transformation corresponds to that of the actual acquired waveform. Obtaining a good FFT transformation therefore always starts with defining the right boundary conditions for the analysis.

Even in the situation, where a set of complete periods is used for the FFT transformation, the reality of the waveform sampling may introduce errors, for example if the sampling is not done coherently or if the signal for the analysis has not reached a stable condition. In the event of insufficient settling time [Ref. 3], the undersettled waveform will exhibit discontinuities at the boundaries of the signal. It is caused by the exponential decay of the signal, for example an AC coupling capacitor. Since the FFT transformation assumes that the signal is repetitive, the end point of the signal in sampling window will be different from that of the start point and the resulting FFT will exhibit spectral leakage, i.e. some of the energy fro the signal will be distributed to other frequency bins, and the FFT will not show a correct spectrum for the actual waveform, although it is repetitive in nature. Since the spectrum of the DUT is interpreted incorrectly in such a situation, there will be frequency components throughout the frequency spectrum that will exhibit a certain element of corruption. Again the problem is to ensure the right electrical signal condition, i.e. make sure that the waveform reaches a steady state condition before the sampling is started or for example applying some precharging of the circuitry. Having the right input condition is preferably rather than manipulating the mathematical conditions of the FFT transformation.

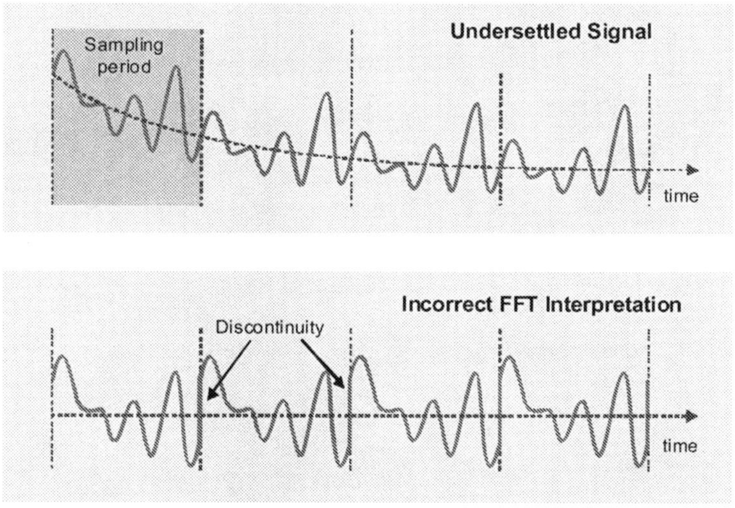

Figure 1-5. Sampling an undersettled signal results in an incorrect FFT interpretation. The upper figure shows an exponential decaying waveform. Although the signal is repetitive in nature, a sampling during the unsettled interval will result in discontinuities, because the FFT transformation wraps the end-point to the starting point of the sampled signal, see lower figure.

The basis for the Fourier Transforms was originally pioneered by Jean-Baptiste Fourier, a French major mathematician, historian and administrator, who developed the theory in his "*Théorie analytique de la chaleur*". The work included the Fourier Series and the Fourier Integral, which serves the basis for transformation of time domain waveforms to the frequency domain and vice-versa. The Fourier Series can be used to describe a waveform, if the waveform is period. Although the approach assumes that the waveform begins at -∞ (minus infinity) and repeats itself to +∞, we can obtains a really good fit by observing the periodicity of a waveform over an observable interval. For example a sine wave observed for a given interval will be assumed turned on at -∞ and continues to +∞. The errors made due to non-ideal conditions of the physical world are normally minor. The Fourier Series can be described as:

$$x(t) = a_o + \sum_{n=1}^{\infty} (a_n \cos(n2\pi f_o t) + b_n \sin(n2\pi f_o t)) \quad (1)$$

where:

$$f_o = \frac{1}{T} \quad \text{(T is the primary period of the signal)} \quad (2)$$

$$a_o = \frac{1}{T} \int_0^T x(t) \; dt \quad (3)$$

$$a_n = \frac{1}{T} \int_0^T x(t) \cos(n2\pi f_o t) \; dt \quad (4)$$

$$b_n = \frac{1}{T} \int_0^T x(t) \sin(n2\pi f_o t) \; dt \quad (5)$$

The coefficient a_o describes the DC values, taken as an average of x(t) over one period, and the coefficients a_n and b_n must be determined for every value of n. The value of n is a positive integer. The Fourier Series describes a completely periodic waveform, and the resulting spectrum analysis includes a number of integer multiples of the basic frequency, the fundamental frequency. Such multiples are referred to as harmonics.

The Fourier Series handles the periodic part of the signal. However, several signals of the real world are not periodic in nature, and a different way of describing this need to be included as well. This is where the Fourier

Integral comes into question. It can be shown [Ref. 3] that the basic equation (1)of the Fourier series can be manipulated to yield:

$$x(t) = \sum_{n=-\infty}^{\infty} c_n e^{j2\pi nf_o t} \tag{6}$$

where:

$$c_n = \frac{1}{T}\int_{-T/2}^{T/2} x(t) e^{-j2\pi nf_o t} dt \tag{7}$$

c_n must be evaluated for each values of n to provide the phase and amplitude of the harmonic component of x(t). The associated frequency is nf_o, and the harmonics are separated by:

$$\Delta f = \frac{1}{T} = f_o \tag{8}$$

Further manipulation of the mathematics, and assuming that $\Delta f \rightarrow 0$, the equation (6) approaches an integral:

$$x(t) = \int_{-\infty}^{\infty} X(f) e^{j2\pi nf_o t} df \tag{9}$$

where the Fourier coefficient X(f) is given by:

$$X(f) = \int_{-\infty}^{\infty} x(t) e^{-j2\pi nf_o t} dt \tag{10}$$

The two integrals of equation (9) and (10) are valid for non-periodic waveforms. The equations include positive and negative frequencies. Furthermore, the magnitudes of real valued signals in the positive and negative frequency domains duplicate each other, whereas the phases in the positive domain have inverted images compared to those of the negative domain, although still duplicated.

If a waveform, generated in the physical domain, is periodic in nature, it is sufficient just to analyse one period of the signal. Such a waveform can be a sine, a square wave, a triangle, a sawtooth or similar type of repetitive signal. In the case of a period signal, the Fourier Series is the most convenient approach for the analysis. The analysis is then based upon the assumption that the signal existed already from minus infinite, before the sampling starts, and will extend forever after the sampling of the waveform stops again. In the analysis, an integer number of waveform periods is used by wrapping the end point of the waveform sample to the start point of the sample. In a subsequent section, the basics of this will be described in more details.

If, instead, the waveform is non-periodic in nature, the most optimal approach is to use the Fourier Integral.

4. FOURIER TRANSFORMS – DFT & FFT

In practical mixed-signal testing, the Fourier transforms are based upon Discrete Fourier Transforms (DFT) and Fast Fourier Transforms (FFT). Coupled with waveform digitising and digital signal processing, the approaches of DFT and FFT have become a most powerful tool for the test engineer. The DFT can be shown to follow the equation:

$$X_d(k\Delta f) = \Delta t \sum_{n=0}^{N-1} x(n\Delta t) e^{-j2\pi k\Delta f\Delta t} \tag{11}$$

where:

- N = Number of samples being applied
- Δt = Sampling interval
- Δf = Sampling interval in frequency domain, equals 1/Δt
- n = Time sample index; values include 0,1,2,….., N-1
- k = Index of computed set of discrete frequency components; values include 0,1,2,…., N-1
-

If Δt=1, and assuming Euler's identify [Ref. 5], equation (11) can be simplified to yield:

$$X_d(k) = \frac{1}{N} \sum x(n) \frac{\cos(2\pi kn)}{N} - jx(n) \frac{\sin(2\pi kn)}{N} \tag{12}$$

Inverse functions also exist for the $X_d(k)$ function.

Since there are N terms to be summed for each N values of k, the DFT algorithm becomes proportional with N^2, because there are N^2 computations to execute. It means that for N=1.024 samples, a total of 1.048.576 computations have to be performed. A more efficient approach is to apply the Fast Fourier Transform, FFT.

The FFT function has been developed during a period where computers were non-existent and computations of DFT became extremely time consuming. The base for the FFT is calculations limited to powers of 2. The algorithm reduces the number of calculations to $N\log_2(N)$ operations. In the example from above where N=1024, the FFT number of computations is reduced to 1.024* $\log_2(1.024)$=10.240, i.e. a reduction in number of calculations of about 100x. The disadvantage of the FFT algorithm is that it only applies to powers of 2, also referred to as modulo 2. In a number of test situations, it is advantageous to apply a number of samples which does not fit into a modulo 2 domain. Instead, a DFT algorithm based on a general concept of prime numbers has been developed. Such a system, also referred to as a Multi-Radix System, breaks down the number system to a system of prime numbers and performs the calculations on the basis of these. As a result, Multi-Radix DFT calculations can now be performed extremely efficiently, consuming only 3-4x the time as that of a true FFT calculation. Given that computers in recent years have seen significant improvements in terms of computational speed, the penalty of applying a Multi-Radix DFT in mixed-signal testing is often insignificant, most of the time being taken up by sampling and transferring of data from the instrumentation to the computer. For that reason, efficient Multi Radix DFT may be preferably due to a number of other advantages. Due to the little difference in performance, such systems are also frequently referred to as FFT.

For a beginner, the FFT and DFT transforms may seem a little awkward, because the transform functions easily get "buried" in mathematics. However, the effect of FFT and DFT can be popularised in accordance to Rosenfeld [Ref. 6]:

> *An FFT can be thought of as a bank of filters. Each filter center frequency is separated from the next by the Fourier frequency (the sampling frequency divided by the number of samples collected). The shape of the filters is determined by the very nature of the FFT. If a signal, which does not exactly fall at the center of one of the analysis frequencies (i.e. a periodic signal not synchronised with the measuring system sample frequency), is filtered using an FFT, the response of each component of the FFT will be dependent upon the shape of the filter.*

In section 6 we shall study the synchronisation aspects further.

5. NON-COHERENT TESTING

If a signal intended for FFT analysis consists of only harmonics of the Fourier frequency (the sampling frequency divided by the number of samples), and if the signal is stable and not undersettled, then the signal endpoints will wrap smoothly around.

However, if a sampled signal has a discontinuity at the endpoints, i.e. that the endpoint does not wrap smoothly from the last sample to the first sample, then the signal is non-coherent. In a non-coherent case, the energy of the individual tones spreads to the surrounding spectral bins and hence leads to a certain corruption of the signal levels of other tones. It means that the Frequency spectrum does not provide a fully correct analysis. In some cases the corruption can lead to rather misleading results.

In real life test applications it is often not practical feasible to apply coherent testing. Reasons can be that the test instrumentation, e.g. functions generators, digitisers and digital I/O functions are not obtaining their test frequencies from a common master clock. Hence, the test system resources are not synchronised to each other and the system itself is consequently not coherent by nature. It may also be that the device under test, DUT, is designed to sample non-coherent waveforms, which may be the example for many devices such as CODECs, ADCs, DACs, etc., and can be difficult to set up for coherent testing.

Given that waveforms in a test situation often are non-coherent, this type of signal in association with FFT analysis is an important characteristic, and we shall return to the problem in section 7.

6. COHERENT TESTING

In mixed-signal testing, a rather fundamental issue of testing is that of employing *coherent* testing. Coherent means that the system under test operates in a repeatable manner where a best trade-off between speed and accuracy can be obtained. It also means that phase and delay measurements can be ensures in a practical manner during production testing. Coherence can best be described as [Ref. 4]:

> *A synchronised mixed-signal system is coherent, if and only if all frequency and timing functions are programmably related in exactly integer ratios.*

This fundamental characteristic can often not be obtained in many real life test applications due to limitations in testing hardware. Instead a principle of applying windowing may have to be applied (see later). However, if feasible, the coherent testing is certainly the preferred way ahead in mixed-signal testing. It ensures a really controlled and repeatable test approach for the mixed-signal circuits.

The coherent testing scheme is a coordinated/synchronised approach offering well devised DSP testing. It allows a synchronisation of the digitiser with each and every clock, signal and distortion component in the system under test. In essence, a coherent test approach, in the mixed-signal domain, is equivalent to that of digital circuit testing, where synchronisation of programmable control of clocks and timings for signals features precise and repeatable results. However, in contrast to the general approach of the digital domain, the use of coherent test is at times less well known in the mixed-signal domain.

6.1 The Mechanical Analogy

Coherent testing can be understood in analogy to the mechanical world. Just think of a mechanical gear consisting of two shafts and two gear wheels. Shaft 1 has a primary frequency of F_1 and shaft 2 has a secondary frequency F_2. The gear wheel on shaft 1 comprises k*M teeth, while the gear wheel on the second shaft has k*N teeth. The constant k is an integer that may be assigned positive integer values (1, 2,....K). In such a coherent system, the relationship between the primary input frequency and the secondary input frequency is:

$$F_2 = \frac{k*M}{k*N} * F_1 = \frac{M}{N} * F_1 \qquad (13)$$

The equation simply states that the secondary frequency F_2 is a proportion of the primary frequency F_1 in the ratio of the number of gear wheel teeth on shaft 1 (M) to the number of gear wheel teeth on shaft 2 (N).

If the common denominator of the gear system k=1, then a count of the teeth will provide an equivalent of the implicit clock cycle within one *primitive period* (P). As the gear wheels rotate, they will pass through the same states periodically, given by the primitive period (t+P), (t+2P), (t+3P) etc.

The mechanical gear system analogy can be expanded to handle many derivatives of the primary frequency.

6.2 The Coherent Mixed-Signal Test Approach

A coherent mixed-signal test system reveals many similarities as to the gear system analogy described above. The systematic approach of relatively prime integers can be seen in the electrical system as well. The basic equations for a coherent system is:

$$f_i = M^* f_{Fourier} \quad (14)$$

$$f_s = N^* f_{Fourier} \quad (15)$$

$$f_{Fourier} = \frac{1}{T_W} \quad (16)$$

where:

- $f_i =$ the input frequency to the digitiser
- $f_s =$ the sample frequency with which the signal is sampled
- $f_{Fourier} = f_{primary} =$ the *Fourier frequency*, also often referred to as the *Primary frequency*.
- M = the number of periods (cycles) of the signal. Must be complete cycles to avoid discontinuities.
- N = the number of samples
- T_W= the sample window width or unit test period (UTP)
- $\frac{M}{N}$: Must be irreducible. Each record shall contain an exact integer number M of cycles, and the number of cycles in a record must be relatively prime to N, the number of samples in the record.

It follows from equations (2) and (3) above that:

$$\frac{f_i}{f_s} = \frac{M}{N} \tag{17}$$

This is in essence the same basic equation as is valid for the mechanical equivalent system (gear box) described earlier.

The requirement that M is a relatively prime to N is essential. If this is not the case, N and M have a common divisor (integer). As a result, the pattern will be repeated, and less than all states (phases) will be tested. Actually, if M and N are not mutually prime and if:

$$\frac{N}{M} = L \tag{18}$$

then only L distinct phases are sampled. This may turn out to be a significant reduction of the resolution in a test. However, if N and M are mutually prime, there are N distinct samples of e.g. an input sine wave. These samples are uniformly distributed between 0 and π. Actually, the use of coherent sampling with N and M prime is the optimum choice of frequency in say an ADC test, since the number of states or phases is maximum.

The restrictions imposed by the mathematics for coherent testing can be illustrated in the examples:

Example 1:

A first approximation for the test a 12 bit ADC may be proposed as:

- $f_i =$ 1.000 Hz
- $f_s =$ 50 kSa/s
- N = 32.000 samples
- From equation (5) it follows that:

$$M = \frac{f_i}{f_s} * N = \frac{1.000}{50.000} * 32.000 = 640$$

However, the requirement of equation 6 shows that:

$$L = \frac{N}{M} = \frac{32.000}{640} = 50$$

Hence, only 50 out of the 2^{12}= 4.096 states of the converter can be tested, since the pattern just repeats itself. A poor test despite the large number of samples.

Example 2

If instead the conditions are changed slightly, e.g. if the number of periods tested are increased to 641, then a more suitable test condition appear

$$f_i = \frac{M}{N} * f_s = \frac{641}{32.000} * 50.000 = 1001{,}5625 Hz$$

A good synthesizer may be able to generate this signal. Furthermore, since N/M =49,92199688 the requirement of N being prime to M is fulfilled, and we are able to test all states of the converter.

From equations (2) and (4) it follows that:

$$T_W = \frac{M}{f_i} = 640ms \text{ n} \qquad \Rightarrow \qquad f_{Fourier} = 1{,}5625 \text{ Hz}$$

The examples show that although mixed-signal testing at first seems quite strait forward, it may eventually prove more complicated and call upon careful planning. As we shall see later on, the testing challenges in mixed-signal can turn out even more complicated than shown above.

7. WINDOWING FUNCTIONS

Testing of mixed-signal devices is typically constraint to a finite set of samples in order to run a subsequent discrete or fast fourier transform. In addition, if the signal is non-coherent, e.g. if a periodic signal is not synchronised with the sampling frequency of the measurement system), we must employ techniques for ensuring that disadvantages due to the non-coherency are suppressed diligently to an acceptable level.

Both situations call upon use of windowing functions. The simplest version of a windowing function is to apply a rectangular window e.g. for sampling as we have already seen in section *2*.

The purpose of the sampling window is to restrict the "infinite" amount of data from a continuous signal (in the time domain) to a finite set of test data. The rectangular window used for the conversion of data provides an accurate result, if the frequency contents of the original signal are harmonically related to the time length of the sampling window (coherent test). However, if this is not the case, we must apply a more advanced windowing function to eliminate or rather reduce the effects of discontinuities at the end-points of the sampling window.

Applying a time-domain window will result in a certain amount of spectral leakage, i.e. spreading of signal energy into incorrect frequency bins. However, it is possible to select a windowing function, although not classical rectangular, that offers low spectral spreading, and at the same time helps smoothening out discontinuity problems at the end points created through non-coherent testing. Selecting the right windowing function remains an engineering virtue, and it is far from always a simple and straightforward choice.

The purpose of the windowing function in FFT is, as was indicated in *3*, to reduce the effects of the discontinuities at the end of the sampling interval, simply because the FFT algorithm assumes periodicity of the waveform beyond the sampling period itself. It is done through repeating the sampling interval infinitely, and the end point of the interval will wrap to the starting point of the interval, and in non-coherent testing these two points will not coincide, hence the discontinuity. We try to eliminate or at least reduce this by using the windowing function.

An FFT transformation can be thought as a filter function as stated in *4*. We will analyse this a little further. We choose a FFT transformation based upon a classical rectangular sample window. A rectangle window in the time domain is transformed to a sin(x)/x in the frequency domain [Ref. 7]. In the transformation we use the magnitude of the transformed signal and possibly also the phase, rather than the signal itself. The signal consists of a main lobe centred on the value of $f=0$ and a train of side lobes, infinite in number, with maxima becoming progressively smaller as the frequency increases. The function reaches its first zero at the Fourier frequency, see Figure 1-6. The side lobes of the transformed signal are only half the width of the main lobe. This is due to the absence of a zero at the value of $f=0$ caused by the fact that :

$$Lim_{x \to 0}(\frac{\sin(x)}{x}) = 1 \quad (19)$$

although sin(x)=0 for x=0 (x is here used in the generic term, although the x-axis in our example is assigned the parameter of frequency).

It should be noted that we could have chosen a different window shape than the rectangle pulse (as well as the equal spaced sampling scheme that is used throughout this article). Had we chosen for example a cosine squared pulse, the result would have been different. This is the novelty, or challenge, that applying different types of windowing functions leads to variations in the results.

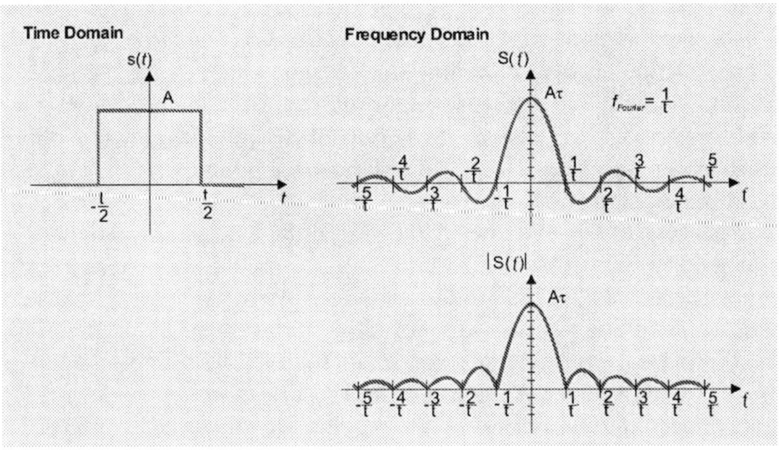

Figure 1-6. A rectangle pulse in the time domain is transformed into a sin(x)/x function in the frequency domain using FFT transforms. The focus of our work is the numerical value of the transformation, i.e. the magnitude of the signal at various points.

The situation can be somewhat simplified by using a unified transformation. Employing a unified amplitude of 1 for the windowing function and selecting e.g. a sample window length of 10 s, we obtain from equation (16):

$$f_{Fourier} = \frac{1}{T_W} = \frac{1}{10} = 0{,}1 Hz$$

The result can be seen from Figure 1-7: It appears that for the choice of a sampling window length of 10 s, we obtain a zero of the |sin(x)/x| function at every multiple of the Fourier frequency, i.e. at 0,1Hz; 0,2Hz; 0,3Hz; 0,4Hz etc. As we shall see later on, see section 7.2 example 3, the position of the zeros as well as the function of the side lobes governs the spectral leakage of the FFT transformation.

It can be seen from Figure 1-6 and Figure 1-7 that the magnitude (numerical value) of the transformed signal remains identical, although the time domain signal in Figure 1-6 is centered around zero, while it has been shifted to only positive values of time Figure 1-7. This is a characteristic of square pulses. However, the phase, which we have not shown here, will vary when the square pulse is shifted relative to time zero.

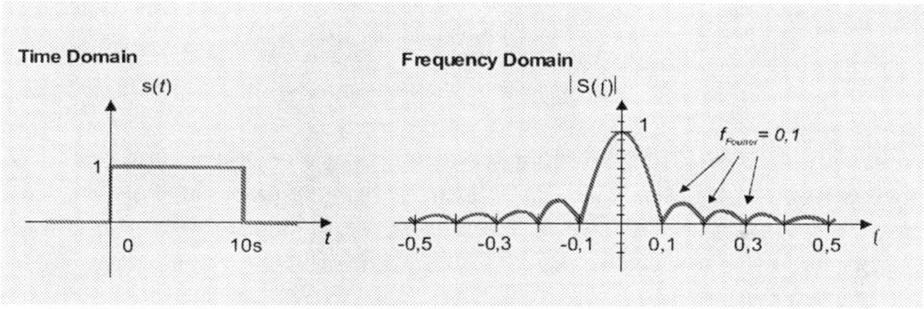

Figure 1-7. The rectangle pulse of 10 s is transformed to a |sin(x)/x| having a Fourier frequency of 0,1 Hz. The transformed signal is zero at the Fourier frequency and at every integer multiples of this frequency. The signal reaches its maximum at f=0 Hz.

The fact that we are using an FFT transformation (Fourier integral) rather than the Fourier series has an impact on the result seen in the frequency domain. If a signal of a given frequency and an amplitude of A is converted to the frequency domain using a Fourier series, then the amplitude in the frequency domain will also be A. However, in the case of a Fourier transform (FFT), a value of the signal exists both as a positive frequency bin and as a negative frequency bin. Due to this, the magnitude in the positive frequency domain will only be ½A except for the DC component (f=0). The value of the frequency component in the negative frequency spectrum duplicates that of the positive spectrum, and the value is also ½A.

The phase of the signal in the positive frequency domain is also mirrored in the negative frequency domain, but the images are inverted.

So far we have seen the effects of the transformation of a windowing function from the time domain to the frequency domain. However, our main interest is not the windowing function but rather the resulting signal spectrum. The windowing of the sampled signal is equivalent to convolving the original signal spectrum with that of the spectrum of the windowing function, i.e. basically multiplying the two spectra.

It should be noted that even if we are not deliberately applying a windowing function, we still face the effects of one. In that situation, we encounter the realities of convolving with a rectangular-shaped window of uniform height, since this is the essential sampling window for sampling in the time-domain and in lack of any other windowing function, this will prevail. If other windowing functions are applied, the sampling window is in essence multiplied by that of the chosen windowing function. Since the situation of "no windowing" still means applying a rectangular window, we sometimes also refer to this situation as applying a uniform or rectangular window.

The effect of the spectrum convolution is that the main lobe(s) of the signal(s) is centred at each frequency component of the original sampled time-domain signal. The side lobes of each signal approach zero at every frequency interval from the main lobe spaced by the Fourier frequency, as we saw it in Figure 1-7. The effects of this are illustrated in Figure 1-8. The result is a discrete frequency spectrum obtained by sampling the continuous, periodic signal spectrum by the FFT, and each frequency line of the FFT is the corresponding value of the continuous convolved spectrum at each frequency line. This is basically equivalent to the sampling of the time domain signal that we apply using the ADC function.

In the case of coherent sampling (integer number of periods in the time domain), the center of the main lobes of the continuous spectrum of the windowing function will coincide with the discrete sampling points given by the Fourier frequency. Hence, the frequency lines of the signal spectrum will fall exactly at the sampling point of Fourier frequency. Furthermore, all side lobes will disappear, since all other leakage elements of the frequency component fall exactly in the zeros of the convolving window spectrum.

However, in the case of a non-coherent sampling, the center of the main lobe of the continuous spectrum of the windowing function (governed by the signal spectrum frequency component) is shifted a fraction from that of the Fourier frequency sampling points. As a result, the frequency component in the FFT transformation will not achieve full amplitude value and the frequency component leakage will also not coincide with the zeros of the side lobes. These effects lead to a certain amplitude error as well as the appearance of frequency components in nearby frequency bins. We shall study this is more detail in Section 7.2, Example 3.

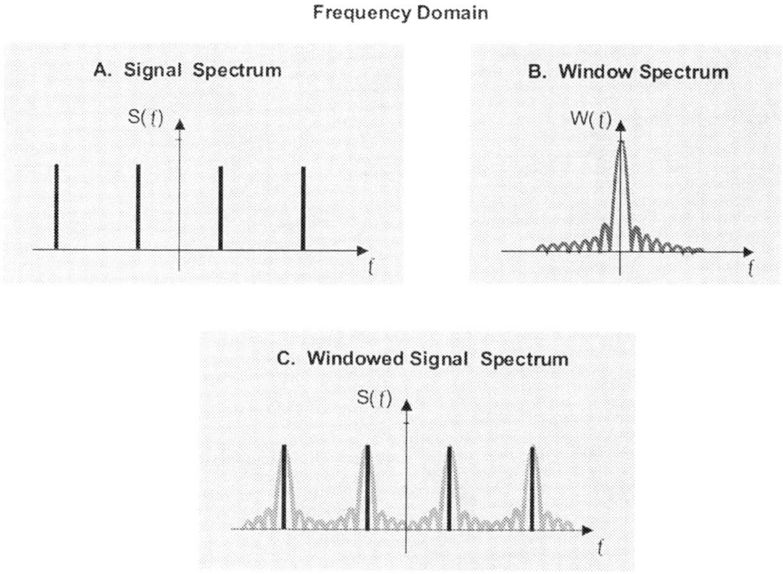

Figure 1-8. The convolving effect of a windowing function in essence "multiplies" the signal spectrum (A) with the window spectrum (B) and results in a windowed signal spectrum (C).

The width of the main lobe also impacts the resulting FFT spectrum. The narrower the main lobes of the windowing function, the easier it will be to separate two closely positioned frequency components of an input signal. However, the narrower the main lobe, the more energy of a frequency component is spread into the side lobes and hence leads to leakage. So we face a trade-off between spectral resolution and leakage suppression.

7.1 Selecting a Proper Window Function

In principle a windowing function aims at setting the values of the derivatives at the boundary ends to zero or near zero. When the windowing data are smoothly set to zero near the boundaries, the period extension of the data in many orders of derivatives.

The operation of windowing is equivalent to multiplying in the time domain the signal waveform with that of the windowing function. This was actually shown in Figure 1-3. Intuitively the "ideal" window should have a weight at the endpoints approaching zero and be one in the signal interval.

This can be obtained using the Rectangle window, where the function can be seen as:

$$h(t)=s(t)*w(t) \tag{20}$$

where:

- s(t) is the function to sampled
- w(t) is the windowing function

and such that :

$$w(t) = \begin{Bmatrix} 1 & for & 0 \le t \le T_W \\ & 0 & otherwise \end{Bmatrix}$$

and h(t) becomes:

$$h(t) = \begin{Bmatrix} s(t) & for & 0 \le t \le T_W \\ & 0 & otherwise \end{Bmatrix}$$

It was used in the previous conversions but as we shall see later on, it suffers from several problems due to lack of reduction of the discontinued function near the end-points.

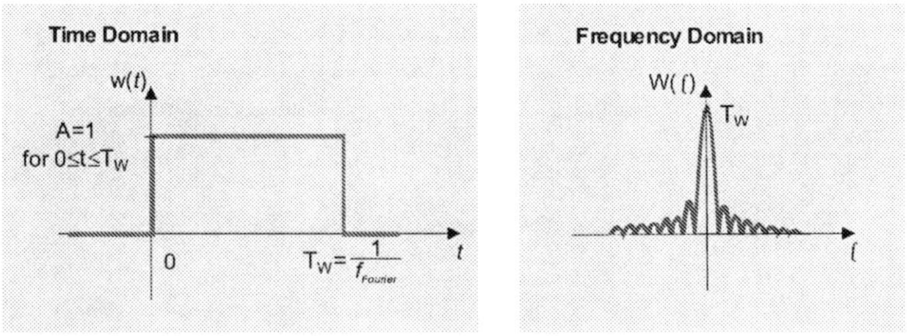

Figure 1-9. The Rectangle function offers an easy conversion formula, and is intuitively the correct windowing function, although not in reality, when non-coherent signals are in question.

For the test engineer, the challenge often becomes one of selecting an optimal windowing function. A large number of windowing functions has been reported over time, but in practical use, only a few are typically used by most mixed-signal engineers. To mention some of the windows:

□ Rectangle	□ Gaussian
□ Triangle	□ Dolph-Chebychew
□ $Cos^2(x)$-Hanning	□ Kaiser-Bessel
□ Hamming	□ Barcilon-Themes
□ Riesz	□ Exact Blackman
□ Rieman	□ Blackman
□ De la Valle-Poussin	□ Minimum 3 sample Blackman-Harris
□ Tukey	□ 61 dB, 3 sample Blackman-Harris
□ Sohman	□ 74 dB, 4 sample Blackman-Harris
□ Poisson	□ 74 dB, 4 sample Kaiser-Bessel
□ Hanning-Poisson	□ Rosenfeld
□ $Cos^4(x)$-	□ Parzen
□ Cauchy	□ Etc.

[Ref. 5] offers a good overview of characteristics of some of the windowing functions and their characteristics such as main-lobe heights, highest side lobes, bandwidth and theoretical Roll-Off. For the user, it is essential to note, that many windowing functions reduce the height (essential signal bin) of the main lobe by –6dB to –8dB compared to the signal using the Rectangle windowing function.

All windows for windowing functions represent compromises. For example, the Hanning window, also referred to as the Cosine Square window, produces a narrower spread of the components near the signal frequency. However it results in higher leakage of the signal frequency into frequency bins further away from the signal frequency [Ref. 6]. Rosenfeld compared the Hanning window with that of the Blackman window. The spread from the Blackman window was twice that of the Hanning window, but most of the leak went into frequency bins near the signal frequency. The leakage into frequency components out-of-band, i.e frequency bins further away, was 20dB lower for the Blackman window than for the Hanning window.

Table 1-1. Some typical Windowing functions [Ref. 5].

Unity Amplitude Window	Equation for Window Shape	Major Lobe Height	Highest Side Lobe [dB]	Band-width (-3dB) [dB]	Theoretical Roll-Off [dB/octave]
Rectangle 	A=1 for $0 \le t \le T_W$	T_W	-13,2	$0{,}86 f_{Fourier}$	6
Extend. cosine 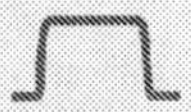	$A=0{,}5(1-\cos(2\pi 5t/T))$ for $0 \le t \le 0{,}1T_W$ & $0{,}9T_W \le t \le T_W$; A=1 for $0{,}1\ T_W \le t \le 0{,}9T_W$	$0{,}9T_W$	-13,5	$0{,}95\ f_{Fourier}$	18 (beyond $5f_{Fourier}$)
Triangle 	$A=2t/T_W$ for $0 \le t \le 0{,}5T_W$; $A=2(1-t/T_W)$ for $0{,}5T_W \le t \le T_W$	$0{,}5T_W$	-26,7	$1{,}27\ f_{Fourier}$	12
Hanning 	$A=0{,}5(1-\cos(2\pi t/T))$ for $0 \le t \le T_W$	$0{,}5T_W$	-31,6	$1{,}39\ f_{Fourier}$	18
Hamming	$A=0{,}08+0{,}46(1-\cos(2\pi t/T))$ for $0 \le t \le T_W$	$0{,}54T_W$	-41,9	$1{,}26\ f_{Fourier}$	6 (beyond $5f_{Fourier}$)
Cosine[4] 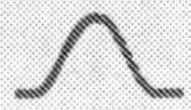	$A=(0{,}5(1-\cos(2\pi t/T)))^2$ for $0 \le t \le T_W$	$0{,}36T_W$	-46,9	$1{,}79\ f_{Fourier}$	30

According to [Ref. 5], the Hamming window is better suited for use with two relatively close signal frequencies than for example the Hanning, Half Cycle Sine[3], Cosine[4] and similar. It is due to the smaller –3dB bandwidth than that of the others.

In other situations, it may be required to use different windowing functions for evaluating different parameters. It may for example be necessary to use another window for analysing the noise performance. Optimising the selection of windows could yield a lower noise floor.

It is worth stressing again that the problems associated with the FFT transformation and the windowing functions are not caused by the input signal itself but rather with the way, we interpret results. It is also due to the fact that we have to consider a finite set of samples of the signal in order to employ our Fourier transformations. So we should be prepared that there will always be some spectral leakage present, when we analyse non-coherent systems.

Below is shown examples of applying windowing functions to sampled signals.

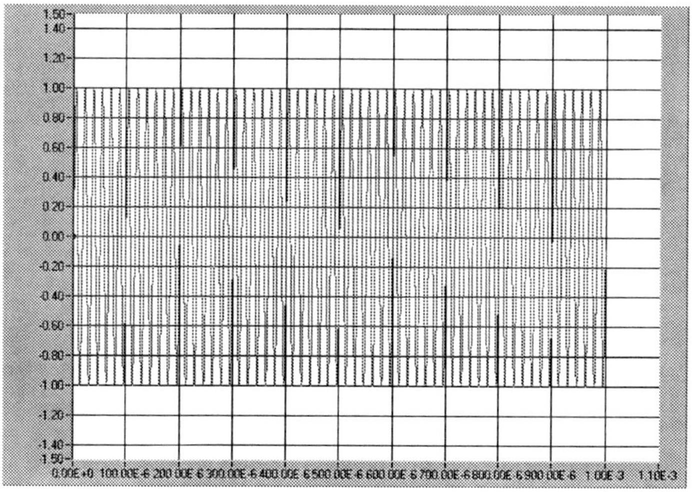

Figure 1-10. Time domain view of a non-coherent signal of 60 kHz, a total of 10.000 samples have been acquired using a sample clock of 10 MSa/s. It follows that the time record is 1 ms.

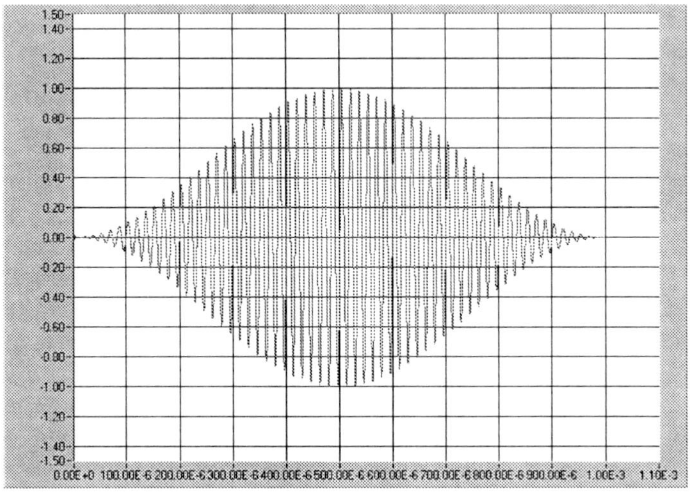

Figure 1-11. Time domain view of the signal of Figure 9 is here processed employing a Hanning (Cosine Square) windowing function. Again the time record is 1 ms, applying 10.000 samples to the 60 kHz signal.

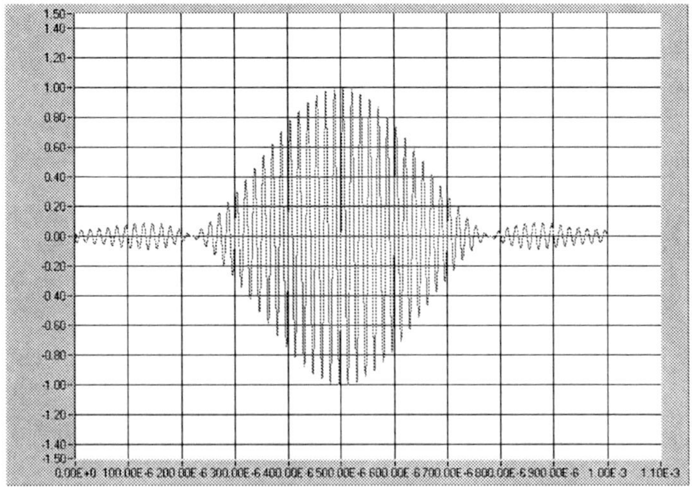

Figure 1-12. The Time domain response of the signal in Figure 9 when exposed to a Flattop windowing function. 10.000 samples are acquired of the 60 kHz signal using a10 MSa/s digitiser.

The conditions for recording is based upon the equation (17):

$$\frac{f_i}{f_s} = \frac{M}{N} \quad \Rightarrow \quad \frac{60 * 10^3 [Hz]}{10 * 10^6 [Hz]} = \frac{60 [periods]}{10 * 10^3 [samples]}$$

Although it at first glance appears as if a whole number of periods are present, the clocks of the digitiser and input generator are independent and a slight deviation creates a minor discontinuity, that we will investigate later on.

7.2 Spectral Leakage

As stated already, lack of coherency in a mixed-signal test system results in spectral leakage. Applying diligent countermeasures such as a careful selection of a windowing function can reduce the effects of spectral leakage, but not really completely remove it.

The lack of coherency spreads some of the energy form the continuous waveform to other frequency bins. It impacts not only the amplitude of a frequency domain signal, but also its phase. Furthermore, the effect is not restricted to the fundamental tone only, but also impacts the harmonics of non sinusoidal repetitive waveform in a way, relative to that of the fundamental signal.

Applying windowing functions influences the energy distribution. It helps significantly in confining the energy of a non-coherent signal to a much narrower range of FFT bins, but if used on a non-coherent signal (non-periodic signals like noise, multi-tone signals, etc.) it unfortunately results in some spectral leakage, due to spreading of energy. Depending on the type of windowing function applied, it is possible to reduce the amount of spectral leakage in such a situation.

Example 3

In order to analyse the effects of spectral leakage, we will analyse and compare a situation of two cosine signals transformed using FFT. We assume cosine input signals that exist prior to t=0 and continues after the end of the windowing period that we apply, see Figure 1-13.

The input cosine has an exact frequency domain line spectrum, because it is exact itself. This applies to Signal A as well as Signal B.

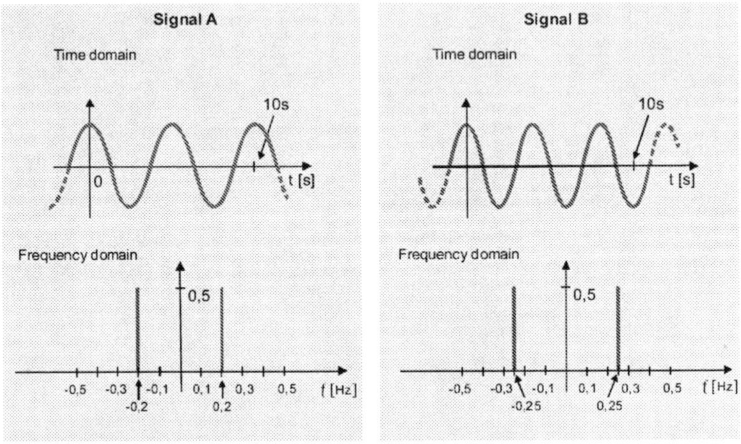

Figure 1-13. Two input signals, Signal A (fi=0,2 Hz) and Signal B (fi=0,25 Hz) and their exact line spectrums in the frequency domains.

To allow for a FFT transformation, we must restrict the signals to a finite time interval, and this is done applying a windowing function. Each signal is sampled over a given window, in our example this is chosen to be a 10 s period. We can think of this as a multiplication of the "pseudo infinite" input signals multiplied by the rectangle windowing functions. The results of the "multiplication" is shown as the time domain signal at top of Signal A and Signal B in Figure 1-14.

Signal A then consists of 2 complete cycles, while Signal B results in 2½ complete cycles during the period of sampling. According to the definitions in sections 5 and 0, the Signal A is coherent, since it assumes periodicity over the window of analysis, while Signal B is non-coherent, because it reveals discontinuity in the end-points. The non-coherency of Signal B is due to the length of the sample window. Had we instead chosen the sample window to be 12 s, Signal B would become coherent while Signal A would become non-coherent, at least for this theoretical study. The coherent test situation occurs, when an integer number of cycles exists in a sampling window, because then each of the harmonics of a signal will also be harmonically related to the window, and hence not produce any leakage effects.

A number of characteristics appear from our selected conditions for the two chosen signals in Figure 1-14. The selection of a 10s Rectangle windowing function results in a f_{Fourie}= 0,1 Hz for the Rectangle window, and it transforms into a |sin(x)/x| conversion (we saw this already in Figure 1-7). It results in a set of zeros for every 0,1 Hz.

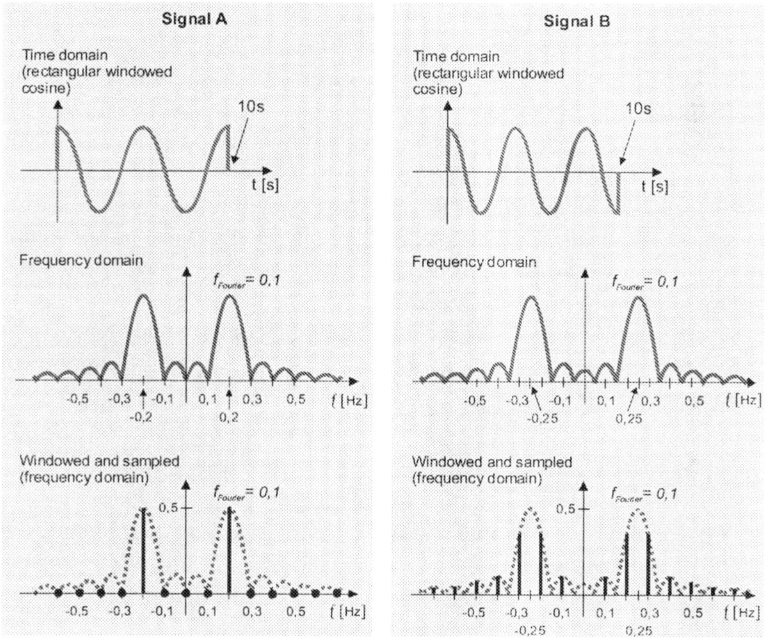

Figure 1-14. Spectral leakage caused by a non-coherent test situation. Signal A operates under coherent test conditions while Signal B is non-coherent due to the discontinuity at the end points of the sample period. The windowing function applies is the rectangle window.

The signal spectrum of Signal A will show a frequency component at 0,2 Hz corresponding to a cycle time of 5 s. The center of the main lobe is at 0,2 Hz and coincides with the signal spectrum. All other frequency components fall exactly at the zeros of the convolving widowing function, separated at 0,1 Hz.

In the case of Signal B, where the cycle time is 4 s and thus should show a signal spectrum frequency component of 0,25 Hz, the main lobe of the windowing convolution function centres at this frequency. However, the discrete frequency components resulting from the FFT are at the frequencies 0,1 Hz, 0,2 Hz, 0,3 Hz, etc. separated by the Fourier frequency. In essence, the main lobe fall between the 0,2 Hz and 0,3 Hz samples. Furthermore, all remaining frequency domain samples no longer fall on the zeros for the main lobes, but actually on the peaks of these, thus causing leakage into other spectral bins.

In the case, where an integer number of cycles of a given frequency is within the sampling window, the higher frequency harmonics of the given signal will also be contained in the FFT transformation and there will be no leakage effects from such signals [Ref. 5]. The same basically applies in a case where non-repetitive signals are completely inside the sampling window. In that event, the FFT transformation assumes that the non-repetitive signal consists of a train of pulses and will handle that correctly. However, if a fraction of the non-repetitive signal is outside of the sampling window, some frequency information will be lost, and the resulting frequency spectrum will contain errors.

In real life applications, a system may contain multi-tones. Some of these may fulfil the coherent requirements, others not. Especially if the waveform is almost periodic, this effect can be seen.

7.3 Some Results Influenced by Spectral Leakage

But how does the unfortunate side effect of spreading of the energy from test tones to adjacent bins influence a typical test situation in mixed-signal testing. In the following we will try to illustrate problems associated with spectral leakage based upon some measurement results.

InFigure 1-16, the signal capture of two non-coherent sine signals are compared. The signal frequencies are approximately 60 kHz, more specifically 60.000 Hz for one of the signals and 59,250 Hz for the second signal. The signals have an amplitude of ±1 Vp and the digitiser used for the sampling is a 14-bit digitiser, sampling at 10 MSa/s and employing a sample of 50.000. Since the basic clock of the digitiser is not synchronised to the basic clock of the input signal, the signals are Non-Coherent. A time domain analysis shows that the 60.000 Hz signal is captured for almost 300 cycles, but not quite, see Figure 1-15, left signal, while the 59.250 Hz signal, Figure 1-15, right signal, is captured for approximately 296¼ signal.

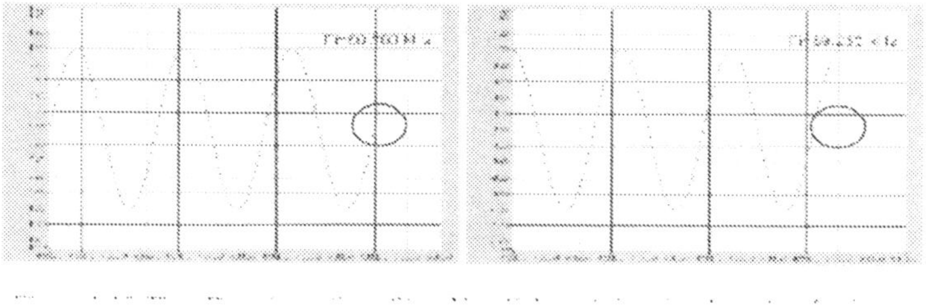

Figure 1-15. Time Domain section of two Non-Coherent sine signals captured using no Windowing. Fs=10 MSa/s, 50.000 samples, Vp=1,0 V. The left signal has fi=60.000 Hz, while the right signal has a slightly different fi=59.250 Hz, otherwise all conditions are identical. The sine signals starts at t=0 and an amplitude=0 V.

The results of performing an FFT transformation of the two signals can be seen in Figure 1-16 A & B, where no windowing is applied in the FFT transformation.Although the two signals are almost identical, the frequency domain appearances are rather different. The difference is due to the fact that the 60.000 Hz signal is close to a coherent signal, and hence only shows little spectral leakage, while the 59.250 Hz signal is "off" by ¼ of cycle, and hence reveals significant spectral leakage. Applying a windowing function like the Hanning window, see Figure 1-16 C & D, remedies the frequency domain appearance rather well, but we notice that large spectral leakage of the 59.250 Hz signal still results in a poorer appearance.

It also appears that the top of the main lobe is reduced, when using the windowing function. In the case of applying no windowing, the top is about –3 dB while in case of using a Hanning window, the top drops to about –9 dB, i.e. 6 dB down from the case of the rectangle window (no windowing). The case of using a Flattop window makes the situation even poorer, dropping to about –14 dB on the main lobe and resulting in a relatively poor FFT spectrum in the case of the 59,250 Hz signal. The noise floor drops correspondingly, so the actual relative difference from signal top to noise floor basically remains constant regardless of the window. Actually, it is possible to scale the values of windowed functions so that this side effect of using a window is balanced out. In the case of the Hanning window, all values of the windowed array should be multiplied by 2 (+6 dB), and in the case of Flat Top, the values should be multiplied by 4,5455 (+13,15 dB).

The Hanning window produces a narrower spread of components near the signal frequency, as was pointed out earlier on. However, when we start using varying sample length in the signal capture, we discover that it has an influence on not only the resolution of the FFT plot, but also the width of the individual lopes, see Figuree 1-18 andFigure 1-19.

using varying sample length in the signal capture, we discover that it has an influence on not only the resolution of the FFT plot, but also the width of the individual lopes, see Figuree 1-18 andFigure 1-19.

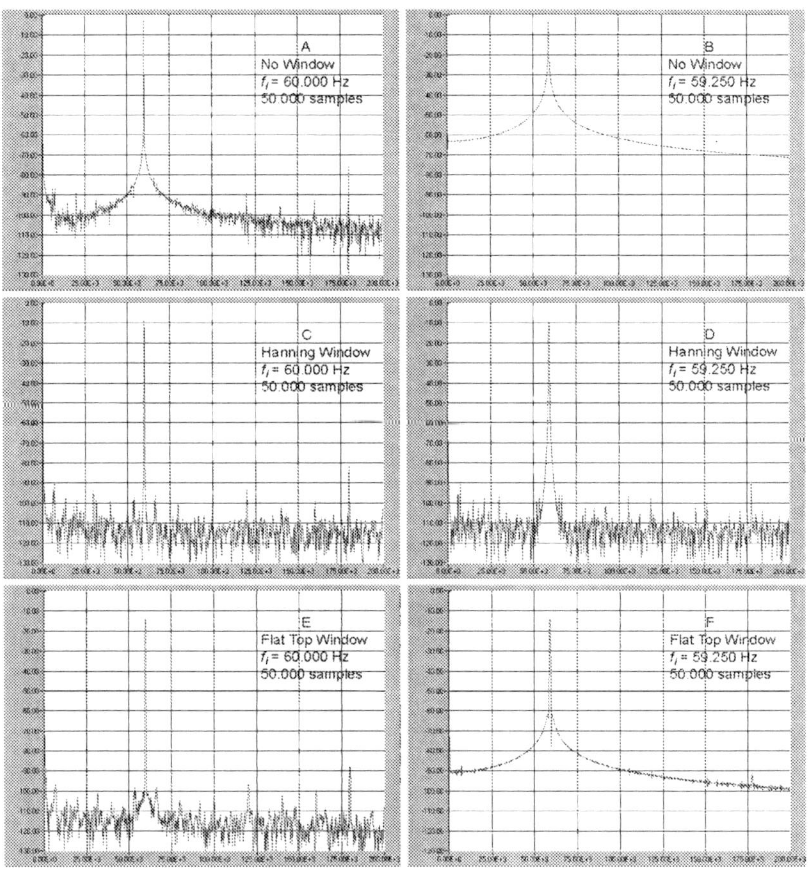

Figure 1-16. FFT of two Non-Coherent signals (see Figure 14) captured using no Windowing(A & B), a Hanning window (C & D) and a Flat Top window (E & F). Fs=10 MSa/s, 50.000 samples, Vp=1,0 V. The left signal has Fi=60.000 Hz, while the right signal has a slightly different Fi=59.250 Hz, otherwise all conditions are identical.

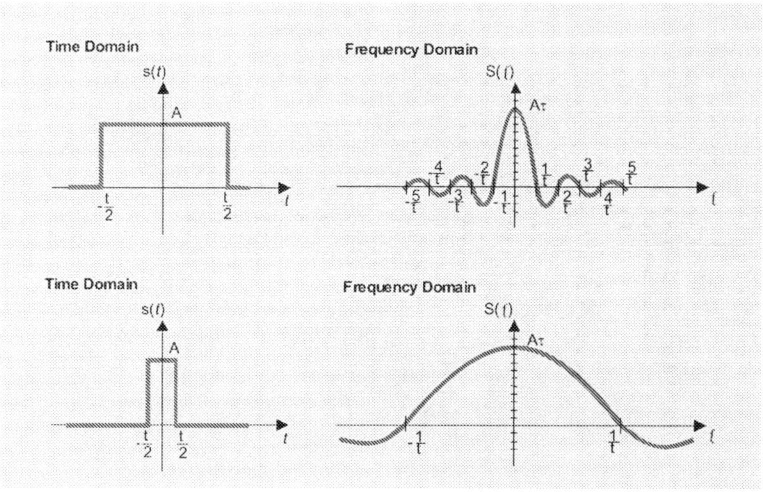

Figure 1-17. The reciprocal nature of the time domain and the frequency domain of FFT transforms also influences the width of a pulse.

What we witness is actually the effects of reciprocity associated with FFT transformations. If a pulse or window in the time domain is decreased, it results in a wider FFT transform in the frequency domain and vice versa, if it increases, seeFigure 1-17. Actually, the reciprocity also applies to transformations from the frequency domain to the time domain.

But this effect allows us to obtain an improved FFT appearance. Take for example the FFT results of Figure 1-19. It appears that if we increase the number of samples from 2.000 to 10.000, the main lobe also becomes 5 times more narrow, and at the same time the effects of the "FFT skirt" is reduced similarly, so that we get a better view of the present noise level. The input frequency, f_i , is constant in the example. Since the sampling frequency, f_s , remains constant in the example, the window sampled over is enlarged by the same amount as the sample size increases. Due to the reciprocity, the width of the FFT transform changes correspondingly, and the sample set is increased or decreased. It is evident both for the Non-Windowed signal as well as for the signal where the Hanning window has been applied.

It means that we can obtain a better appearance of the FFT result by increasing the sample size. As the FFT skirt diminishes, so does it for the skirts of other frequencies present in the spectrum. Notice for example the

3rd harmonic at 177.750 Hz inFigure 1-19, B, D, F and H, i.e. the FFT transformations, where a Hanning window has been applied. The top is about –90 dB in all four figures, but at low samples, the side lobe is quite broad. In the case of No Windowing, i.e. in Figure 1-19 A, C, E and G, this is hardly visible except in G figure due to the high spectral leakage present.

Comparing the FFT results of Figure 1-18 and Figure 1-19 clearly shows the influence of the synchronisation, especially in the case where no windowing is applied. In the case of an almost fit to coherent test conditions, i.e. where the f_i =60.000 Hz, the FFT transformation even in the case of no windowing, yields reasonable results.

The obvious disadvantage of increasing the sample size in a test situation, especially in production related test, is the time required for the sampling. A few thousand samples, or may be even below 1000 samples, may be required to offer an efficient throughput in production testing. The luxury of applying samples different than modulo 2 samples (powers of 2) may also be prohibited, because the DFT runs typically up to 3x slower than the FFT based upon modulo 2.

Another issue is that of sampling speed. This is illustrated in Figure 1-20. Two different sampling speeds are applied. In Figure 1-20 A and B, the signal is sampled using 10 MSa/s, while Figure 1-20 C& D use 50 MSa/s. Since the latter samples 5x faster, the number of cycles of the input signal included in the FFT transformation drops from about 296 ¼ cycles to about 59 ¼ cycles.

The shorter duration in the time domain means a wider transformation in the frequency domain. This is also obvious from the figure. For example the main lobe width at –90 dB in Figure 1-20 B is about 5,3 kHz while the width at – 90 dB in Figure 1-20 D is about 27 kHz, which is quite precisely a difference of 5x.

Had we instead of making the number of samples constant chosen to make the length of the sampling window constant, we would need in the case of f_s =50 MHz a sample length N=250.000 samples to fill up the 5 ms time it takes to run 50.000 samples at f_s =10 MHz. In that situation our FFT results would appear to look identical.

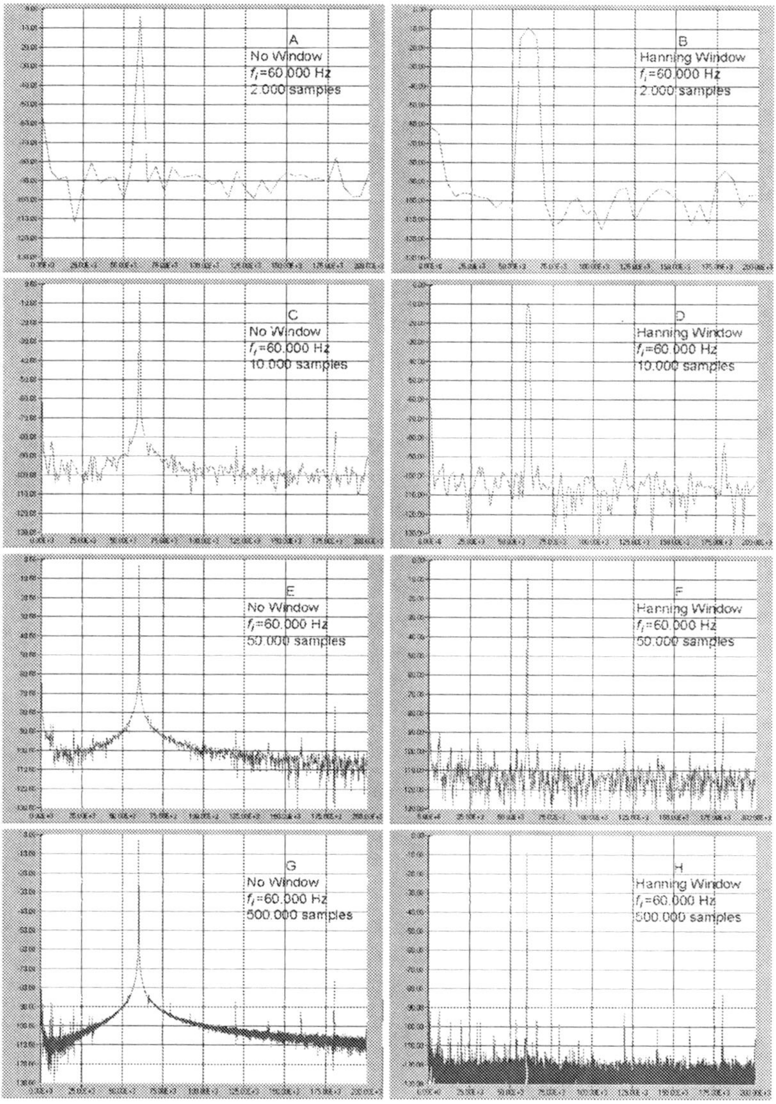

Figure 1-18. See previous page. The Figure shows the FFT results in the frequency domain of a Non-Coherent signal captured using Fs=10 MSa/s, and a signal amplitude of Vp=1,0 V. The Fi=60.000 Hz applies to all figures A-H. The left figures (A, C, E, G) are transformed using no windowing function except the rectangular sampling window, while the figures to the right (B, D, F, H) have been transformed using a Hanning window.

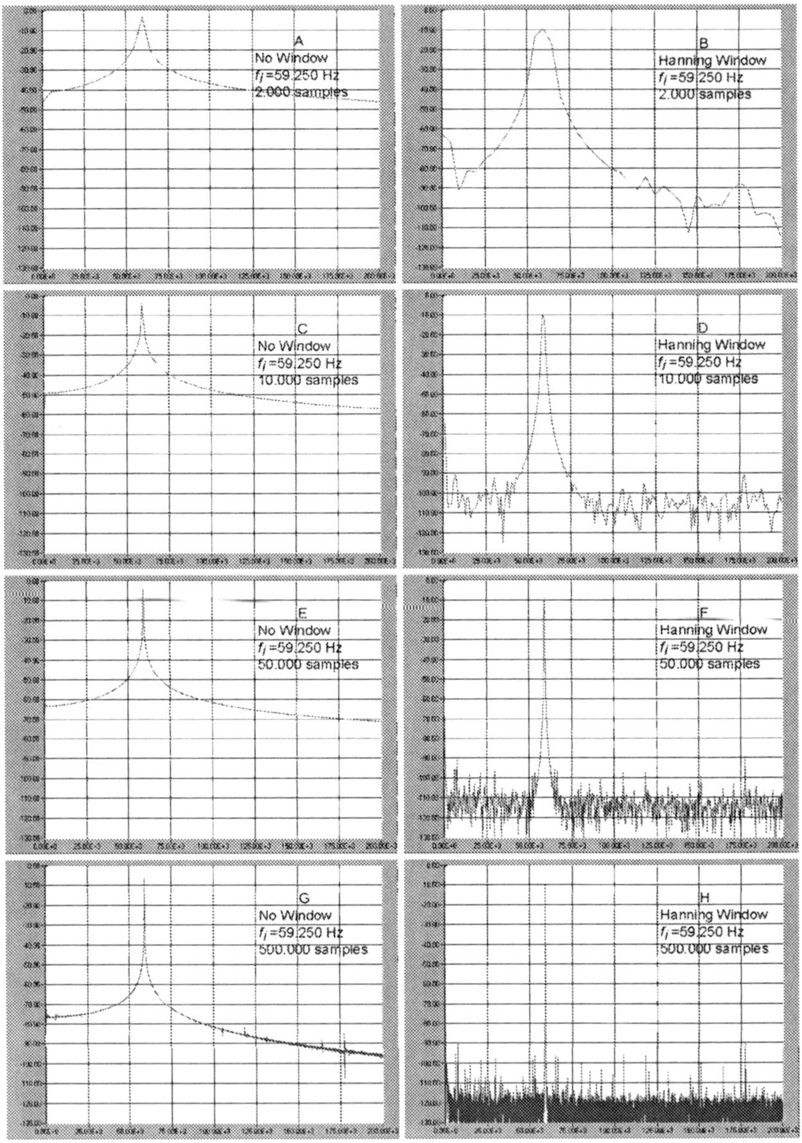

Figure 1-19. See previous page. The Figure shows the FFT results in the frequency domain of a Non-Coherent signal captured using Fs=10 MSa/s, and a signal amplitude of Vp=1,0 V. The Fi=59.250 Hz applies to all figures A-H. The left figures (A, C, E, G) are transformed using no windowing function except the rectangular sampling window, while the figures to the right (B, D, F, H) have been transformed using a Hanning window.

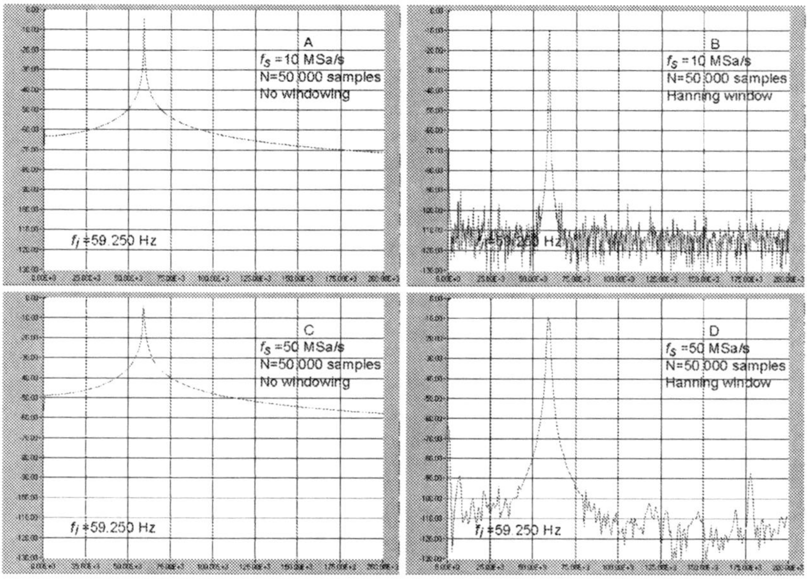

Figure 1-20. The effect of sampling speed. A signal of fi=59.250 Hz is sampled at 10 MSa/s in figure A and B, and at 50 MSa/s in figure C and D. In figure A and C no window is applied, while figure B and D use a Hanning window.

8. CONCLUSIONS

Applying FFT analysis is a highly powerful approach in modern mixed-signal testing. To easy the use, efficient tools can be built into the test suite approach. However, disregard the ease of use, a basic understanding of a number of fundamentals are require in order to ensure reliable and efficient testing. The present paper describes a number of fundamental problems, but quite a few other also exists. For the user, it is essential to become familiar with the issues to make sure that good mixed-signal test results can be obtained.

9. REFERENCES

[Ref.1] "OPTIMISTIC: An Approach for Optimising Mixed-Signal Design Validation" Birger Schneider,
International Mixed-Signal Test Workshop, Montreux, CH, June 2002, IEEE

[Ref. 2] "Production Testing of PCM (Digital)Audio Circuits"
Mark Landry
International Test Conference 1983, pages 767-770, IEEE

[Ref. 3] "Improving DSP-Based Measurements with Spectral Interpolation"
Mark Burns
International Test Conference 1995, pages 355-362, IEEE

[Ref. 4] "DSP-Based Testing of Analog and Mixed-Signal Circuits"
Matthew Mahoney
IEEE Computer Society O.N. 785, ISBN 0-8186-0785-8, 1987

[Ref. 5] "Tutorial No. 3 - Mixed-Signal Testing"
ETC91, European Test Conference, IEEE et al.

[Ref. 6] "DSP Measurement of Frequency"
Eric Rosenfeld
International Test Conference 1986, pages 981-986, IEEE

[Ref. 7]: "Telecommunications"
J. Brown & E.V.D. Glazier
Chapman & Hall Ltd. And Science Paperbacks, 1966, SBN412 20420 7, pages 38-44

Chapter 2

ANALOG AND MIXED-SIGNAL TEST BUS: IEEE 1149.4 TEST STANDARD

Florence Azaïs, Pascal Nouet
LIRMM, CNRS / University Montpellier II

Abstract: This chapter aims to introduce the basis of the IEEE 1149.4 mixed-signal test bus standard, reviewing its architecture, its instruction set and a test methodology using the standard.

Key words: Mixed-signal test standard, Mixed-signal test buses

1. MOTIVATION

In the early nineties, the IEEE Standard Test Access Port and Boundary-Scan Architecture [1] was introduce as a low-cost alternative to bed-of-nails tester. This so-called IEEE 1149.1 Standard provides facilities for application and monitoring of digital test signals through the use of a test bus residing on board and some DFT hardware residing on chip. It allows to scan the boundaries of any compliant Integrated Circuit (IC) from a single test port of the Printed Circuit Board (PCB) connected to an Automatic Test Equipment (ATE). It is then possible to monitor internal logic states of the digital system during the mission mode or to perform an integrity test verifying the interconnections between chips. Moreover, it is also possible to test the internal IC circuitry by applying test stimuli and collecting test responses via test resources or initiating a Built-In Self Test (BIST) procedure and collecting the result directly from the board connector.

J.L. Huertas (ed.), Test and Design-for-Testability in Mixed-Signal Integrated Circuits, 45–71.

However, the electronic world is no longer a pure digital world but a Mixed-Signal world where analog and digital signals are exchanged on-chip and between chips. In the late nineties, the IEEE 1149.4 Standard [2] was proposed to address problems of board-level testing when analog or mixed-signal ICs are involved. This standard extends the capabilities of the so-called boundary-scan to analog or mixed-signal ICs and electronic systems. Nowadays, this standard is a major concern for designing and testing systems-on-a-chip (SOC) that contain analog elements, digital logic and memory resources, each with their specific testing requirements.

2. MIXED SIGNAL TEST BUS STANDARD OVERVIEW

Basically, the IEEE 1149.4 standard specifies the use of an analog test bus which will reside on board, some additional DFT hardware integrated on-chip, and the protocol associated with these new features.

2.1 General Architecture

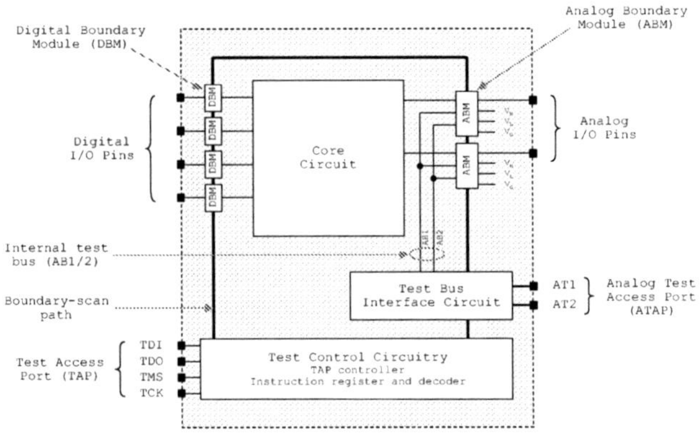

Figure 2-1. On-chip test architecture of an 1149.4-conformant IC

The general architecture of an 1149.4-conformant IC is illustrated in Figure 2-1, which shows the main mandatory elements of the standard. To comply with the IEEE 1149.1 standard, it comprises a dedicated Test Access Port (TAP) composed of four required pins *TDI*, *TDO*, *TMS* and *TCK*, a collection of Digital Boundary Modules (DBM) associated with every digital function pin, and a test control circuitry composed of a TAP controller, an instruction register and an instruction decoder. These features permit loading and unloading of both instructions and test data and provide access to the core circuitry for application and monitoring of digital test signals. The 1149.4 standard adds an Analog Test Access Port (ATAP), a Test Bus Interface Circuit (TBIC), Analog Boundary Modules (ABM) associated to every analog function pin and a two-wire internal analog test bus (*AB1/2*).

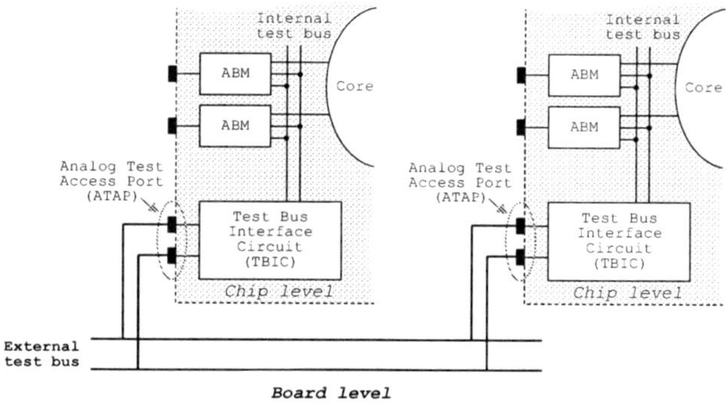

Figure 2-2. General architecture of an 1149.4-compliant board

Facilities for application and monitoring of analog test signals are illustrated in Figure 2-2. At chip level, the internal test bus lines are distributed to every ABM. Connection of the internal test bus lines to the ATAP pins is controlled by the TBIC. At board level, a two-wire external test bus is specified to carry signals on the board between the test equipment and the ATAP pins.

2.2 Register Architecture

The overall register structure of a 1149.4-conformant chip is shown in Figure 2-3. It basically consists of a number of shift-registers connected in parallel between *TDI* and *TDO* pins. Test equipment is able to serially load any of these registers through *TDI*, and to serially inspect the contents of any register through *TDO*.

This architecture is purely digital and essentially identical to that defined in Standard 1149.1. It comprises the following mandatory elements:

- an instruction register,
- a bypass register,

– a boundary-scan register.

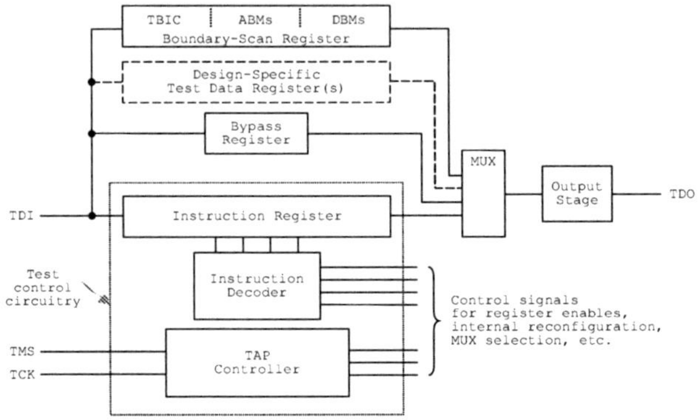

Figure 2-3. Test register structure

Optionally, additional registers may be included by the designer to provide extra user-defined testability features. The only difference in the test register structure between standards 1149.1 and 1149.4 resides in the boundary-scan chain composition. The boundary-scan register not only includes the register stages of all the DBMs but also the control stages of all the ABMs and the control stages of the TBIC. All these stages may be concatenated in any order.

Besides, it should be noted a specificity of mixed-signal devices comprising both analog and digital cores. In addition to the boundary modules associated with the function pins (both analog and digital), further DBMs may be included at the interface between the analog and digital cores. Depending whether these internal modules are incorporated or not, the

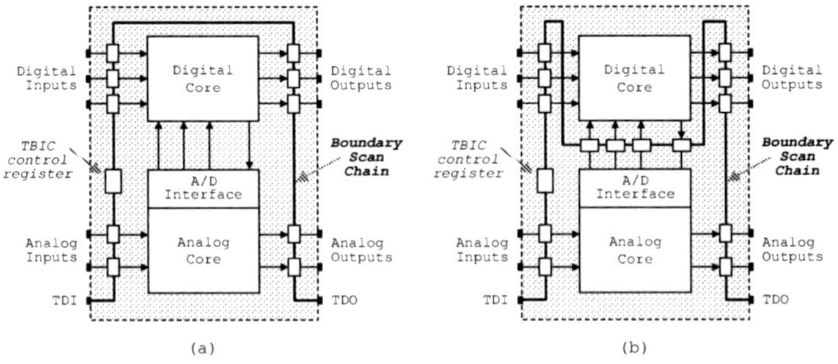

Figure 2-4. Alternative forms of the boundary-scan register

boundary-scan register can take one or the other of the two alternative forms depicted in Figure 2-4.

2.3 Control of the test architecture

The overall control of the testability structure is managed by the test control circuitry composed of a TAP controller, an instruction register, and an instruction decoder (see Figure 2-3). It provides the control signals required for the transfer of data and for the choice of the test register that operates between *TDI* and *TDO* pins. The TAP controller is a synchronous finite state machine that operates according to the state diagram shown in Figure 2-5, where state transitions occur on the rising edge of *TCK* according to the value present on *TMS*. It is identical to that specified in standard 1149.1.

2.4 Test Capabilities

The major role of the test structures described in the 1149.4 standard is to provide electronic access to the individual chip pins of a mixed-signal board. Test features defined in the standard permit to achieve test access without mechanical contacts and allow to target various types of defects.

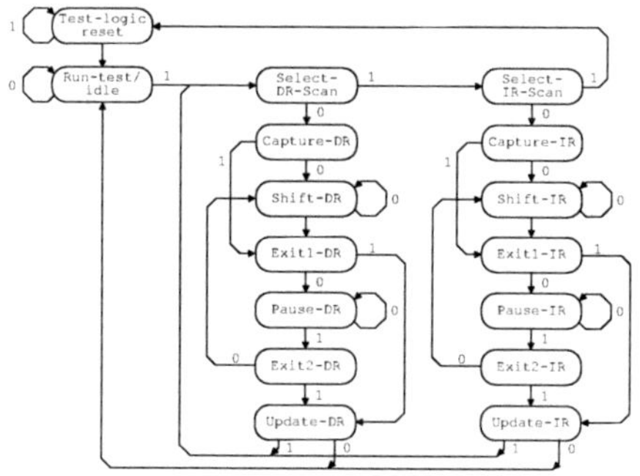

Figure 2-5. TAP controller state transition diagram

As illustrated in Figure 2-6, defects likely to occur on a mixed-signal board include shorts, opens, and components that are out of their tolerance range. It should be noted that opens can occur either on analog or digital lines, and shorts can occur between analog pins, digital pins, analog and digital pins, or across discrete components in the interconnect. The 1149.4 standard addresses these various defects through three classes of tests:

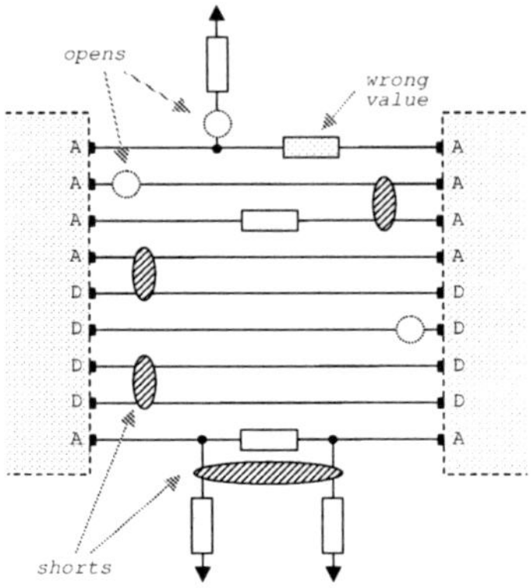

Figure 2-6. Overview of targeted defects on a mixed-signal board

- **Interconnect test**: testing for shorts and opens among the interconnections in a board whether digital or analog. This aspect of the standard offers the ability to perform mixed-signal interconnect testing at the same time and using the same procedure as digital interconnect testing since full compatibility with 1149.1 is ensured.
- **Parametric test**: performing analog characterization measurements via the application of analog stimuli and analog response monitoring. This aspect of the standard permits to handle the test of discrete components often interposed between chips in a mixed-signal board (level shifting, passive filtering, A.C. coupling, ...).
- **Internal test**: testing the internal circuitry of the chip whether in isolation or while mounted on a substrate. This aspect of the standard is optional since it may require the inclusion of internal test structures impacting cost and performance of the circuit.

3. IEEE 1149.4 INSTRUCTION SET

The instruction set for a 1149.4-conformant IC is almost identical to that for a 1149.1-conformant IC. It must include the mandatory instructions from standard 1149.1, namely instructions *BYPASS*, *SAMPLE/PRELOAD*, and *EXTEST*. It may also include optional instructions defined in standard 1149.1 as well as user-defined instructions. The only mandatory instruction that must be added in standard 1149.4 is the *PROBE* instruction designed to support analog testing. However, interpretation of many 1149.1 instructions is specific for analog pins. This section briefly reviews the different instructions, starting with the mandatory ones followed by the optional ones.

3.1 Mandatory Instructions

Every IC that conforms to the standard must support the following instructions, which consist of dedicated bit-patterns loaded in the instruction register and decoded by the instruction decoder. Coding of instructions can be freely chosen by the designer except explicitly specified.

3.1.1 BYPASS Instruction

The definition of this instruction is identical to that for the standard 1149.1. It can be selected by loading "all-1's" in the instruction register (mandatory code), or optionally other codes if designed for. Its main function is to select the *BYPASS* register as the serial path between *TDI* and *TDO* in the *shift-DR* controller state. It also ensures that the chip operates in normal function mode, i.e. all function pins are connected to the core circuitry and are isolated from all test resources.

BYPASS can be used to allow efficient test access to a specific IC on a board. As an example, let us consider a board having 100 chips connected in series into a single scan chain. Without this instruction, test data must pass through all the chips when testing one specific IC in the middle of the chain. To reduce the length of the scan path, all chips except that specific IC can be put into the *BYPASS* mode, where only one single storage cell lies between the *TDI* and *TDO* ports of these chips. The length of the bit stream input to the serial path is therefore considerably reduced.

3.1.2 SAMPLE/PRELOAD Instruction

This mandatory instruction allows test data to be shifted into and out the boundary-scan register during the normal operation of the chip. This instruction selects the boundary-scan register to be connected for serial access between *TDI* and *TDO* and ensures that the test resources have no effect on the normal functional mode of the chip.

This instruction has actually two functions. The first (*SAMPLE*) is to use the boundary-scan register for capturing a snapshot of the normal operation of the component. This snapshot is taken on the rising edge of *TCK* in the

Capture-DR controller state, and can be examined serially by shifting out through *TDO*. The instruction operates according standard 1149.1 for digital pins, bus has an expanded meaning for analog pins: the snapshot of the signal corresponds to the digitized value of the analog voltage appearing at the pin, which is captured by a digitizer in the ABM and stored as a one-bit result in the boundary-scan register.

The second function of this instruction (*PRELOAD*) is to allow test data to be loaded into the latched parallel outputs of the boundary-scan register in preparation for applying another instruction. In this way, a digital data pattern can be specified for each ABM by loading the appropriate values in its control register, which is part of the boundary-scan register. The loading is performed on the falling edge of *TCK* in the *Update-DR* controller state.

3.1.3 EXTEST Instruction

The *EXTEST* instruction is the central feature of the standard. It is the instruction dedicated to the test of the board interconnect, either simple interconnect or extended interconnect with discrete components included in the interconnection pathway. It can be selected by loading "all-0's" in the instruction register (mandatory code), or optionally other codes if designed for. When this instruction is executed, the boundary-scan register is selected as the serial path between *TDI* and *TDO* in the *shift-DR* controller state. All function output pins are isolated from the core circuitry so that signals driven on those pins solely depend on the content of their boundary modules. Function input pins are monitored by their boundary modules and may be isolated from the core circuitry. Although core isolation facility is optional, the analog circuitry must be designed to be immune to test voltages applied on the pin and its interconnect.

As in standard 1149.1, the *EXTEST* instruction is used to test simple interconnect by applying logic levels to all pins. Standard 1149.4 defines some extended facilities to deal with the test of extended interconnect. In particular, analog measurements of discrete components connected to the pins are allowed by injecting test stimuli and monitoring test responses at the pins using both the internal and external analog test busses.

3.1.4 PROBE Instruction

The *PROBE* instruction is the only new mandatory instruction in 1149.4 relative to 1149.1. This instruction is analogous to the *SAMPLE* instruction but allows continuous-time access to pin signals during the normal operation of the chip instead of a simple snapshot of the normal operation of the chip.

When the *PROBE* instruction is active, the boundary-scan register is connected between *TDI* and *TDO* and all function pins are connected to the core circuitry. In addition, one or both of the ATAP pins are connected to the internal analog test bus lines as determined by the contents of the TBIC control register, and the internal analog test bus lines are connected to individual function pins as determined by the contents of the corresponding ABM control registers. The *PRELOAD* phase of the *SAMPLE/PRELOAD* instruction can be used to set up the contents of the ABM and TBIC control registers.

In effect, the basic function of the *PROBE* instruction is to connect a pin to an analog test bus while the system functions normally, so that some characteristics of the signal at the pin can be measured. This electronic probe is extremely useful for in-situ measurements in analog and mixed-signal systems. Note that only one analog pin signal can be accessed at a time.

3.2 Optional Instructions

These instructions are proposed to chip designers in order to maximize the efficiency of the mixed-signal test bus. For example, if a BISTed memory is embedded in a chip, the *RUNBIST* instruction should be supported in order to check the memory at the board level.

3.2.1 INTEST Instruction

The *INTEST* instruction is dedicated to the test of the internal IC circuitry. This instruction is used to apply test stimuli and collect test responses via the test resources. Note that if *INTEST* (or *RUNBIST*) is supported, the boundary-scan register must include digital boundary modules at the interface between the analog and digital portions of the on-chip circuitry, as shown in Figure 4. This is necessary to ensure that the analog core will not interact with the digital core during test and vice-versa.

When the *INTEST* instruction is executed, the boundary-scan register is selected for connection between *TDI* and *TDO*, all digital pins (except clocks) are disconnected from the core and solely controlled by their boundary modules, while all analog pins remain connected to the analog core. The test of the digital part of the core proceeds exactly as with standard 1149.1: test data are supplied by the boundary-scan register to every input of the digital core and test responses on every output of the digital core are captured in the boundary-scan register. In addition, the presence of the analog test structure permits some testing of the analog core.

In particular, any analog pin signal can be probed via the analog buses to allow signal injection or signal monitoring, as determined by the contents of the ABM and TBIC control registers. However at a given time, only one analog pin can be stimulated and only one observed since there is only one pair of analog test bus lines. Consequently, all other analog input pins must have signals applied to them from off-chip for a meaningful 1149.4 test.

3.2.2 HIGHZ Instruction

The *HIGHZ* instruction selects the *BYPASS* register for connection between *TDI* and *TDO* and disconnects all pins, including ATAP pins, from the core and from all test circuitry. Disabling all the pins allows the use of a bed-of-nails tester to test board-level interconnect without conflict with the chip.

3.2.3 CLAMP Instruction

The *CLAMP* instruction selects the *BYPASS* register for connection between *TDI* and *TDO* and disconnects all pins from core. The state of the pins is solely controlled by the contents of the boundary-scan register. The *CLAMP* instruction is therefore a combination of the *BYPASS* and *EXTEST* instruction.

As the *EXTEST* instruction, it can be used to perform analog measurements by injecting test stimuli and monitoring test responses at the pins via the analog test busses. However, since data shifted into *TDI* bypasses the boundary-scan register and goes directly to *TDO*, the state of the pins does not change while *CLAMP* is executed. This can be used to keep parts of the circuit quiescent while measuring voltages and current in other parts.

As the *BYPASS* instruction, the *CLAMP* instruction can also be used to allow a rapid access to any IC in a board, but maintaining the ICs in test mode rather than in functional mode.

3.2.4 RUNBIST Instruction

The *RUNBIST* instruction allows for the execution of a self-test procedure of the chip. It is initiated when the TAP controller enters the *Run-Test/Idle* state and leaves in the test data register connected between *TDI* and *TDO* a test result signature. This instruction actually behaves according to standard 1149.1, with analog pins either controlled by the boundary-scan register or in high-impedance state.

4. IEEE 1149.4 ARCHITECTURE

As illustrated in Figure 2-1, the TBIC and the ABM are the main new features defined in the "dot 4" extension of the boundary-scan standard. These elements will be described in details in this section.

4.1 Test Bus Interface Circuit

The TBIC is the interface between the analog external test bus and the analog internal test bus. It provides a link between signals carried from and to the ATE and signals carried within the component from and to the analog function pins. It is therefore a key element in the testability infrastructure. Its main functions are:

- to connect each of the internal analog test bus lines to the ATAP pins in order to support parametric testing,
- to apply or capture logic values on the ATAP pins in order to test the external analog test bus lines with respect to opens, shorts and stuck-at faults,
- to perform characterization of the internal analog test bus lines in order to improve the accuracy of analog measurements,
- to isolate the internal analog test bus lines from the ATAP in order to avoid interference and cross-coupling during the normal operation of the chip.

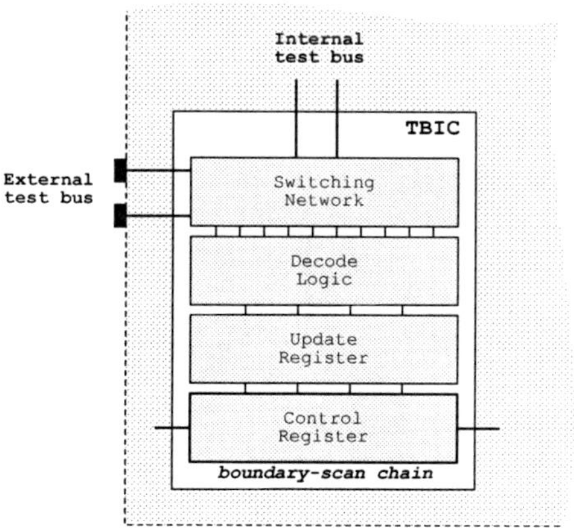

Figure 2-7. TBIC general architecture

This functionality is implemented by means of a switching network connected between the internal test bus lines and the ATAP pins. The control of the switching network is realized through a control register, an update register and a decode logic as described in Figure 2-7. It is worth noting that the control register takes part in the boundary-scan chain. Hence, it can be loaded and unloaded through the TAP as for the classical "dot 1" boundary-scan cells.

4.1.1 TBIC Switching Network

The minimal switching infrastructure required in the TBIC is derived from the mandatory capabilities defined in the standard.

- In functional mode, the internal analog test bus lines must be isolated from the ATAP. In that case, *AB1* and *AB2* must not interfere with the normal operation of the chip. Typically, this can be achieved by connecting *AB1* and *AB2* to a good reference voltage (e.g., ground). In a more general way, it is recommended to connect the internal test bus lines to a low-noise DC source (V_{clamp} in the standard) whenever they are not in use.
- During both functional and test modes, the TBIC must permit to transmit a current from *AT1* to *AB1* and/or transmit a voltage from *AB2* to *AT2*. Further recommendations are given in the standard to support both current and voltage transmission modes and to permit bi-directional transmission. In this purpose, the use of real switches such as CMOS transmission gates is encouraged rather than current and voltage followers.
- During test mode, the TBIC must permit to apply one among two reference voltages (V_H and V_L) to the ATAP pins. In that case, ATAP pins should be disconnected from the internal analog test bus.
- During test mode, the TBIC must permit to convey a current from *AT1* to *AT2* via *AB1* and convey a voltage from *AT1* to *AT2* via *AB2*.

A possible implementation of all these capabilities consists of a set of ten switches as shown in Figure 2-8. Mandatory digitizers are intended to monitor the voltage present on each of the ATAP pins with respect to a threshold reference voltage. They deliver the result as a logic bit stating whether the pin voltage is higher or lower than the reference voltage. These digitizers will be discussed later in this section. In the presented network, the clamp option is implemented and the role of switches can be summarized as follows:

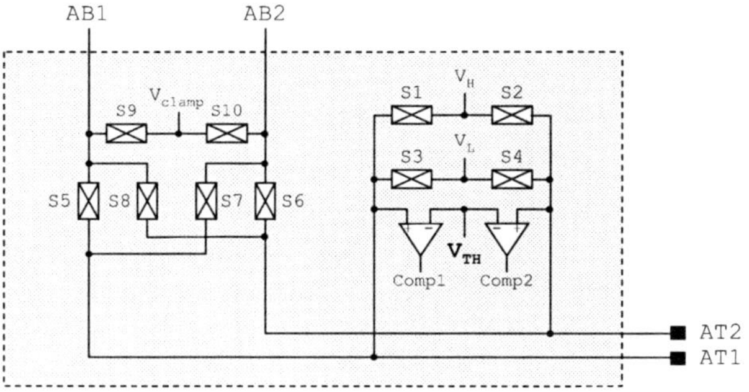

Figure 2-8. TBIC switching network

- Switches S1 to S4 permit to drive each one of the ATAP pins either to V_H or V_L. These switches are used (together with digitizers) during simple interconnect test.
- Switches S5 to S8 interface the ATAP pins to the internal test bus lines. Switches S5 and S6 permit to connect or isolate *AT1* and *AT2* to/from *AB1* and *AB2* respectively. They provide current and voltage transmission facilities used during parametric test. Switches S7 and S8 are used to connect *AT1* to *AT2* via one of the internal test bus lines in order to perform the characterization of the *AT1*/*AT2* path.

Table 2-1. Minimal switching configurations of the TBIC

Conf	Closed Switch (es)	Connections	Function
C0	None	Isolate *AT1/2*	Mission mode
C1	S6	Connect *AB2* to *AT2*	Analog measurements in *PROBE, INTEST & EXTEST*
C2	S5	Connect *AB1* to *AT1*	
C3	S5 and S6	Connect *AB1/2* to *AT1/2*	
C4	S6 and S7	Drive 00 on *AT1/2*	Digital interconnect test in *EXTEST*
C5	S2 and S3	Drive 01 on *AT1/2*	
C6	S1 and S4	Drive 10 on *AT1/2*	
C7	S1 and S2	Drive 11 on *AT1/2*	
C8	S6 and S7	Connect *AT1* to *AT2* via *AB2*	Characterization in *EXTEST*
C9	S5 and S8	Connect *AT1* to *AT2* via *AB1*	

- Switches S9 and S10 clamp the internal test bus lines to an internal voltage source V_{Clamp} in order to eliminate undesirable noise when they are not in use. Note that these two switches can be controlled by the complemented control signal of S5 and S6. In this way, S9 is closed when S5 is open (*AB1* not in use) and S10 is closed when S6 is open (*AB2* not in use).

Table 2-1 summarizes the different useful configurations of the TBIC switching network together with their associated functionality. It is implicitly assumed that *AB1* and *AB2* are clamped when not in use. For example in configuration C0, switches S9 and S10 are closed since switches S5 and S6 are opened.

Regarding the practical design of the switching network, we can draw several comments. Two DC voltage sources V_L and V_H are used for acting as logic levels for simple interconnect testing. These voltage levels can be chosen at the discretion of the chip designer but they should be easily identified as a low or a high logic level. They may be identical for both pins (e.g., using power rail voltages) or specific to each pin.

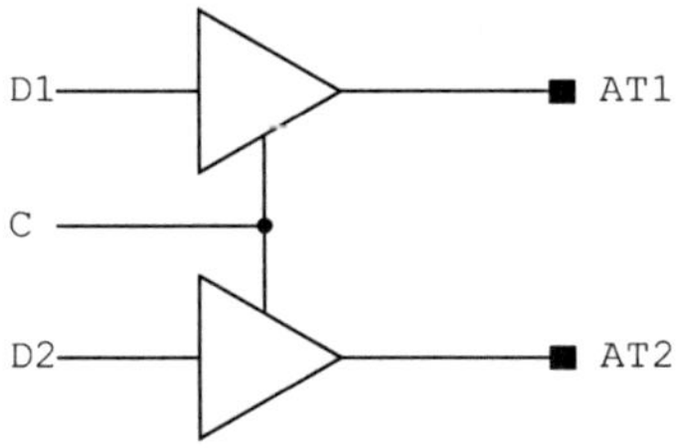

Figure 2-9. Alternative implementation of V_L/V_H application facility

In particular, an alternative implementation of the S1 to S4 part of the switching network consists in using a pair of logic drivers controlled by three logic signals to provide V_L and V_H levels. As illustrated in Figure 2-9, *C* acts as a control signal while *D1* and *D2* are defining the level to be driven on each pin:

- when *C=0*, both drivers are in a high-impedance state condition thus corresponding to the open state of S1 to S4 conceptual switches,
- when *C=1*, ATAP pins are driven to a logic level that may be high or low depending on the logic value of *D1* and *D2*; the values of V_L or V_H being only determined by the two drivers, they are not necessarily the same.

Digitization of the analog signal at the pin is another important feature for performing simple interconnect testing in accordance with the digital boundary-scan standard. Digitization is realized with respect to a threshold voltage V_{TH} that may correspond to an actual voltage source or to an intrinsic feature of a simple logic gate. This reference voltage is typically mid-range between V_H and V_L and is strictly defined in the standard to be in the following range:

$$\frac{V_H + 3 \cdot V_L}{4} \leq V_{TH} \leq \frac{3 \cdot V_H + V_L}{4} \tag{1}$$

Several simplifications may be obtained when using simple V_H and V_L corresponding to the power rail voltages:

- for single power supply circuits, it comes:

$$\frac{V_{CC}}{4} \leq V_{TH} \leq \frac{3 \cdot V_{CC}}{4} \tag{2}$$

- for symmetrical power supply circuits, it comes:

$$\frac{-V_{CC}}{2} \leq V_{TH} \leq \frac{V_{CC}}{2} \tag{3}$$

In most of the cases, digitization will be practically implemented using a simple inverter or preferably using two cascaded inverters to obtain a non-inverting behavior. In order to detect short-circuits in an unambiguous manner it is recommended to use a specifically designed logic gate with a threshold different from the one of standard digital buffers. Alternatively, a voltage trigger may be used when the threshold must be adjusted and/or externally provided to implement user-specified features for example.

Finally, note that although Figure 8 shows V_{TH} and V_{clamp} as a single value associated with both the ATAP pins, they could be pin-specific.

4.1.2 TBIC Control

The TBIC control circuitry allows to select one configuration of the switching network. Figure 2-10 shows the minimum control infrastructure required in the standard, which consists of:

- A 4-bit scan register, which is a part of the boundary-scan chain. The four bits *Calibrate* (*Ca*), *Control* (*Co*), *Data1* (*D1*) and *Data2* (*D2*) can be serially loaded and unloaded through *TDI* and *TDO*. Two multiplexers are included allowing the digitized voltages of *AT1* and *AT2* to be captured in cells *Data1* and *Data2*, respectively.

- A 4-bit update register, which is parallel loaded from the corresponding stages of the control register.
- A decode logic that provides the control signals for each switch from the contents of the update register together with two mode signals supplied by the TAP controller.

To select a given configuration of the switching network, appropriate control codes have to be loaded in the update register in conjunction with the two mode signals delivered by the TAP controller. These mode signals are generated according to the active instruction in the instruction decoder. Table 2-2 summarizes the mandatory assignments between the six control inputs of the decode logic and the ten mandatory configurations presented in Table 2-1.

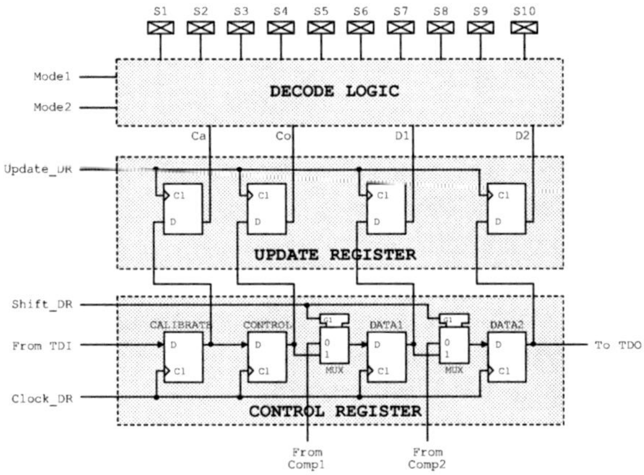

Figure 2-10. TBIC control structure

Note that in normal mission mode as well as in *BYPASS, SAMPLE/PRELOAD* and *HIGHZ*, *Mode2* signal is set to the low logic level and is sufficient to put the switching network in the C0 configuration: internal analog bus lines are disconnected from the ATAP and clamped whatever the control code loaded in the update register. At the contrary, when *Mode2* is high, the switching network is configured with respect to the control codes loaded in the update register. In essence, the *Co*-bit acts as an enable signal for applying logic levels to the pins:

- when *Co*=1, the two bits *D1* and *D2* are defining which of the logic value (V_L or V_H) is applied on *AT1* and *AT2*, respectively.
- when *Co*=0, the two bits *D1* and *D2* control the *AT1-AB1* connection (switch S5) and the *AT2-AB2* connection (switch S6), respectively.

Finally, the *Ca*-bit enables the characterization of the *AT1-AB1-AT2* and *AT1-AB2-AT2* paths.

Table 2-2. Assignments between TBIC decode logic inputs and TBIC switching configurations

M_1M_2	$C_aC_oD_1D_2$	Conf	Connections	Function
0 0 (*BYPASS, SAM/PRE*)	XXXX	C0	*AB1/2* disconnected from *AT1/2* and clamped	Mission mode
1 0 (*HIGHZ*)	XXXX	C0	*AB1/2* disconnected from *AT1/2* and clamped	HIGHZ state
0 1 (*PROBE, INTEST*)	0 0 0 0	C0	*AB1/2* disconnected from *AT1/2* and clamped	Apply and/or monitor analog signals during functional mode
	0 0 0 1	C1	*AB2* connected to *AT2*, *AB1* clamped	
	0 0 1 0	C2	*AB1* connected to *AT1*, *AB2* clamped	
	0 0 1 1	C3	*AB1* connected to *AT1*; *AB2* connected to *AT2*	
1 1 (*EXTEST, CLAMP, RUNBIST*)	0 0 0 0	C0	*AB1/2* disconnected from *AT1/2* and clamped	Apply and/or monitor analog signals during test mode (extended interconnect test)
	0 0 0 1	C1	*AB2* connected to *AT2*, *AB1* clamped	
	0 0 1 0	C2	*AB1* connected to *AT1*, *AB2* clamped	
	0 0 1 1	C3	*AB1* connected to *AT1*; *AB2* connected to *AT2*	
	0 1 0 0	C4	V_L *on AT1/2*; *AB1/2* clamped	Apply logic signals during test mode (simple interconnect test)
	0 1 0 1	C5	V_L on *AT1*, V_H on *AT2*; *AB1/2* clamped	
	0 1 1 0	C6	V_H on *AT1*, V_L on *AT2*; *AB1/2* clamped	
	0 1 1 1	C7	V_H on *AT1/2*; *AB1/2* clamped	
	1 0 0 0	C0	*AB1/2* disconnected from *AT1/2* and clamped	Characterization during test mode
	1 0 0 1	C8	*AT1* connected to *AT2* via *AB2*; *AB1* clamped	
	1 0 1 0	C9	*AT1* connected to *AT2* via *AB1*; *AB2* clamped	

4.2 Analog Boundary Module

As specified in the standard, every analog function pin is required to have an Analog Boundary Module associated with it. This ABM interfaces analog primary I/O's with the core circuitry. During test execution, an ABM can react in the same manner than a DBM by:
- delivering either a high or a low logic level to the pin,
- monitoring the logic level at the pin by capturing a one-bit representation of the signal,
- putting the pin in a high impedance state.

In addition, several analog features are added to the ABM to extend the test capabilities:
- a current and/or voltage may be transmitted from the internal test bus to the pin and reciprocally,
- the pin may be connected to a highly stable reference voltage (generally the ground).

This functionality is basically implemented by means of a switching network allowing the pin to be put in the high impedance state and to be connected to dc voltages or to the analog internal test bus. Similarly to the TBIC architecture, the control of the switching network is realized through a control register that is a part of the boundary-scan chain, an update register and a decode logic. This general architecture is described in Figure 2-11.

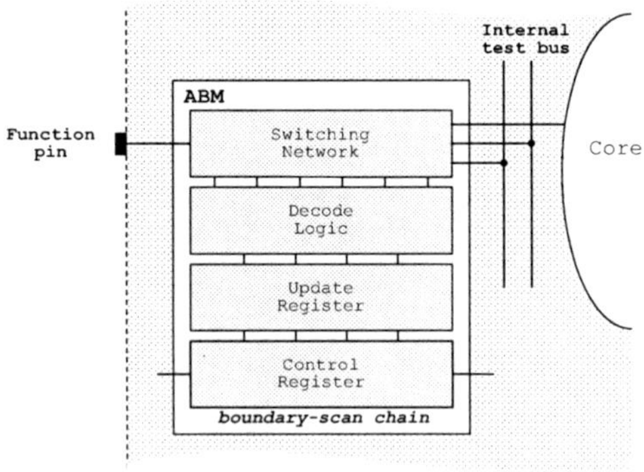

Figure 2-11. ABM general architecture

It is worth noting that as an ABM exhibits all the features of a DBM, it is possible to connect a digital function pin to an ABM. This is compliant with the standard and the way is then opened for the analog characterization of a digital signal or for the measurement of impedance, fan-out, pull-up resistor, etc. ...

4.2.1 ABM Switching Network

From the set of mandatory functions defined in the standard, a possible implementation of the ABM switching network consists of a set of six conceptual switches:

- Switches SH and SL permit to drive the pin either to V_H or V_L. These two switches together with the digitizer that monitor the pin's voltage allow to perform simple interconnect test.
- Switch SG allows the pin to be connected to the high quality reference voltage V_G. This facility will be used in extended interconnect test as discussed in the following section.
- Switches SB1 and SB2 allow the pin to the connected to the internal analog test bus lines *AB1* and *AB2*. Electronic access to the pin is then provided (through the TBIC) so that analog measurements can be performed without any physical probe. This facility can be used in both functional and test modes, addressing internal test or extended interconnect test respectively.
- Switch SD is the core disconnect facility allowing the pin to be isolated from the core while external test is taking place.

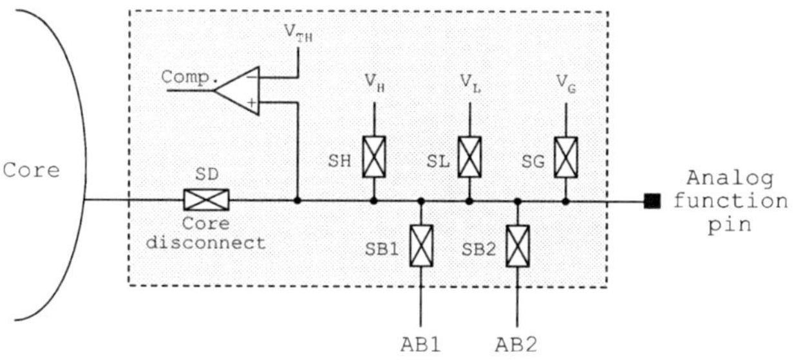

Figure 2-12. ABM general architecture

Regarding the practical implementation of the switching network, the core disconnect facility is one of the more controversial aspect of the standard. On one hand, isolation of the pin from the core during external testing is necessary to avoid interference between the core and test signals. On the other hand, the mission of the analog core shall not be affected by the core disconnect facility and one knows how difficult it is to ask a designer to add an extra component on the signal path. Finally, it comes that the core disconnect facility is mandatory for an ABM attached to an output pin, but optional for an ABM attached to an input pin. In this latter case, it is the responsibility of the designer to ensure that activity in the core does not affect measurements at the pin and vice versa.

Concerning output pins, it is strongly recommended whenever possible to integrate part of the ABM functionality into the pin driver. In that way, dedicated pin drivers for functional analog outputs may be designed with the following features:

- in functional mode, the pin driver connects the core to the pin with respect to the functional mode specifications,
- in test mode, the pin driver may drive the pin to any of the reference voltage sources (V_H, V_L or V_G), or alternatively may be placed in a high-impedance state to comply with the core disconnect facility.

It is worth noting that V_G may be one of the logic level (V_H or V_L) if their stability is sufficient to avoid errors during parametric testing. In addition, all considerations presented in the previous section regarding digitizers for the TBIC are still active for the ABM. In the same manner, voltage level of the reference sources can be chosen by the designer as identical of all pins or specific to each pin.

Table 2-3 summarizes the different useful configurations of the ABM switching network together with their associated functionality.

4.2.2 ABM Control

The ABM control circuitry allows to select one configuration of the switching network. The minimal control infrastructure required in the standard has a similar architecture to the TBIC one. As depicted in Figure 2-13, it comprises:

- A 4-bit scan register, which is a part of the boundary-scan chain. The four bits are denoted *Data* (*D*), *Control* (*C*), *Bus1* (*B1*) and *Bus2* (*B2*). The *Data* cell can capture the digitized representation of the voltage at the pin depending on the *shift_DR* signal.
- A 4-bit update register, which is parallel loaded from the corresponding stages of the control register.
- A decode logic providing the control signals for each switch from the contents of the update register together with the two mode signals supplied by the TAP controller.

Table 2-3. Minimal switching configurations of the ABM

Conf	Closed Switch (es)	Connections	Function
C0	None	Completely isolated	CD state
C1	SD	Connected to core;	Mission mode
C2	SD and SB2	Connected to core and *AB2*	Analog measurements in *PROBE* and *INTEST*
C3	SD and SB1	Connected to core and *AB1*	
C4	SD, SB1 and SB2	Connected to core and *AB1/2*	
C5	SB2	Connected to *AB2*	Analog measurements in *EXTEST*
C6	SB1	Connected to *AB1*	
C7	SB1 and SB2	Connected to *AB1/2*	
C8	SG	Connected to V_G	
C9	SL	Connected to V_L	Digital interconnect test in *EXTEST*
C10	SH	Connected to V_H	
C11	SG and SB2	Connected to V_G and *AB2*	Parametric measurements of voltage sources (V_G, V_L, V_H) in *EXTEST*
C12	SG and SB1	Connected to V_G and *AB1*	
C13	SG, SB1 and SB2	Connected to V_G and *AB1/2*	
C14	SL and SB2	Connected to V_L and *AB2*	
C15	SL and SB1	Connected to V_L and *AB1*	
C16	SL, SB1 and SB2	Connected to V_L and *AB1/2*	
C17	SH and SB2	Connected to V_H and *AB2*	
C18	SH and SB1	Connected to V_H and *AB1*	
C19	SH, SB1 and SB2	Connected to V_H and *AB1/2*	

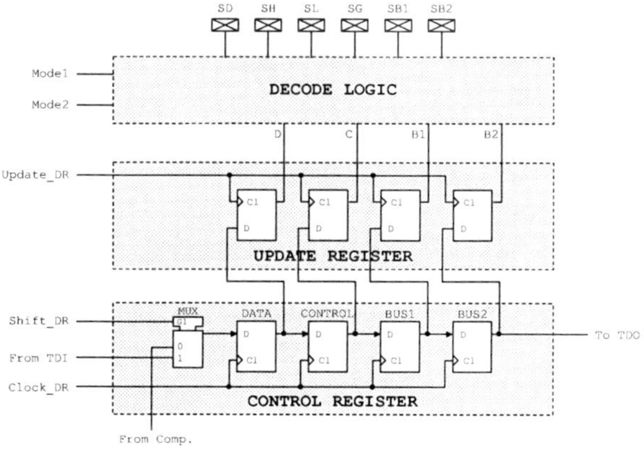

Figure 2-13. ABM control structure

The selected configuration of the switching network is determined from the contents of the update register along with the Instruction Register command. In normal mission mode as well as in *BYPASS* and *SAMPLE/PRELOAD*, the mode signals generated by the TAP controller force the pin to be connected to core and disconnected from all test circuitry whatever the control code loaded in the update register is. In the same way, the *HIGHZ* instruction constrains the pin to be fully isolated whatever the control codes loaded in the update register. For all other instructions, the contents of the update register is used by the decode logic to select the active configuration. In all cases, the *B1*-bit controls the connection of *AB1* to the pin and the *B2*-bit controls the connection of *AB2* to the pin. In test mode, the *C*-bit acts as an enable signal for applying logic levels. When *C*=1, the *D*-bit determines the logic value (V_L or V_H) applied to the pin. When *C*=0, the *D*-bit determines whether V_G is connected to the pin. Table 2-4 summarizes the mandatory assignments between the six control inputs of the decoder logic and the twenty ABM configurations.

5. TEST METHODOLOGIES USING IEEE 1149.4

This section introduces system test methodologies using the IEEE 1149.4 Mixed-Signal Test Bus Standard. The usage of the 1149.4 test infrastructure is illustrated through applications in interconnect test, parametric test and internal test. Note that before the 1149.4 hardware can be used to test any interconnect, external impedance or core, it have to be verified. A detailed description of how to validate this hardware by structural test methods is given in [3].

5.1 Interconnect Test

The first class of tests addressed by the standard is simple interconnect testing. A simple interconnect basically consists in a conductor connecting two or more pins without external component inserted in the interconnect pathway. Note that this conductor may connect digital pins, analog pins or analog and digital pins. The IEEE 1149.4 standard provides facilities and instructions for testing shorts, opens and stuck-at faults in simple interconnects, covering both analog and digital pins.

Table 2-4. Assignments between ABM decode logic inputs and ABM switching configurations

M_1M_2	DCB_1B_2	Conf	Connections	Function
0 0 (*BYPASS, SAM/PRE*)	XXXX	C1	Connected to core; Isolated from test circuits	Mission mode
1 0 (*HIGHZ*)	XXXX	C0	Completely isolated	*HIGHZ* state
0 1 (*PROBE, INTEST*)	0 0 0 0	C0	Completely isolated	Apply and/or monitor analog signals during functional mode
	0 0 0 1	C2	Connected to core and *AB2*	
	0 0 1 0	C3	Connected to core and *AB1*	
	0 0 1 1	C4	Connected to core *AB1/2*	
1 1 (*EXTEST, CLAMP, RUNBIST*)	0 0 0 0	C0	Completely isolated	Apply and/or monitor analog signals or apply reference voltage V_G during test mode (extended interconnect test)
	0 0 0 1	C5	Connected to *AB2*	
	0 0 1 0	C6	Connected to *AB1*	
	0 0 1 1	C7	Connected to *AB1/2*	
	0 1 0 0	C8	Connected to V_G	
	1 0 0 0	C9	Connected to V_L	Apply logic signals in test mode (simple interconnect test)
	1 1 0 0	C10	Connected to V_H	
	0 1 0 1	C11	Connected to V_G and *AB2*	Characterization of reference voltages (V_G, V_L, V_H) during test mode
	0 1 1 0	C12	Connected to V_G and *AB1*	
	0 1 1 1	C13	Connected to V_G and *AB1/2*	
	1 0 0 1	C14	Connected to V_L and *AB2*	
	1 0 1 0	C15	Connected to V_L and *AB1*	
	1 0 1 1	C16	Connected to V_L and *AB1/2*	
	1 1 0 1	C17	Connected to V_H and *AB2*	
	1 1 1 0	C18	Connected to V_H and *AB1*	
	1 1 1 1	C19	Connected to V_H and *AB1/2*	

The basic principle for interconnect testing consists in driving a logic signal on one pin of the interconnect (the driving node) and verifying the logic signal received at another pin of the same interconnect (the sensing node). With the 1149.4 infrastructure, interconnect testing is performed by loading test data into the boundary modules associated with driving nodes and capturing test responses into the boundary modules associated to sensing nodes. In case of digital pins, the test data is directly loaded into the DBMs while in case of analog pins, test data loaded in the ABMs consists of control codes for driving V_H and V_L voltage levels to the pin. In the same way, test response captured into the DBMs directly correspond to logic signal present at the pins while test response captured into the ABMs corresponds to the

1-bit digitization of the analog signal present at the pin. Despite these differences, the test procedure is strictly identical to that used in 1149.1.

5.2 Parametric Test

The second class of tests addressed by the standard is parametric test, which is the main additional feature of the dot 4 extension in terms of capability. Indeed, simple interconnect test is not sufficient for mixed-signal applications where discrete passive components such as resistance, capacitance and inductance may be placed in the signal path between analog pins. To avoid the use of an in-circuit tester, the IEEE 1149.4 introduces the possibility of performing accurate impedance measurements. Fundamentally, the idea is to measure the impedance of an element by injecting a current through this element to a DC voltage and measuring the voltages at the two points of the element.

To illustrate this approach, let us assume a discrete component (Z_{cut}) connected between two analog pins (P1 and P2) belonging to the same chip, or more generally to a couple of chips that comply with the dot 4 standard. The test setup for measuring the impedance Z_{cut} is illustrated in Figure 2-14, where only the TBIC and ABM switches involved in the test are represented:

- an ATE is connected to the board connector in order to apply a current on *AT1* and to monitor a voltage on *AT2*,
- the current is carried out to P1 via a path composed of switch S5 of the TBIC, the internal bus line *AB1* and finally switch SB1 of the ABM connected to P1,
- the analog pin P2 is connected to the quality reference voltage V_G via the switch SG of its associated ABM.

The measurement itself is then a two step procedure. Indeed, both analog pins P1 and P2 can be connected to *AT2*, but only one at a time. The voltage measurement path flowing through switch SB2, the analog internal test bus line *AB2* and finally switch S6 in the TBIC is then activated successively in both chips to measure V_{P1} and V_{P2} (Figure 2-14.a and b respectively). Assuming that the internal impedance of the voltage measurement resource is greater than the impedance to be measured and the impedance of the switches, both voltage readings effectively correspond to the actual voltage on each analog pin. Following the same assumption, the complete current is flowing through the impedance under test and finally it comes:

$$Z_{CUT} = \frac{V_{P1} - V_{P2}}{I_t} \tag{4}$$

Similar procedures may be established for measuring:

An impedance directly connected between an analog pin and the ground. In that case V_{P2} is equal to zero and must not be measured.

All elements of a more complex network such as an attenuator, a filter and so on. In that case, a multi-step procedure allows to calculate all the elements of the network by solving a set of equations with the same number of unknowns.

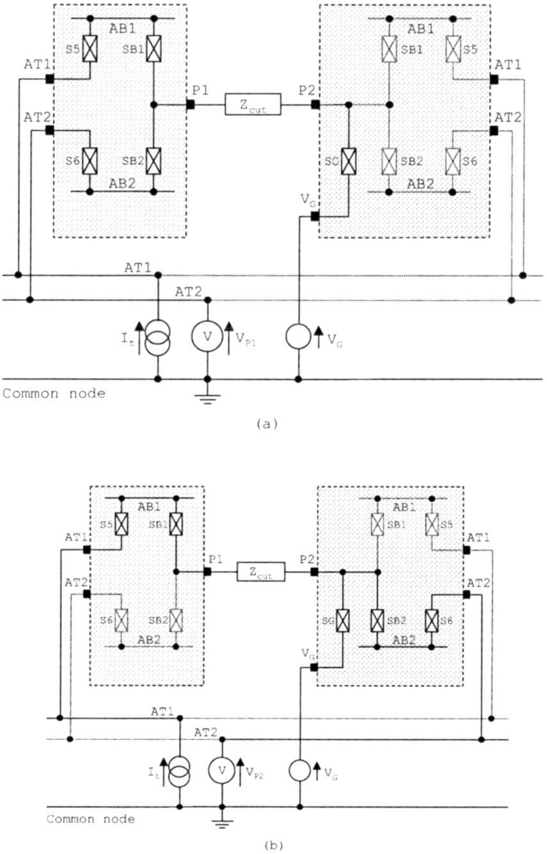

Figure 2-14. Measurement of external impedance using 1149.4 (active path in solid line)

5.3 Internal Test

Finally, the last class of tests addressed by the standard is internal test. The IEEE 1149.4 standard allows to test the internal circuitry of a digital core as it would be the case with a 1149.1-compliant IC, but also adds some facilities for the test of an analog core. In particular, the analog test bus can be used to supply analog stimulus to input pins and monitor the analog voltage present on output pins.

Figure 2-15 shows a typical analog core test. Since there is normally only one pair of analog test bus lines, only one of the analog input pins is supplied through AT1 and only one of the analog output pins is fully monitored through AT2. Other analog input pins remain connected to the external circuitry on the board and should be held quiescent for a meaningful test. In addition, input signals normally supplied by the digital core are determined by DBMs fitted at the interface between the digital and analog cores. Regarding test response monitoring although only one analog output is fully observed, useful information can be obtained from the digital test results captured in the DBMs at the analog/digital interface and the digitized results captured in the control register of all ABMs. If further information is required, the test can be repeated for the same stimulus conditions but monitoring each analog output in turn.

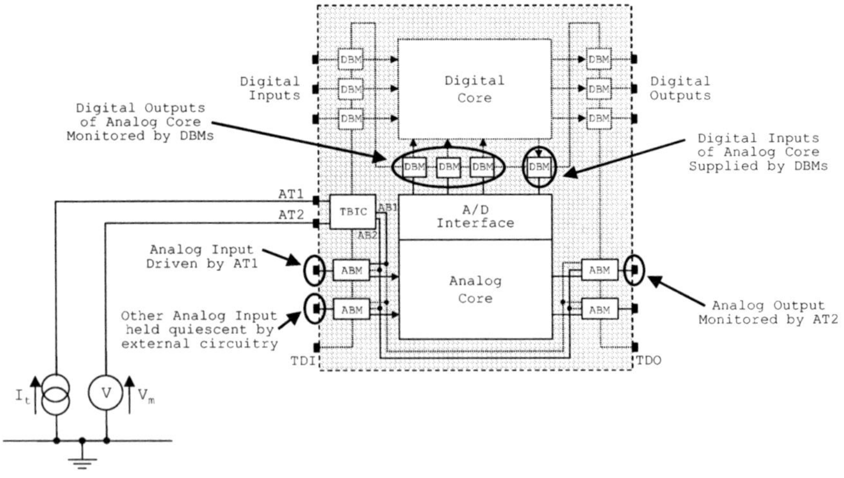

Figure 2-15. Analog core test using 1149.4 (active path in solid line)

BIBLIOGRAPHY

1. "IEEE Standard Test Access Port and Boundary-Scan Architecture", IEEE Standard 1149.1-1990. *IEEE Standards Board*, 345 East 47th St. New York, NY 10017, 1990.
2. "IEEE Standard for a Mixed Signal Test Bus", IEEE Standard 1149.4-1999. *IEEE Standards Board*, 345 East 47th St. New York, NY 10017, 1999.
3. "Analog and Mixed-Signal Boundary-Scan - A Guide to the IEEE 1149.4 Test Standard", Edited by Adam Osseiran, *Kluwer Academic Publishers*, Boston, 1999.

Chapter 3

TEST OF A/D CONVERTERS

From converter characteristics to built-in self-test proposals

Andreas Lechner & Andrew Richardson
University of Lancaster, UK

Abstract: Converter testing is a complex process. Test specifications are application dependent and in most cases require sophisticated stimulus generation and, response analysis equipment. The processing of sampled data to generate specification measurements also requires complex DSP algorithms. Furthermore, test complexity increases with speed, resolution and technology advances. This chapter summaries the key test specifications, conventional methods of verifying these specifications and potential built-in self-test solutions.

1. INTROCUTION

Analogue-to-digital (A/D) converters are one of the most frequently used mixed-signal functions creating an interface between sensing and actuation devices in the industrial control, transportation, consumer electronics and instrumentation industries and conversion of analogue voice and video data in the computing and communications communities. In control applications, the trend is towards medium speed (10-100 kHz) and high resolution (>16 bits) with test requirements focused towards linearity testing. In communications applications, trends are similar, however, dynamic performance tends to be critical, especially in voice processing applications. Consumer goods are another important application where high conversion speed (up to 100's of MHz) and low-medium resolution (8-12 bits) is the norm.

Of interest therefore for the test engineer is the optimization of test programs towards verification of the key application specifications. The

J.L. Huertas (ed.), Test and Design-for-Testability in Mixed-Signal Integrated Circuits, 73–98.

problems include the range of specifications requiring verification and the very different implementations used (e.g. ΣΔ vers. flash). Today, most test strategies are based on DSP testing where a known stimulus is injected into the device and its digital output processed using, for example, fast Fourier transform techniques to extract dynamic specifications, such as total harmonic distortion and inter-modulation distortion. The cost of implementing these measurements is becoming excessive. As an example, to measure signal-to-noise and distortion ratio of a converter with a 90 dB resolution will typically require around 8,000 samples which translates to around 0.1 to 0.2 seconds of test time when acquisition and processing tasks are taken into account. Total harmonic distortion measurements to an 80 dB resolution will require double this. Taking into account that in many cases these measurements must be carried out for two or more gain settings and possible input signal amplitudes, test time can rapidly grow to several seconds. These estimates should be considered in the context of total test time for a system-on-chip or mixed signal IC which ideally needs to be below a second. Note also that the converter will generally occupy only a small area of the device.

These issues illustrate the importance of utilizing the best possible methods available for converter testing. This chapter will present not only the conventional techniques for testing converters but a selection of new ideas and test implementations that target both test time reduction and the demand for high cost analogue test equipment.

2. A/D CONVERSION

The process of A/D conversion is a transformation of a continuous analog input waveform into a sequence of discrete digital words. An A/D converter usually requires a sample-and-hold operation at the front end of the signal path to ensure that the analog signal remains constant during the conversion process. The conversion process itself varies between different A/D converter architectures, that include serial, folding, ΣΔ and interpolating A/D converters [1, 2].

The converter's transfer characteristics are commonly represented by a staircase function. Figure 3-1 illustrates the transfer function of an ideal N-bit A/D converter, where N is the converter's resolution. This corresponds to the number of digitized bits. For linear converters the full-scale (FS) input range from V_{min} to V_{max} is segmented into 2^N equally sized bins (so called code bins) of nominal width Q, also frequently referred to as 1 LSB (least significant bit):

$$Q = \mathrm{FS}/2^N = (V_{\max} - V_{\min})/2^N = 1\,\mathrm{LSB} \quad (1)$$

The ideal code bin width Q, usually given in volts, may also be given as a percentage of the full-scale range. By standard convention, the first code bin starts at voltage V_{min} and is numbered 0, followed by the first code transition level $T[1]$ to code bin 1, up the last code transition level $T[2^N\text{-}1]$ to the highest code bin $[2^N\text{-}1]$ which reaches the maximum converter input voltage V_{max} [3]. In the ideal case, all code bin centers fall onto a straight line with equidistant code transition levels, as illustrated in Figure 3-1. The analog equivalent of a digital A/D converter output code k corresponds to the location of the particular ideal code bin center V_k on the horizontal axis.

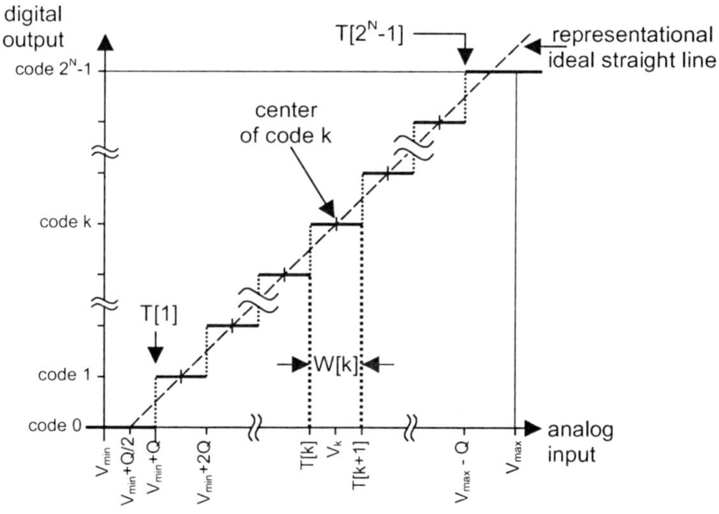

Figure 3-1. Ideal *N*-bit A/D converter transfer function

The quantization process itself introduces an error corresponding to the difference between the A/D converter's analog input and the equivalent analog value of its output, which is depicted over the full-scale range in Figure 3-2. With an rms value of $Q/\sqrt{(12)}$ for a uniform probability distribution between $Q/2$ and $-Q/2$ and an rms value of $FS/(2\sqrt{(2)})$ for a full-scale input sine wave, the ideal or theoretical signal-to-noise ratio (SNR) for an N-bit converter can be given in decibels as:

$$\mathrm{SNR}_{ideal}(\mathrm{dB}) = 10\log_{10}\left[\frac{\left(FS/2\sqrt{2}\right)^2}{\left(Q/\sqrt{12}\right)^2}\right] = 10\log_{10}\left[\left(2^N\sqrt{\frac{12}{8}}\right)^2\right]$$

$$= 20\log_{10}\left[2^N\right] + 20\log_{10}\left[\sqrt{\frac{12}{8}}\right] = 6.02N + 1.76 \qquad (2)$$

For real A/D converters, further errors affect the conversion accuracy and converter performance. The following sections will introduce the main static and dynamic performance parameters that are usually verified to meet the specifications in production testing. Standardized performance parameters associated with the A/D converter transient response and frequency response can be found in [3].

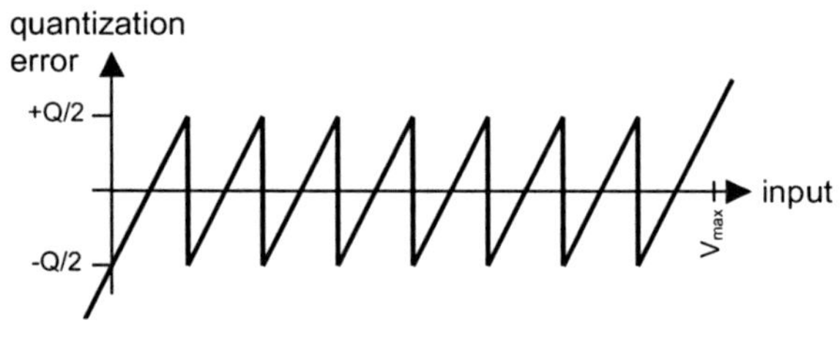

Figure 3-2. A/D conversion quantization error

2.1 Static A/D Converter Performance Parameters

Apart from the systematic quantization error due to finite converter resolution, A/D converters have further static errors mainly due to deviations in transition levels from the ideal case and are affected by internally and externally generated noise. One of the characteristic parameters which can indicate conversion errors is the real code widths. A particular code bin width, *W[k]*, can be determined from its adjacent code transition levels *T[k]* and *T[k+1]*, as indicated in Figure 3-1:

$$W[k] = T[k+1] - T[k] \quad \text{for } 1 \le k \le 2^N - 2 \quad (3)$$

where code transition level $T[k]$ corresponds to the analog input voltage where half the digital outputs are greater or equal to code k while the other half are below code k.

In addition to assessment of converter performance from transition levels and code bin widths, the real transfer function may also be approximated by a straight line for comparison with the ideal case. The straight line can be determined though a linear regression computation where the regions close to the upper and lower end of the transfer function are ignored to avoid data corruption due to overdriving the converter (input voltage exceeds the real full-scale range). The following main static performance parameters are introduced and described below: gain and offset, differential non-linearity and integral non-linearity.

The basic effect of *offset* in A/D converters is frequently described as a uniform lateral displacement of the transfer function, while a deviation from ideal *gain* corresponds to a difference in the transfer function's slope after offset compensation. With regard to performance verification and test, offset and gain can be defined as two parameters, V_{OS} and G, in a straight line fit for the real code transition levels, as given on the left hand side in equation (4) [3, 4]. The values for offset and gain can be determined through an optimization procedure aiming at minimum matching error $\varepsilon(k)$ between gain and offset adjusted real transition levels and the ideal values (right side of equation (4)).

$$G \times T[k] + V_{OS} + \varepsilon[k] = Q \times (k-1) + T[1]_{ideal} \quad \text{for } 1 \le k \le 2^N - 1 \quad (4)$$

where G is the gain, V_{OS} the offset, Q the ideal code bin width, $T[1]_{ideal}$ the ideal first transition level, and $T[k]$ the real transition level between codes k and k-1. The value for V_{OS} corresponds to the analogue equivalent of the offset effect observed at the output.

However, different optimization techniques yield slightly different values for offset, gain and the remaining matching error. For example the matching may be achieved through mean squared value minimization for $\varepsilon(k)$ for all k [3], alternatively, the maximum of the matching errors may be reduced. Simpler offset and gain measurements are often based on targeting an exact match in equation (4) for the first and last code transition levels, $T[1]$ and $T[2^N-1]$ ($\varepsilon(1)$ and $\varepsilon(2^N-1)$ equal to zero) referred to as *terminal based* offset and gain. An example for this case is illustrated in Figure 3-3. An alternative methodology is to employ the straight line approximation of the real transfer function mentioned above. Offset and gain values are then determined through matching this real straight line with the ideal straight line, which again can deviate slightly from the optimization process results [3].

Differential nonlinearity (DNL) is a measure of the deviation of the gain and offset corrected real code widths from the ideal value. DNL values are given in LSBs for the codes 1 to (2^N-2) as a function of k as:

$$\mathrm{DNL}[k](\mathrm{LSB}) = \frac{(W[k]-Q)}{Q} \qquad \text{for } 1 \le k \le 2^N - 2 \quad (5)$$

where $W[k]$ is the width of code k determined from the gain and offset corrected code transition levels as given in equation (3), and Q the ideal code bin width. Note that neither the real code bin widths nor the ideal value are defined at either end of the transfer function. As an example, a differential nonlinearity of approximately +¼ LSB in code m is included Figure 3-3. The *absolute* or *maximum DNL* corresponds to the maximum value of |DNL[k]| over the range of k given in equation (5). A value of -1 for DNL[k] corresponds to a *missing code*.

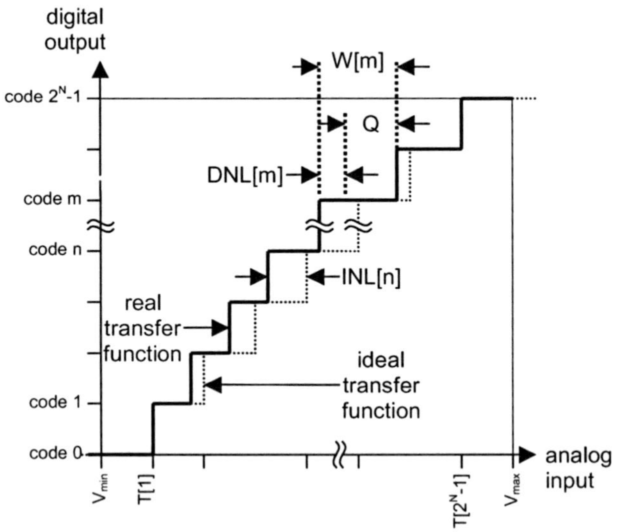

Figure 3-3. Terminal-based DNL and INL in A/D converter transfer function

Integral nonlinearity (INL) quantifies the absolute deviation of a gain and offset compensated transfer curve from the ideal case. INL values are given in LSBs at the code transition levels as a function of k:

$$\mathrm{INL}[k](\mathrm{LSB}) = \frac{\varepsilon[k]}{Q} \qquad \text{for } 1 \le k \le 2^N - 2 \qquad (6)$$

where $\varepsilon(k)$ is the matching error from equation (4) and Q the ideal code bin width, both given in volts. Alternatively, INL[k] values can also be given as a percentage of the full-scale input range. As an example, Figure 3-3 also depicts a integral nonlinearity of approximately +⅔ LSB in code n. Plots of INL[k] values over k provide useful information on the converter performance, as the overall shape of the INL[k] curve enables some initial conclusions on the predominance of even- or odd-order harmonics [5]. However, the exact values for the INL do depend on the type of gain and offset correction methodology applied, which should be documented. The *absolute* or *maximum INL*, usually provided as an A/D converter specification, corresponds to the maximum value of |INL[k]| over the range of k given in equation (6).

Some additional performance characteristics are defined for A/D converters. An A/D converter is said to be *monotonic*, if the output is either consistently increasing or decreasing with an either consistently increasing or decreasing input. If the change at input and output are of the same direction, the converter is *non-inverting*. When the change at the input and output are of opposite direction, the converter is *inverting*. An A/D converter can also be affected by *hysteresis*. This is a condition where the computation of the transfer function yields different results for an increasing and a decreasing input stimulus that are beyond normal measurement uncertainties. For more details see [3].

2.2 Dynamic A/D Converter Performance Parameters

A/D converter performance is also expressed in the frequency domain. This section introduces the main dynamic performance parameters associated with the converter's output spectrum, while the determination of their values in converter testing described in section 3.4.

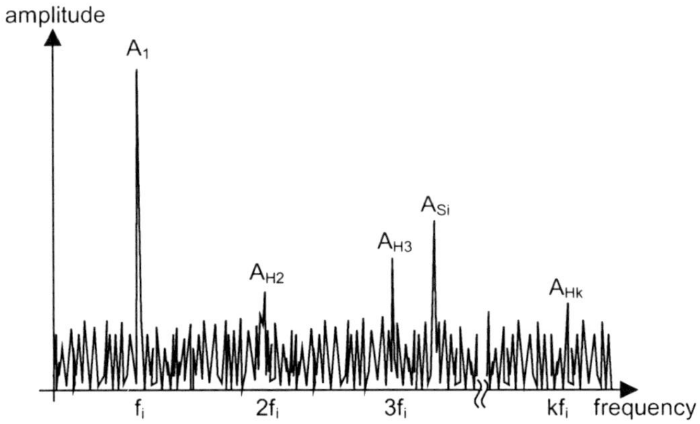

Figure 3-4. A/D converter output spectrum

Figure 3-4 illustrates an A/D converter output spectrum, a plot of frequency component magnitude over a range of frequency bins. Such a spectrum can be obtained from a spectrum analyzer or a discrete Fourier transform (DFT) [6] through analysis of the A/D converter response to a spectrally pure sine wave input of frequency f_i. The original input signal can be identified as the fundamental of amplitude A_1. The second to k^{th} harmonic distortion component, A_{H2} to A_{Hk}, occur at non-aliased frequencies that are integer multiples of f_i. Additionally, non-harmonic or spurious components, such as A_{Si} in Figure 3-4, can be seen at other than the input signal or harmonics frequencies. The main dynamic performance parameters given below can be extracted from the output spectrum in the form of ratios of rms amplitudes of particular spectral components, which also relates to signal power ratios. The calculation of these from the results of a DFT is outlined in section 3.4.1. Note that the input signal frequency and amplitude, and in some cases the sampling frequency, have an impact on the actual performance parameter value, and have to be provided with test results and performance specifications.

The *signal-to-noise and distortion ratio* (SINAD) relates the input signal to the noise including harmonics. The SINAD can be determined from rms values for the input signal and the total noise (including harmonics), which also relates to the power, P, carried in the corresponding signal component. The SINAD is given in decibels as:

$$\mathrm{SINAD(dB)} = 20\log_{10}\left[\frac{\mathrm{rms}\left(signal\right)}{\mathrm{rms}\left(totalnoise\right)}\right] = 10\log_{10}\left[\frac{P_{signal}}{P_{totalnoise}}\right] \quad (7)$$

The *effective number of bits* (ENOB) compares the performance of a real A/D converter to the ideal case with regard to noise [7]. The ENOB is determined through:

$$\mathrm{ENOB} = N - \log_2\left[\frac{\mathrm{rms}\left(totalnoise\right)}{\mathrm{rms}\left(idealnoise\right)}\right] \quad (8)$$

where N is the number of bits of the real converter. In other words, an ideal A/D converter with a resolution equal to the ENOB determined for a real A/D converter will have the same rms noise level for the specified input signal amplitude and frequency. The ENOB and SINAD performance parameters can be correlated to each other as analyzed in [3].

Total harmonic distortion (THD) is a measure of the total output signal power contained in the second to k^{th} harmonic component, where k is usually in the range from 5 to 10 (depending on the ratio of the particular harmonic distortion power to the random noise power) [8]. The THD can be determined from rms values of the input signal and the harmonic components and is commonly expressed as the ratio of the powers in decibels:

$$\mathrm{THD(dB)} = 20\log_{10}\left[\sqrt{\sum_{i=2}^{k} A_{Hi(rms)}^2} \Big/ A_{1(rms)}\right] = 10\log_{10}\left[\frac{P_{harmonic}}{P_{input}}\right] \quad (9)$$

where $A_{1(rms)}$ is the rms for the signal and $A_{Hi(rms)}$ the rms for the ith harmonic. THD is given in decibels and usually with respect to a full-scale input (dBFS). Where the THD is given in dBc, the unit is in decibels with respect to a carrier signal of specified amplitude.

The *spurious free dynamic range* (SFDR) quantifies the available dynamic range as a ratio of the fundamentals amplitude to the amplitude of the largest harmonic or spurious component and is given in decibels:

$$\mathrm{SFDR(dB)} = 20\log_{10}\left[\frac{A_1}{\max\left\{A_{H(\max)}, A_{S(\max)}\right\}}\right] \quad (10)$$

where $A_{H(max)}$ and $A_{S(max)}$ are the amplitudes of the largest harmonic component and spurious component, respectively.

While the dynamic performance parameters introduced above are essential for an understanding of A/D converter test methodologies (section 3), an entire range of further performance parameters is included in the IEEE standard 1241 [3], such as various signal-to-noise ratios specified for particular bandwidths or for particular noise components. Furthermore, some performance parameters are defined to assess intermodulation distortion in A/D converters with a two-tone or multiple-tone sine wave input.

3. A/D CONVERTER TEST APPROACHES

This section introduces A/D converter test methodologies, for static and dynamic performance parameter testing. The basic test setup and other prerequisites are briefly described in the next section. For further reference, an introduction to production test of integrated circuits, ranging from test methodologies and design-for-test basics to aspects relating to automatic test equipment can be found in [9]. Details on DSP-based testing of analog and mixed-signal circuits are provided in [10, 11].

3.1 Setup for A/D Converter Test

The generic test setup is illustrated in Figure 3-5. In generic terms, a suitable stimulus supplied by a test source is applied to the A/D converter under test via some type of test access mechanism. The test stimulus generator block corresponds to one or more sine wave, arbitrary waveform or pulse generator(s) depending on the type of test to be executed. Generally, the response is captured for processing in a test sink again facilitating a test access mechanism.

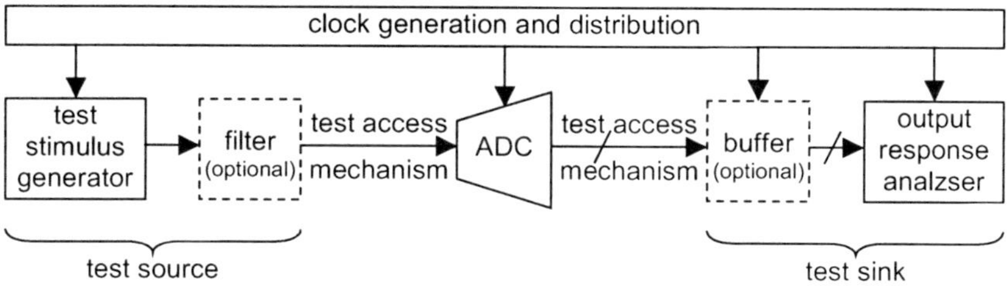

Figure 3-5. A/D converter test setup

In a conventional A/D converter test setup, test source and test sink are part of the external automatic test equipment (ATE) and are centrally controlled. The ATE interfaces with the IC via a device interface board; functional IC input/output pins and IC-internal interconnects may be facilitated as a test access mechanism. However, in the majority of cases some other means of test access has to be incorporated in the early stages of IC design due to access restrictions, such as limited pin count or converters being deeply embedded in a complex system-on-chip (SOC). Systematic design methodologies which increase test access, referred to as design-for-testability (DfT), are standardized at various system levels. The IEEE standard 1149.1, also known as boundary scan, supports digital IC and board level tests [12]. Its extension to analog and mixed-signal systems, IEEE standard 1149.4, adds an analog test bus to increase access to analog IC pins and internal nodes [13]. For SOC implementations, an IEEE standard for interfacing to IC-internal subsystems, so-called embedded cores, and documentation of their test requirements is expected to be approved in the near future [14].

3.2 Capturing the Test Response

The action of collecting a set of A/D converter output samples and transferring it to the output response analyzer is commonly referred to as taking a data record. The aim is to accumulate consecutive samples, however, for high-speed converters interfacing restrictions may require output decimation [3]. This is a process in which only every i^{th} sample of a consecutive sequence is recorded at a lower speed than the A/D converter's sampling speed.

On the other hand, the A/D converter maximum sampling frequency restricts the rate at which a waveform can be digitized and therefore the measurement bandwidth. When sampling periodic waveforms, it is generally desirable to record an integer number of waveform periods while not re-sampling identical points at different cycles. This can be assured by applying coherent sampling, equation (11), where additionally the number of samples in the record, *M*, and the number of cycles, *N*, are in the ratio of relative prime numbers [10].

$$f_i = f_s * N/M \tag{11}$$

Where f_s is the sampling frequency and f_i the input waveform frequency.

With such a sampling scheme it is possible to systematically rearrange consecutive samples taken over multiple periods to an equivalent data record for a single period taken at higher speed. This technique is called *equivalent time sampling* or time shuffling and is illustrated in the Figure 3-6.

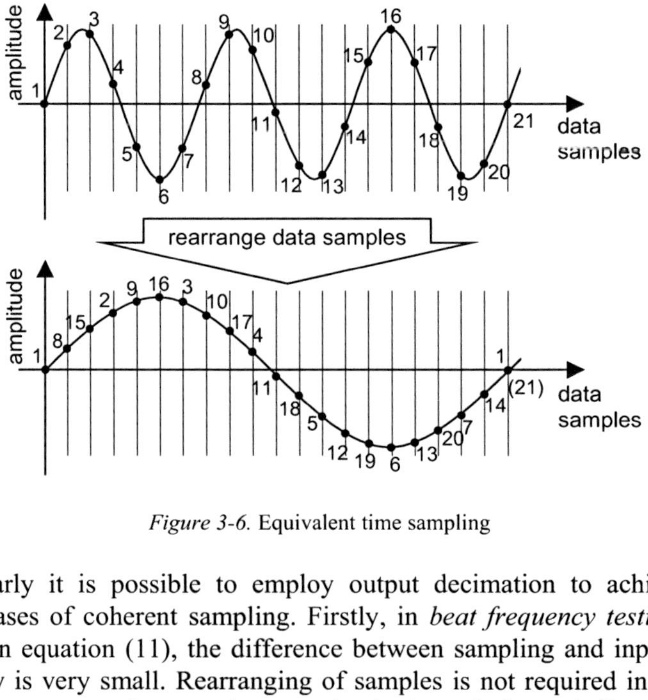

Figure 3-6. Equivalent time sampling

Similarly it is possible to employ output decimation to achieve two special cases of coherent sampling. Firstly, in *beat frequency testing*, with $N=M+1$ in equation (11), the difference between sampling and input signal frequency is very small. Rearranging of samples is not required in the data record where successive samples step slowly through the periodic waveform as illustrated in Figure 3-7a. Secondly, in *envelope testing*, where $N=1+M/2$ and M is a multiple of 4, the sampling frequency is nearly twice the input frequency stepping sequentially through both halves of the input waveform phases as illustrated in Figure 3-7b.While the latter sampling schemes allow a quick visualization of the waveforms shape, the sampling techniques introduced can be employed in the test methodologies described in the next sections.

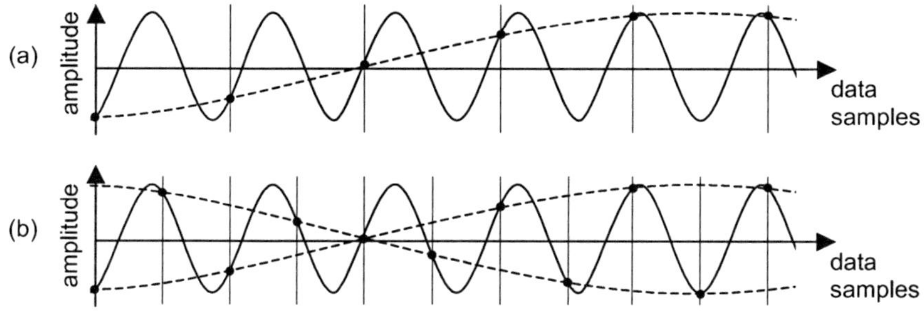

Figure 3-7. (a) Beat frequency testing and (b) envelope testing

3.3 Static Performance Parameter Test

A/D converter performance can be verified in terms of static performance parameters, introduced in section 2.1, through assessment of the transfer function. One way to compute the transfer function is to apply a DC voltage to the A/D converter that is stepping through an analog voltage range slightly larger than the full-scale input range. For each step a number of data pairs (input voltage / output code) have to be computed. The transfer function can then be approximated as the curve of the mean output code over corresponding input voltage. The input voltage step size and the number of output codes averaged for each input voltage step depends on the ideal code bin width, the level of random noise and the required measurement accuracy, which can be assessed through computation of the standard deviation of the output.

In static performance production test, however, the use of continuous signals is more desirable. The following sections introduce two A/D converter test methodologies widely in use today which measure code transition levels, namely feedback-loop testing and histogram testing [3, 10, 11, 15]. From those values, static performance parameters are determined as described in section 2.1 which can then be compared to the test thresholds.

3.3.1 Feedback-loop Test Methodology

In 1975 Corcoran et al. published a test methodology for A/D converters that incorporates a feedback loop to force the A/D converter input voltage to oscillate around a desired code transition level [16]. On the test source side, an analog integrator is employed which continuously integrates either a positive or a negative reference voltage, V_{ref+} and V_{ref-}, for stimulus generation (Figure 3-8a). The reference voltage to integrate is toggled depending on a comparison result between the A/D converter's output code, C, and a set desired output code, D, after each conversion. If $C < D$, the positive reference voltage is connected to the analogue integrator to set a positive slope in the test stimulus. If $C > D$, the negative reference voltage is chosen to obtain a negative slope in the test stimulus. Once the input stimulus has reached the desired code transition level $T[D]$, the feedback from the digital comparator enforces oscillation around $T[D]$ at the converter input. Measuring the average voltage at the A/D converter input yields the value of the particular code transition level.

Several further adaptations of this technique, also referred to as servo testing, have been described in literature and were also included in IEEE standards [3, 17]. For example, the test source can be based on D/A conversion of a counter output, as illustrated in Figure 3-8b, where the D/A converter resolution is larger than the A/D converter resolution. Instead of toggling an analog reference voltage, the content of the counter can be incremented or decremented to obtain a positive or negative slope in the test stimulus. Alternatively, an accumulator may be chosen instead of a counter which allows increasing or decreasing the D/A converter input by an adjustable amount [18]. In either case, additional low-pass filtering may be incorporated to increase the effective test stimulus resolution. The test evaluation may be performed by measuring the average A/D converter input voltage, as above, or by extrapolating that voltage from the average D/A converter input through digital processing.

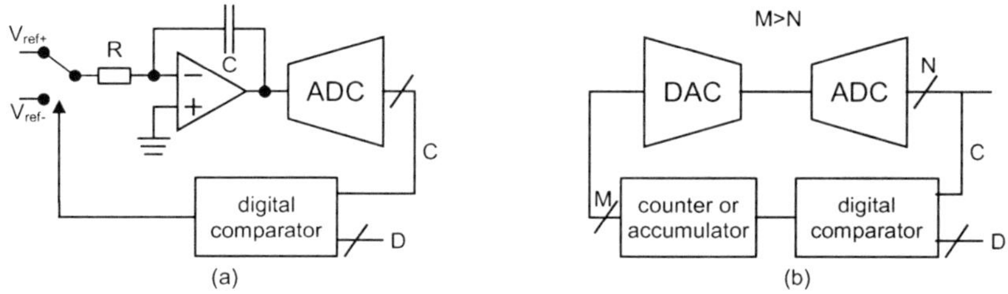

Figure 3-8. Feedback-loop configurations, (a) integrator (b) D/A conversion

The test can be automated easily. Starting at the first code transition level with $D = 1$, the desired output code D is sequentially incremented each time the average A/D converter input corresponding to $T[D]$ has been determined. The loop is continued until the last transition level with $D = 2^N-1$ is computed. If test time restrictions prevent measurement of all transition levels, a basic linearity test may be performed for code words formed by a set number of most significant bits while DNL testing is only applied to a limited number of codes [19].

In feedback-loop A/D converter testing, stimulus accuracy is crucial; the positive and negative slope rates must especially be equal, as any mismatch directly affects the ratio of code occurrences at either side of the code transition level. Also, the effect of test stimulus slope (and step size in D/A conversion) on the dynamics of the feedback loop must be analyzed in more detail [5, 15, 18]. Regarding test application time, the number of conversions required to reach a particular code transition level and to settle the feedback loop has to be assessed. This is especially true for high-speed A/D converters, where the conversion delay may be smaller than the delay through the feedback loop. Stable oscillation around a code transition level may not be achievable in this case.

3.3.2 Histogram Test Methodology

In histogram testing, A/D converter code transition levels are not measured directly but determined through statistical analysis of converter activity [20]. For a known periodic input stimulus, the histogram of code occurrences (code counts) is computed over an integer number of input waveform periods. There are two types of histograms employing test stimuli of different characteristic [21]. Firstly, there is the *ramp histogram*, also called linear histogram, computed for a linear - typically triangular – waveform . Secondly, the *sine wave histogram*, frequently referred to as

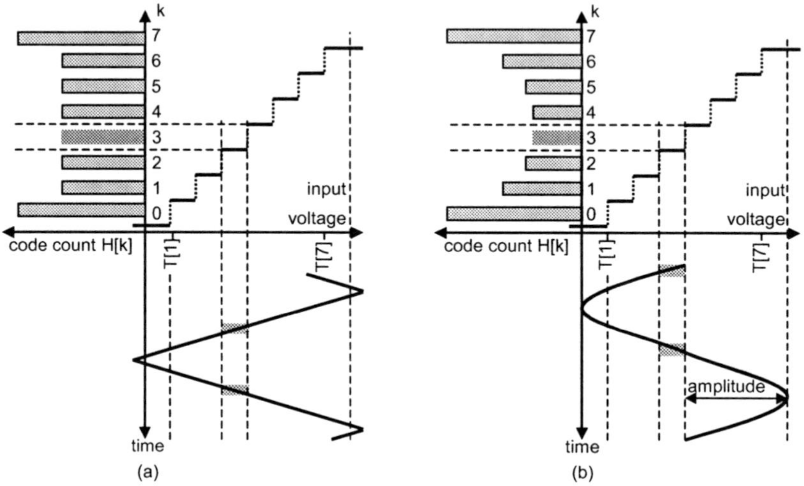

Figure 3-9. Histogram generation (a) linear (b) sine wave

dynamic histogram, is collected for a sinusoidal input waveform. The computation is illustrated for both types for an ideal 3-bit converter in Figure 3-9. Note that in the illustration the triangular waveform overdrives the A/D converter causing higher code counts in *H[0]* and *H[7]*, while the sinusoidal wave only touches the boundaries of the full-scale range.

Generally, histograms support analysis of the converter's static performance parameters. A missing code *m* is easily identified as the corresponding code count *H[m]* is equal to zero. Also offset is easily identified as a shift in the code counts and gain directly relates to average code count. However, the converter linearity is assessed via the determination of code transition levels.

For ramp histograms, where ideal values for *H[2]* to $H[2^N\text{-}2]$ are equal, code transition levels can be given as in the first part of equation (12), where *C* is an offset component and *A* a gain factor which is multiplied with the accumulated code count up to the transition to code *k* [3]. As the widths of the extreme codes, 0 and 2^N-1, cannot be defined, their code counts are usually set to zero ($H[0] = H[2^N\text{-}1] = 0$). In these cases, *C* and *A* can be determined as shown in equation (12), where the first code transition level, *T[1]*, is interpreted as the offset component. The gain factor defines the proportion of the full-scale input range for a single sample in the histogram, where H_{tot} is the total number of samples in the entire histogram.

$$T[k] = C + A\sum_{i=0}^{k-1} H[i] = T[1] + \frac{T[2^N - 1] - T[1]}{H_{tot}} \sum_{i=0}^{k-1} H[i] \tag{12}$$

For sine wave histograms, code transition levels have to be computed differently as ideal values for *H[2]* to *H[2^N-2]* are not equal [22]. With a stimulus $v[t] = A*\sin[\omega t+\phi]+C$, the transition levels can be given as [3, 4]:

$$T[k] = C - A\cos\left[\left(\pi\sum_{i=0}^{k-1} H[i]\right) \Big/ H_{tot}\right] \tag{13}$$

where offset component *C* and gain factor *G* correspond to the input sine wave's offset and amplitude.

For either type of histogram, high stimulus accuracy is essential as most deviations (ramp linearity, sine wave distortion) have a direct impact on the test result. For high frequency stimuli, tests may also detect dynamic converter failure. The choice of sine wave histograms can be advantageous, as stimulus verification and high frequency signal generation are easier to achieve [11]. An advantage of ramp histograms is that generally a lower number of samples is required due to constant ideal code counts. The number of code counts is an important test parameter, as it is directly proportional to test application time and depends on required test accuracy and confidence level. In [23] an equation is derived for the number of samples required for random sampling of a full-scale sine wave histogram. This number can be reduced through controlled sampling and overdriving of the converter; a relationship is derived in [4].

A shortcoming of histogram testing in general is the loss of information associated with the accumulation of code counts only and not their order of occurrence. Imagine a situation where code bins were swapped, leading to a non-monotonic transfer function. There will be no effect on a ramp histogram and detection for a sine wave histogram depends on the code locations. A more realistic converter failure escaping histogram testing, is the occurrence of code sparkles. Usually this is a dynamic effect where an output code of unexpected difference to its neighbors occurs. However, such effects can become detectable via accumulation of each code's indices (locations) in a so-called *weight* array, which can be computed in addition to the histogram accumulated in a *tally* array [10].

3.4 Dynamic Performance Parameter Test

Generally, the aim in dynamic performance parameter testing is to identify the signal's components at the A/D converter output, such as the converted input signal, harmonics, and random noise, and to compute performance parameters introduced in section 2.2. For the majority of these parameters and determination of signal components, a transformation from time domain to frequency domain is required. A/D converter testing employing discrete Fourier transformation is described in the next section. However, some dynamic performance parameters can also be determined in the time domain from an A/D converter model generated to match a data record taken from a real converter. So-called sine-fit testing is introduced in section 3.4.2. For either technique, it is assumed that a single tone sine wave stimulus is applied to the A/D converter.

3.4.1 Frequency Domain Test Methodology

This section focuses on the application of frequency domain test to A/D converters. It is beyond the scope of this chapter to provide an introduction to Fourier transformation [6] or to discuss general aspects of DSP-based testing and spectrum analysis [10, 11] in great detail.

A signal can be described in time or frequency domain where the Fourier analysis is employed to move from one domain to the other without loss of information. For coherent sampling of periodic signals with the number of samples taken in the time domain being a power of two, the discrete Fourier transformation can be computed more efficiently through fast Fourier transformation (FFT) algorithms. If coherent sampling of all signal components cannot be guaranteed, a periodic repetition of the sampled waveform section can lead to discontinuities at either end of the sampling interval causing spectral leakage. In such cases windowing has to be applied, a processing step in which the sampled waveform section is mathematically manipulated to converge to zero amplitude towards the interval boundaries, effectively removing discontinuities [24]. In either case, the A/D converter output signal is decomposed into its individual frequency components for performance analysis. The frequency range covered by the spectrum analysis depends on the rate of A/C converter output code sampling, f_S. The number of discrete frequency points, also referred to as frequency bins, is determined by the number of samples, N, processed in the FFT. While accounting for aliasing, signal and sampling frequencies have to be chosen to allow sufficient spacing between the harmonics and the fundamental component.

The graphical presentation of the spectrum obtained from the analysis, frequently referred to as FFT plot, illustrates the particular signal component amplitude with its frequency on the x-axis (Figure 3-10). The number of frequency bins is equal to $N/2$ and their widths are equal to f_s/N.

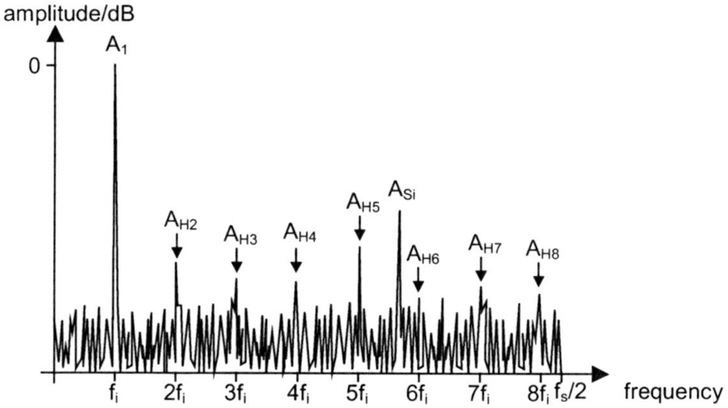

Figure 3-10. A/D converter output spectrum

The following spectrum features can be identified in Figure 3-10: Firstly the fundamental component, A_1, corresponding to the input signal, secondly the harmonic distortion components, A_{H2} to A_{H8}, thirdly large spurious components, such as A_{Si}, and finally the remaining noise floor representing quantization and random noise. Dynamic performance parameters, such as SINAD, THD and SFDR, can be calculated from the particular signal components' real amplitudes (not in decibels) or the power contained in them, as given in section 2.2 and described in [25] including multi-tone testing.

3.4.2 Sine Wave Fitting Test Methodology

Fitting a sine wave function to the data recorded when a sinusoidal stimulus is applied to the A/D converter allows assessment of the general performance. In [3], sampling of at least five stimulus cycles is quoted as a rule of thumb, where it also has to be born mind that conversion errors may be concealed with increasing record size due to averaging effects. Again, sampling and input stimulus frequency should be selected to record data that is uniformly spaced over a single period of the waveform. However, sampling of a non-integer number of cycles does not cause problems which have been mentioned for spectrum analysis above. In any case, the mathematical model of the fitted sine wave relates to the data y_n sampled at time instants t_n as:

$$y_n = A\cos(\omega t_n + \theta) + C \qquad \text{for } n = 1,\ldots,M \tag{14}$$

where A is the amplitude, θ the phase and C the offset of the fitted sine wave of angular frequency ω. When the frequencies of the input stimulus and the sampling and therefore parameter ω of the fitted function are known, the remaining three sine wave parameters can be calculated through minimization of the rms error between the data record and the model (*three-parameter least square fit*, [3]). When the frequencies are unknown or not stable, then the *four-parameter least square fit* has to be employed. Here, an iteration process beginning with an initial estimate for the angular frequency surrounds the least square minimization process. The value for ω is updated between loops until the change in obtained sine wave parameters remains small. The three parameter and four parameter fitting process is derived and described in far more detail in [17].

The performance parameter that is usually computed for the fitted A/D converter model is the ENOB [7], equation (8). Some further performance analysis can be achieved by test execution under different conditions, such as various input stimulus amplitudes or frequencies as described in [26]. A potential problem is that this test methodology does not verify the converter performance over its entire full-scale input range, as the test stimulus amplitude has to be chosen to avoid clipping. Localized conversion error, affecting a very small section of the transfer function, may also escape unnoticed due to the averaging effect of the fitting process.

4. A/D CONVERTER BUILT-IN SELF-TEST

Built-in self-test (BIST) for analogue and mixed signal components has been identified as one of the major requirements for future economic deep sub-micron IC test [27, 28]. The main advantage of BIST is to reduce test access requirements. At the same time the growing performance gap between the circuit under test and the external tester is addressed by the migration of tester functions onto the chip. In addition, parasitics induced from the external tester and the demands on the tester can be reduced. Finally, analogue BIST is expected to eventually enable the use of cheaper, digital only or so-called DfT-testers that will help with the integration of analogue virtual components - including BIST - for digital SOC applications. Here the aim is to enable the SOC integrator to avoid the use of expensive mixed signal test equipment. Also, for multi-chip modules, on-chip test support hardware helps to migrate the test of analogue circuitry to the wafer level. It is expected that the re-use of BIST structures will significantly reduce escalating test generation costs, test time and time-to-market for a range of devices. Full BIST has to include circuitry to implement both test stimulus generator (TSG) and output response analyzer (ORA). This section briefly summarizes BIST solutions that have been proposed for performance parameter testing of A/D converters, some of which have been commercialized.

Most BIST approaches for A/D converter testing aim to implement one of the converter testing techniques described in section 3. In [29] it is proposed to accumulate a converter histogram in an on-chip RAM while the test stimulus is generated externally. The accumulated code counts can be compared against test thresholds on-chip to test for DNL; further test analysis has to be performed off-chip. This test solution can be extended towards a full BIST by including an on-chip triangular waveform generator [30]. In a similar approach, the histogram-based analog BIST (HABIST™), additional memory and ORA circuitry can be integrated to store a reference histogram on-chip for more complete static performance parameter testing of A/D converters [31]. This commercialized approach [32] also allows the use of the tested A/D converter with the BIST circuitry to apply histogram-based testing to other analogue blocks included in the same IC. As illustrated in Figure 3-11, the on-chip integration of a sine wave or saw tooth TSG is optional. The histogram is accumulated in a RAM where the converter output provides the address and a read-modify-write cycle updates the corresponding code count. The response analysis is performed after test data accumulation and subtraction of a golden reference histogram. As for the TSG, on-chip implementation of the full ORA is optional.

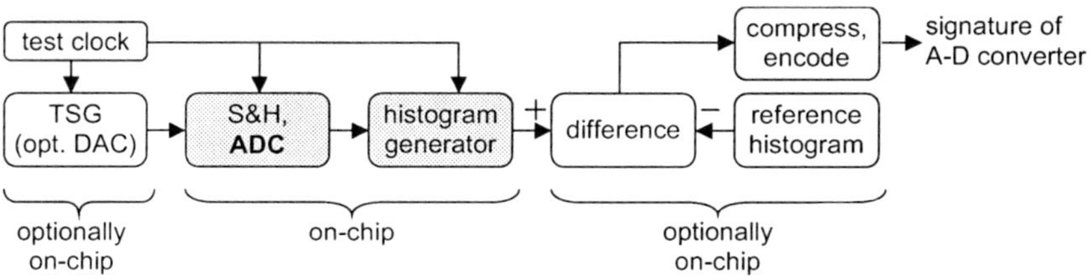

Figure 3-11. HABIST™ scheme applied to A/D converter (ADC)

Also the feedback-loop test methodology has been considered for a straightforward BIST implementation [33]. The oscillating input signal is generated through the charging or discharging of a capacitor with a positive or a negative reference current *I*, generated on-chip (Figure 3-12). Testing for differential and integral nonlinearity is based on the measurement of the oscillation frequency on the switch control line (*ctrl*) similarly to feedback-loop testing (section 3.3.1).

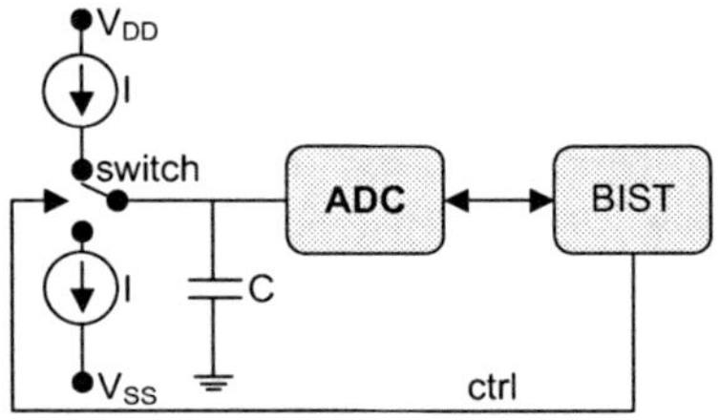

Figure 3-12. Oscillation BIST applied to A/D converter (ADC)

The dynamic performance of an A/D converter can be assessed through spectrum analysis or sine wave curve fitting as described above. It has been proposed to add an on-chip TSG and to facilitate an available DSP core to implement a BIST scheme for either of these test techniques [34]. Similarly, a BIST approach for mixed signal ICs containing a DSP, memory, and both, A/D and D/A converters has been proposed which computes an FFT and evaluates test results on-chip [35].

A BIST scheme to test an A/D-D/A converter chain for dynamic performance without the availability of a DSP is proposed in [36]. For high resolution A/D converter testing, the response analysis is conducted by integrating evenly distributed $n/4$ samples for each quarter of a ramp response (S_0 to S_4 in Figure 3-13a/b). The coefficients of a fitted third order polynomial can be calculated from these four sums and relate to DC offset and gain, and 2nd and 3rd order harmonic distortion expected for a sine wave input. While the integration is conducted by BIST circuitry, the on-chip extraction of coefficients, performance parameters and the comparison to test thresholds is optional. The expected ramp input stimulus can be approximated by a low pass filtered four-segment staircase-like stimulus, as illustrated in Figure 3-13c [37]. The BIST circuitry generates a pulse-width modulated waveform with five different duty cycles. These are applied to an off-chip low pass filter in turn to generate a rising and a falling step for each quarter of the converter's input range. The integration process is conducted during the rising / falling edges of the exponential (approximately 17% of the step width, as illustrated by shaded regions in Figure 3-13c) to achieve a relatively linear output code distribution.

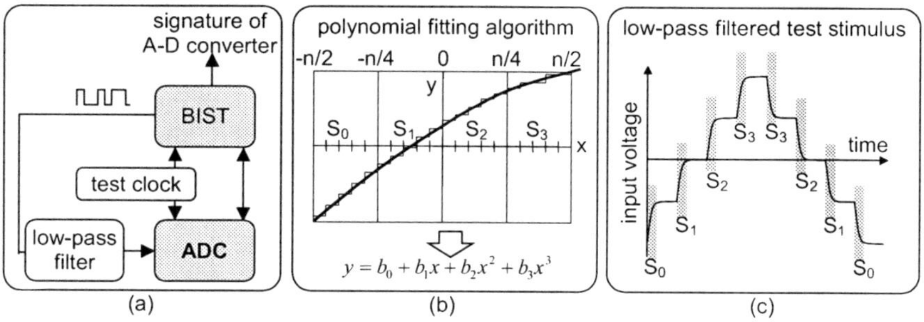

Figure 3-13. A/D converter BIST employing polynomial fitting algorithm

While the advantages of analog and mixed-signal BIST solutions are clear, drawbacks due to limited test sets, excessive area overhead or a low confidence in test results have hampered wide industrial use. BIST techniques summarized above are mostly limited to particular A/D converter architectures. Histogram-based BIST, for example, may result in excessive area overhead for high-resolution converters. The polynomial fitting algorithm BIST scheme is aimed at high-resolution converter testing, but relies on the assumption that a third order polynomial accurately fits the test response.

More work may be required to identify converter performance parameters crucial for testing. Test requirements and realistic failure modes will depend on particular converter architectures. An example study can be found in [38].

5. SUMMARY AND CONCLUSIONS

This chapter has discussed the key parameters and specifications normally targeted in A/D converter testing, methods for extracting these performance parameters and potential solutions for either implementing full self-test or migrating test resources from external test equipment to the device under test. Table 1 provides a summary of the advantages and limitations of five of the main test methods used in A/D converter testing.

Table 3-1. Summary of A/D converter test approaches

Technique	Performance parameters tested	Major advantages	Main limitations
Histogram-based	Static performance (offset and gain error, DNL, INL, missing codes, etc.)	Well-established, Complete linearity test	Long test time, amount of data, No test for dynamic performance, Test stimulus accuracy
Servo-loop	Static performance (offset and gain error, DNL, INL)	Accurate measurement of transition edges (not based on statistics)	Test stimulus accuracy Measurement accuracy
Sine-wave curve-fitting	DNL, INL, missing codes, aperture uncertainty, noise	Tests for dynamic performance	Input frequency is a sub multiple of sample frequency, Lack of convergence of algorithm, Measurement accuracy
Beat frequency testing	Dynamic characteristic	Quick and simple visual demonstration of ADC failures	No accurate test
FFT-based	Dynamic performance (THD, SINAD, SNR, ENOB)	Tests for dynamic performance, Well-established	No tests for linearity

The field now faces major new challenges, as the demand for higher resolution devices becomes the norm. The concept of design reuse in the form of integrating third party designs is also having a major impact on the test requirements as in many cases system integrators wishing to utilize a high performance converter functions will not normally have the engineering or production test equipment required to test these devices. The concept of being able to supply an A/D converter with an embedded test solution that requires only digital external test equipment is hence a major goal.

In the case of on-chip test solutions proposed or available commercially, limitations need to be understood before investing design effort. Histogram testing, for example, will require a large amount of data to be stored and evaluated on-chip while requiring long test times. For servo-loop based solutions, the oscillation around a single transition level may be difficult to achieve under realistic noise levels. Sine-wave fitting will require some significant area overhead for the on-chip computation, as do FFT-based solutions, and may still not achieve satisfying measurement accuracy and resolution. Further work is therefore required to quantify test times, associated cost and measurement accuracies and generate quality test quality metrics.

REFERENCES

1. R. van de Plassche: Integrated analog-to-digital and digital-to-analog converters, Kluwer Academic Publishers, 1994
2. R. L. Geiger, P. E. Allen & N. R. Strader: VLSI design techniques for analog and digital circuits, McGraw-Hill, New York, 1990
3. IEEE Standard 1241-2000: IEEE standard for terminology and test methods for analog-to-digital converters, Institute of Electrical and Electronics Engineers Inc., 2000
4. J. Blair: Histogram measurement of ADC nonlinearities using sine waves, IEEE Transactions on Instrumentation and Measurement, 1994, vol. 43, no. 3, pp. 373-383
5. Maxim Application Note A177: INL/DNL measurements for high-speed analog-to-digital converters (ADCs), Maxim Integrated Products, September 2000
6. A. V. Oppenheim & R. W. Schafer: Discrete-time signal processing, Prentice Hall, 1989
7. T. Linnenbrink: Effective bits: is that all there is?, IEEE Transactions on Instrumentation and Measurement, 1984, vol. IM-33, no. 3, pp. 184-187
8. T. C. Hofner: Dynamic ADC testing part I. Defining and testing dynamic ADC parameters, Microwaves & RF, 2000, vol. 39, no. 11, pp. 75-84,162
9. A. Grochowski, D. Bhattacharya, T. R. Viswanathan & K. Laker: Integrated circuit testing, IEEE Transactions on Circuits and Systems II: Analog and Digital Signal Processing, 1997, vol. 44, no. 8, pp. 610-633
10. M. Mahoney: DSP-based testing of analog and mixed-signal circuits, IEEE Computer Society, 1987
11. M. Burns & G. W. Roberts: An introduction to mixed-signal IC test and measurement, Oxford University Press, 2001
12. IEEE Standard 1149.1-2001: IEEE standard test access port and boundary-scan architecture 2001, Institute of Electrical and Electronics Engineers Inc., 2001
13. IEEE Standard 1149.4-1999: IEEE Standard for a Mixed-Signal Test Bus, Institute of Electrical and Electronics Engineers Inc., 1999
14. IEEE P1500 Working Group on a Standard for Embedded Core Test (SECT), web site: http://grouper.ieee.org/groups/1500/, 2003
15. S. Max: Testing high speed high accuracy analog to digital converters embedded in systems on a chip, IEEE International Test Conference, 28-30th Sept. 1999, Atlantic City, NJ, USA, pp. 763-771
16. J. J. Corcoran, T. Hornak & P. B. Skov: A high resolution error plotter for analog-to-digital converters, IEEE Transactions on Instrumentation and Measurement, 1975, vol. 24, no. 4, pp. 370-374
17. IEEE Standard 1057-1994 (R2001): IEEE standard for digitizing waveform recorders, Institute of Electrical and Electronics Engineers Inc., 2001
18. S. Max: Optimum measurement ADC transitions using a feedback loop, 16th IEEE Instrumentation and Measurement Technology Conference, 24-26th May 1999, Venice, Italy, pp. 1415-1420
19. T. M. Sounders & R. D. Flach: An NBS calibration service for A/D and D/A converters, IEEE International Test Conference, 27-29th Oct. 1981, Philadelphia, PA, USA, pp. 290-303
20. G. Pretzl: Dynamic testing of high speed aid converters, IEEE Journal of Solid-State Circuits, 1978, vol. 13, pp. 368-371

21. O. J. Downing & P. T. Johnson: A method for assessment of the performance of high speed analog / digital converters, Electronics Letters, 1978, vol. 14, no. 8, pp. 238-240
22. M. Van den Bossche, J. Schoukens & J. Renneboog: Dynamic testing and diagnosis of A/D converters, IEEE Transactions on Circuits and Systems, 1986, vol. 33, no. 8, pp. 775-785
23. J. Doernberg, H. S. Lee and D. A. Hodges: Full-speed testing of A/D converters, IEEE Journal of Solid-State Circuits, 1984, vol. 19, no. 6, pp. 820-827
24. F. J. Harris: On the use of windows for harmonic analysis with the discrete Fourier transform, Proceedings of the IEEE, 1978, vol. 66, no. 1, pp. 51-83
25. T. C. Hofner: Dynamic ADC testing part 2. Measuring and evaluating dynamic line parameters, Microwaves & RF, 2000, vol. 39, no. 13, pp. 78-94
26. B. E. Peetz: Dynamic testing of waveform recorders, IEEE Transactions on Instrumentation and Measurement, 1983, vol. 32, no. 1, pp. 12-17
27 Semiconductor Industry Association: International Technology Roadmap for Semiconductors, 2001 Edition
28. S. Sunter: Mini Tutorial: Mixed signal test, 7th IEEE International Mixed-Signal Testing Workshop, 13-15th June 2001, Atlanta, GA, USA
29. R. Bobba & B. Stevens: Fast embedded A/D converter testing using the microcontroller's resources, IEEE International Test Conference, 10-14th Sept. 1990, Washington, DC, USA, pp. 598-604
30. J. Raczkowycz & S. Allott: Embedded ADC characterization techniques using a BIST structure, an ADC model and histogram data, Microelectronics Journal, 1996, vol. 27, no. 6, pp. 539-549
31. A. Frisch & T. Almy: HABIST: histogram-based analog built in self test, IEEE International Test Conference, 3-5th Nov. 1997, Washington, DC, USA, pp. 760-767
32. Fluence Technology Incorporated: BISTMaxx^{TM} product catalog, 2000
33. K. Arabi & B. Kaminska: Oscillation built-in self test (OBIST) scheme for functional and structural testing of analog and mixed-signal circuits, IEEE International Test Conference, 3-5th Nov. 1997, Washington, DC, USA, pp. 786-795
34. M. F. Toner & G. W. Roberts: A BIST scheme for a SNR, gain tracking, and frequency response test of a sigma-delta ADC, IEEE Transactions on Circuits and Systems II: Analog and Digital Signal Processing, 1995, vol. 42, no. 1, pp. 1-15
35. E. Teraoka, T. Kengaku, I. Yasui, K. Ishikawa, T. Matsuo, H. Wakada, N. Sakashita, Y. Shimazu & T. Tokada: A built-in self-test for ADC and DAC in a single chip speech CODEC, IEEE International Test Conference, 17-21st Oct. 1993, Baltimore, MD, USA, pp. 791-796
36. S. K. Sunter & N. Nagi: A simplified polynomial-fitting algorithm for DAC and ADC BIST, IEEE International Test Conference, 3-5th Nov. 1997, Washington, DC, USA, pp. 389-395
37 A. Roy, S. Sunter, A. Fudoli & D. Appello: High accuracy stimulus generation for A/D converter BIST, IEEE International Test Conference, 8-10th Oct. 2002, Baltimore, MD, USA, pp. 1031-1039
38 A. Lechner, A. Richardson & B. Hermes: Short circuit faults in state-of-the-art ADC's - are they hard or soft, 10th Asian Test Conference, 19-21st Nov. 2001, Kyoto, Japan, pp. 417-422

Chapter 4

PHASE LOCKED LOOP TEST METHODOLOGIES

Current Characterisation and Production Test Practices

Martin John Burbidge & Andrew Richardson
Lancaster University.

1. INTRODUCTION: PLL OPERATION AND TEST MOTIVATIONS.

Phase locked loops are incorporated into almost every large-scale mixed signal and digital system on chip (SOC). Various types of PLL architectures exist including fully analogue, fully digital, semi-digital, and software based. Currently the most commonly used PLL architecture for SOC environments and chipset applications is the Charge-Pump (CP) semi-digital type. This architecture is commonly used for clock synthesis applications, such as the supply of a high frequency on-chip clock, which is derived from a low frequency board level clock. In addition, CP-PLL architectures are now frequently used for demanding RF (Radio Frequency) synthesis, and data synchronization applications. On chip system blocks that rely on correct PLL operation may include third party IP cores, ADCs, DACs and user defined logic (UDL). Basically, any on-chip function that requires a stable clock will be reliant on correct PLL operation. As a direct consequence it is essential that the PLL function is reliably verified during both the design and debug phase and through production testing.

This chapter focuses on test approaches related to embedded CP-PLLs used for the purpose of clock generation for SOC. However, methods discussed will generally apply to CP-PLLs used for other applications.

J.L. Huertas (ed.), Test and Design-for-Testability in Mixed-Signal Integrated Circuits, 99–136.

1.1 Phase Locked Loop key elements operation and test issues.

The charge-pump PLL architecture of figure 1 consists of a phase

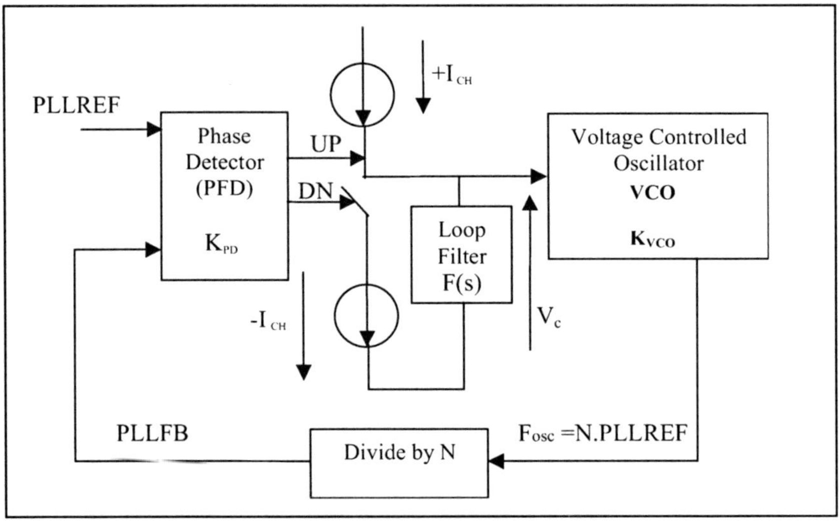

Figure 1 Block diagram of typical CP-PLL configuration

detector, a charge pump, a loop filter (LF), a voltage controlled oscillator (VCO) and a feed back divider (÷N). The phase frequency detector (PFD) senses the relative timing differences between the edges of the reference clock and VCO clock (feedback clock) and applies charge-up or charge down pulses to the charge pump that are proportional to the timing difference. The pulses are most commonly used to switch current sources, which charge or discharge a capacitor in the loop filter. The voltage at the output of the loop filter is applied to the input of the VCO, which changes oscillation frequency as a function of its input voltage. Note that ideally when the feedback and reference clocks are equal, that is they are both phase and frequency aligned, the charge pump transistors will operate in such a way as to maintain the loop filter voltage at a constant value. In this condition, the PLL is "locked" which implies that the output signal phase and frequency is aligned to the input within a certain limit. Note that the division block up converts the VCO output frequency to an integer multiple of the frequency present on its reference input (PLLREF). It follows that when the PLL is in its locked state:

$$F_{out} = N \cdot PLLREF$$

Equation 1

In figure 1, the following conversion gains are used for the respective blocks.

K_{PD} = Phase detector gain = $I_{ch}/2\pi$ $(A\ r^{-1})$
F(s) = Loop filter transfer function
Kvco = VCO gain $(r\, s^{-1}\, v^{-1})$

Using feedback theory the generalized transfer equation in the Laplace domain for the system depicted in figure 1 is.

$$H(s) = \frac{\theta_o(s)}{\theta_i(s)} = \frac{N \cdot K_{PD} \cdot K_{VCO} \cdot F(s)}{sN + K_{PD} \cdot K_{VCO} \cdot F(s)}$$

Equation 2

Note that by substituting suitable values for N and F(s) equation 1 will generally apply to any order PLL system. [1]. Specific transfer equations are provided as part of the loop filter description.

It must be noted that even for the case of a CP-PLL the implementation details for the blocks may vary widely, however in many applications designers attempt to design the PLL to exhibit the response of a second order system. This is due to the fact that second order systems can be characterized using well-established techniques. The response of a second order CP-PLL will be generally considered in this chapter. [2] [3] [4].
A brief description of each of the blocks now follows. Further, basic principles of CP-PLL operation are given in [1][3][5][6].

Phase Frequency Detector:

The phase detector most commonly used in CP-PLL implementations is the type 4 edge sensitive PFD. The PFD may be designed to operate on rising or falling edges, for the purpose of this discussion it will be assumed that the PFD is rising edge sensitive A schematic of this type of PFD is shown below. In Figure 2, PFDUP and PFDDN represent the control signals for the up and down current sources respectively. When considering the operation of the PFD it is also useful to also consider the change in VCO output frequency. Considering phase alignment of the PFD input signals;
θ_{REF} will be used to designate the instantaneous phase of PLLREF and θ_{FB} will be used to designate the instantaneous phase of the PLLFB signal. Using

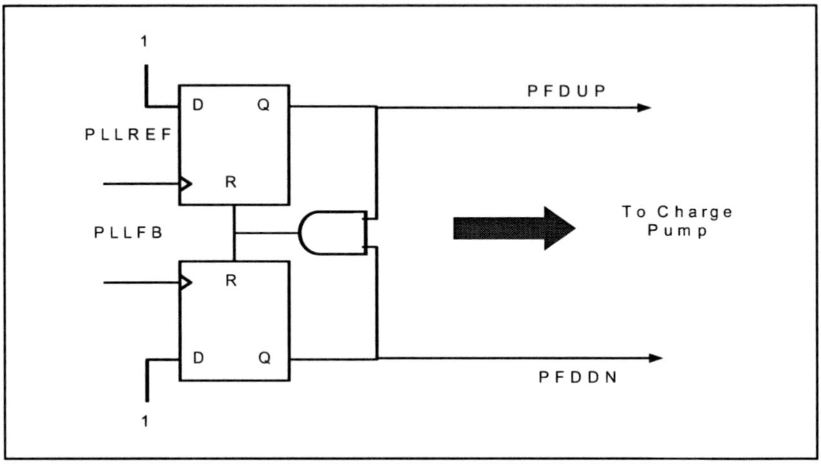

Figure 2 Typical implementation of type 4 PFD

this convention and with reference to figure 2 and figure3 the PFD operation is now explained.

(1) <u>$\theta_{FB}(t)$ leads $\theta_i(t)$</u> => LF voltage falls and VCO frequency falls to try and reduce the difference between $\theta_i(t)$ and θ_{FB} (t).

(2) <u>$\theta_i(t)$ leads $\theta_{FB}(t)$</u> => LF voltage rises and VCO frequency rises to try and reduce the difference between $\theta_i(t)$ and θ_{FB} (t)

(3) <u>$\theta_i(t)$ coincident with $\theta_{FB}(t)$</u> => The PLL is locked and in its stable state.

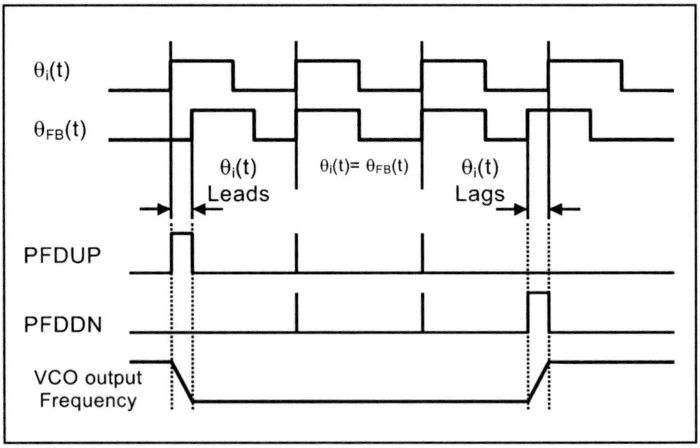

Figure 3 Operation of PFD and associated increase in VCO output frequency

Typical configurations for CP and F(s):

As above, the loop filter is designed to ensure that the whole system exhibits a second order response. A typical loop filter and charge pump configuration used for fully embedded CP-PLLs is illustrated in Figure 4.

The Laplace domain transfer function for this circuit is:

$$F(s) = R1 + \frac{1}{sC1}$$

Equation 3

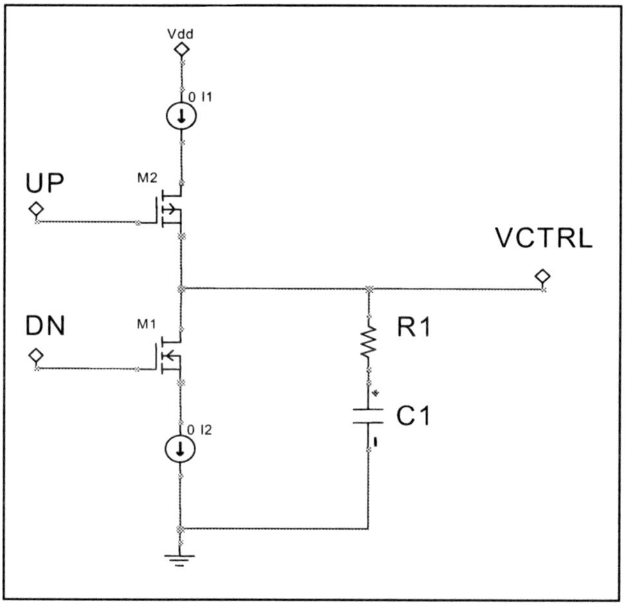

Figure 4 Typical Charge Pump and Loop filter configuration

Where we define following [3] : τ_1=R1C1

Now if the above loop filter transfer function is substituted into equation (2) for F(s). The following equation can be derived:

$$H(s) = \frac{\theta_O(s)}{\theta_i(s)} = \frac{2\zeta\omega_n s + \omega_n^2}{s^2 + 2\zeta\omega_n s + \omega_n^2}$$

Equation 4

Where:

$$\omega_n = \sqrt{\left(\frac{K_O I_P}{N 2\pi C_1}\right)} \qquad \zeta = \frac{\tau_2}{2}\left[\frac{K_O I_P}{N 2\pi C_1}\right]$$

Equation 5 Equation 6

It must be mentioned that for CP-PLLs in general and for embedded CP_PLLs specifically, that the loop filter node can be considered as the critical controlling node of the PLL. Any noise coupled into this node will generally manifest itself as a direct instantaneous alteration of the VCO output frequency, this action will be observed as PLL output jitter. Consequently PLL designers usually spend a great deal of design effort in screening this node. In addition, correct loop filter operation is essential if the PLL is to function properly over all desired operational ranges. Embedded loop filters usually include one or more large area MOSFET capacitors. These structures may be sensitive to spot defects, such as gate oxide shorts [7].

Matching of the Charge-Pump currents is also a critical part of PLL design. Leakage and mismatch in the Charge pump will lead to deterministic jitter on the PLL output.

Voltage Controlled Oscillator:

For embedded CP-PLL configurations the VCO is usually constructed as a current starved ring oscillator structure. This is primarily due to the ease of implementation in CMOS technologies. The structure may be single ended or differential, with differential configurations being preferred, due to their superior noise rejection capabilities. A typical single ended current starved ring oscillator structure is illustrated in Figure 5.

In this circuit, VCTRL is the input control voltage taken from the loop filter node and Fout is the VCO output signal. Note that to prevent excessive loading of the VCO, its output is usually connected to buffer stages.

The transfer gain of the VCO is found from the ratio of output frequency deviation to a corresponding change in control voltage. That is:

$$Kvco = \frac{F2 - F1}{V2 - V1}(MHzV^{-1})or(rs^{-1}V^{-1})$$

Equation 7

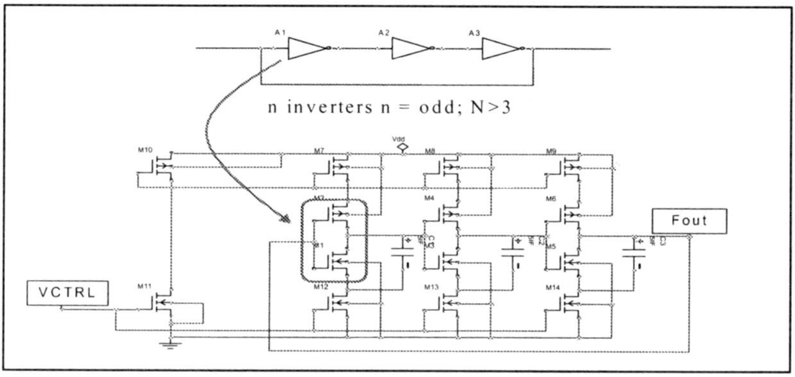

Figure 5 Single ended current starved ring oscillator structure.

Where F2 is the output frequency corresponding to V2 and F1 is the output frequency corresponding to V1. An example of experimental measurement of the VCO gain is given in section 2.1.1.4.

Digital structures:

The digital blocks of the CP-PLL are generally constructed from standard digital structures. In some cases feedback dividers and PFDs may be constructed from standard cells. However, in many situations, to meet stringent timing requirements, the digital structures are constructed using full custom layout techniques. Digital structures of the PLL are generally less sensitive than the analogue structures and they are often modified to ease testing of the PLL.

1.2 Typical CP-PLL Test Specifications.

Important functional characteristics that are often stated for CP-PLL performance are listed below:

- 1) Lock time / and ability to achieve lock from system startup.
- 2) Capture Range and Lock range.
- 3) Phase / Frequency step response time.
- 4) Overshoot.
- 5) Loop bandwidth (ω3dB).
- 6) Output jitter.

All of the above parameters are interrelated to a certain extent. For example the loop bandwidth will have an effect on the PLL output jitter. However, loop bandwidth, lock time, overshoot and step response time are also directly related to the natural frequency and damping of the system. It must be mentioned that, certain non-idealities or faults may attribute to further jitter on the PLL output, or increased lock time. Examples of typical measurements for these parameters are provided in later sections.
Table 1 provides an initial analysis of testing issues for the PLL sub-blocks. Fault models that are suggested for use in fault coverage calculations for each of the blocks. Further research and justification for the use of fault models in the key PLL sub-blocks are given in [7][8][9][10][11][12][13].

PLL Block	**Structures**		**Direct access / modification**	**At speed testing required**	**Commonly Suggested Fault Models**
	A	**D**			
(1) PFD		√	Yes	Yes	Single stuck at faults
(2) CP	√		No	Yes	MOS Transistor faults *
(3) LF	√		No	Yes	“”
(4) OSC	√	√	Yes	Yes	“”
(5) DIV		√	Yes	Yes	Single stuck at faults
* Mos transistor catastrophic faults: Gate to drain shorts; gate to source shorts; drain opens; source opens.					

Table 1 Common examples of test accessibility.

Note also that the fault models suggested in table 1 can also be used to assess the fault coverage of BIST (Built In Self Test) techniques. It should be noted however that many fault types are related to the structure realization of the PLL hence these guidelines should be used with care. Faults that may be implementation dependant include:

- Problems with interconnect due to pinholes or resistive vias.
- Coupling faults.

Obviously a high performance PLL will be routed in such a way so as to attempt to minimize the probability of these faults occurring.

1.2.1 Jitter Overview.

Jitter has been mentioned several times with respect to the PLL sub block descriptions. Non-Idealities, faults, or bad design practices, such as poor matching of structures, poor layout and insufficient de-coupling between critical functions can lead to excessive jitter in the PLL output. Jitter may be divided into two main classes as follows:

- Random Jitter (Gaussian Jitter): Is caused by non-deterministic events, such as coupling of electrical noise into the charge pump structures, the VCO control input, or the PLLs digital structures.
- Deterministic or correlated jitter: Can be viewed as changes in the PLL outputs instantaneous phase or frequency that can be directly correlated to changes in PLL operation or changes in the host system operation. Typical examples may be phase spurs due to charge pump mismatch, or a notable change in output jitter when a system output driver is switched.

For clock synthesis based applications jitter can be regarded as the amount of time variation that is present on the periodic output signal.

The diagram below shows a typical spread or bell shaped curve that would be used to represent truly random or Gaussian type jitter, superimposed upon this plot is a non Gausian shape that may occur from deterministic jitter sources.

As a general approximation, the maximum random signal jitter will increase

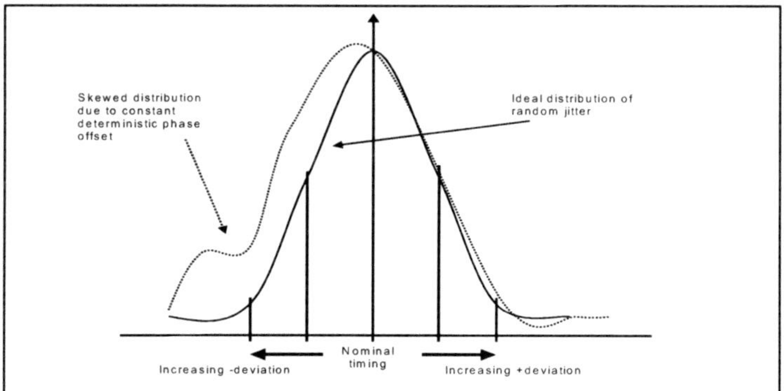

Figure 6 Ideal and non-ideal jitter probability density curves

to a peak-to-peak value of 6 standard deviations from the ideal case. However it, must be noted that due to non-deterministic effects, that the actual measured case may vary significantly from this. This statistical technique of viewing jitter is often used in conjunction with jitter histogram measurements. Obviously the confidence levels of the measurements will increase when an increasing number of measurements are taken. An example of this technique is given in later sections (see section 2.2.1).
A measurement often quoted that relates to non-deterministic jitter is that of RMS_jitter. The expression for RMS_jitter is given below.

$$RMS\_jitter = \sqrt{\frac{1}{N-1} \cdot \sum_{i=1}^{N} (T_i - \overline{T})^2}$$

Equation 8

Where $\overline{T}$ is the mean of the measured time intervals and is defined as.

$$\overline{T} = \frac{1}{N} \cdot \sum_{i=1}^{N} T_i$$

Equation 9

In both of the above equations N represents the total number of samples taken and T_i represents the time dispersion of each individual sample.

For clock signals, jitter measurements are often classified in terms of short-term jitter and long term jitter. These terms are further described below:

Short term Jitter: This covers short-term variations in the clock signal output period. Commonly used terms include:

- **Period Jitter:** Is defined as the maximum or minimum deviation (whichever is the greatest) of the output period from the ideal period.
- **Cycle to Cycle Jitter:** Is defined as the period difference between consecutive clock cycles. That is Cycle-to-Cycle Jitter = [period (n) – period (n-1)]. It must be noted that Cycle-to-Cycle Jitter represents the upper bound for the period jitter.
- **Duty Cycle distortion Jitter:** Is the change in the duty cycle relative to the ideal duty cycle. The relationship often quoted for duty cycle is given below.

$$Duty\_cycle = \frac{High_{period}}{High_{period} + Low_{period}} \cdot 100 \quad (\%)$$

Equation 10

Where $High_{period}$ is time duration the signal is high during one cycle of the waveform and Low_{period} is the time duration the signal is low over one period of the measured waveform. In an ideal situation the duty cycle will be 50%, the duty cycle distortion jitter will measure the deviation of the output waveform duty cycle from the ideal position. Typical requirements for duty cycle jitter is that it should be within 45 – 55 % [14] [15].

The above jitter parameters are often quoted as being measured in terms of degrees deviation with respect to an ideal waveform. Another metric often encountered is that of a UI (Unit Interval), where one UI is equivalent to 360°. A graphical representation of a unit interval is given below.

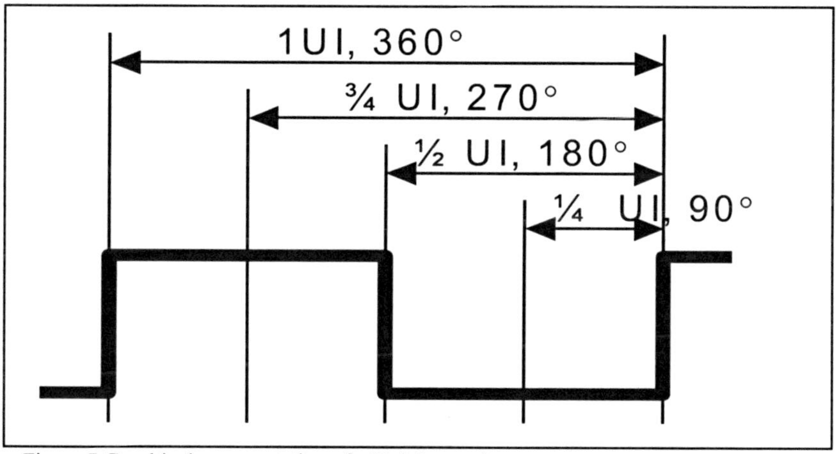

Figure 7 Graphical representation of a Unit Interval

<u>Long term Jitter:</u> Provides a measure of the long-term stability of the PLL output. That is, it represents the drift of the clock signal over time. It is usually specified over a certain time interval (usually a second) and expressed in ppm. For example, a long-term jitter specification of 1ppm would mean that a signal edge is allowed to drift by 1us from the ideal position in 1 second.

A graphical representation of all of the mentioned forms of jitter is provided below.

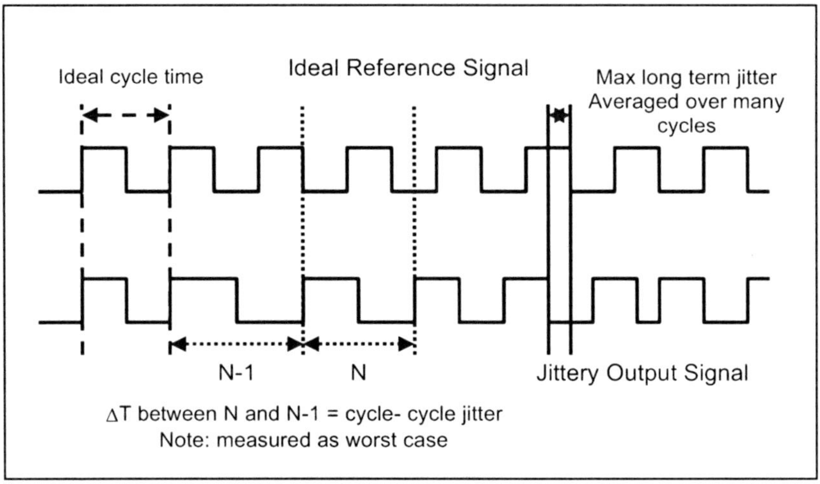

Figure 8 Pictorial summary of various types of jitter.

All of the above measurements, both long term and short term rely on either being able to discern the timing fluctuations of the signal when compared to an ideal reference or discern the timing difference between consecutive cycles. Absolute Jitter specifications will be dependant upon the specific application and will relate to the maximum output frequency of the PLL.
For example, in a 622Mbps SONET (Synchronous Optical Network) PLL [16] the maximum pk-pk generated jitter , is stated as 2.6 mUI. For this application. one unit interval is calculated as:

$$1UI = \frac{1}{622.08MHz} = 1.60ns$$

Equation 11

Therefore, 2.6mUI is equivalent to

$$2.6mUI = \frac{1UI}{1000} \cdot 2.6 = \frac{1.608ns}{1000} \cdot 2.6 = 4.1808\,ps$$

Equation 12

Further examples of jitter specifications and allowable tolerances are given in [17].
It can be seen from the above example that measurement of jitter requires highly accurate and jitter free reference signals that are a much higher frequency than that of the device being measured

.

2. TRADITIONAL TEST TECHNIQUES.

This section will explain traditional or commonly employed CP-PLL test techniques that are used for the evaluation of PLL operation. Many of the techniques will be applicable for an analogue or semi digital type of PLL or CP-PLL, however it must be recognized that although the basic principles may hold, the test stimuli may have to undergo slight modification for the fully analog case. The section is subdivided into two subsections, focusing on characterization and production test techniques respectively.

2.1 Characterization Focused Tests.

In this subsection we will review typical techniques that are used to characterize a PLL system. Characterization in this context will refer to measurements made by the PLL designer upon an initial test die for the purpose of verifying correct circuit functionality, and to allow generation of the device or data sheet [18]. Characterization based tests usually cover a greater number of operational parameters than that carried out for production test. Also they can be carried out using specific special function test equipment and hardware, as appose to general-purpose production test equipment.

2.1.1 Operational Parameter Based Measurements.

The key parameter based measurements employed for CP-PLL verification generally consist of the following tests:

- 1) Lock Range and capture range.
- 2) Transient response:

  - Correct operation from power up of the system incorporating the PLL. This test is often ascertained using a frequency lock test (FLT).
  - Correct step response of the system, when the PLL is switched between two frequencies or phases.

- 3) Phase transfer function (or jitter transfer function monitoring):
  - To ascertain correct 3dB bandwidth of the PLL system.
  - To ascertain correct phase response of the PLL system.

The above tests can be considered to be full functionality tests as they are carried out upon the whole PLL system. It must also be mentioned that for second order systems, both of the above techniques can be used to extract defining parameters such as ω_n (natural frequency) and ζ (damping).

Further tests are often carried out that are based upon structural decomposition of the PLL into its separate building blocks. These techniques are often used to enhance production test techniques [19][20]. Typical tests are;

- Charge pump current monitoring is used to ascertain the charge pump gain and hence the phase detector transfer characteristic. It is also used to monitor for excessive charge pump mismatch.
- 4) Direct control of the VCO is used to allow estimation of the VCO transfer characteristic.

The decomposition tests are also often coupled with some form of noise immunity test, that allow the designer ascertain the sensitivity of the VCO or charge pump structures to noise on the PLLs supply rails. As the CP currents and VCO control inputs are critical controlling nodes of the PLL, and determine the instantaneous output frequency of the PLL , any coupling of noise onto these nodes will cause jitter on the PLL output. Thus noise immunity tests are particularly important in the initial characterization phases. A more detailed discussion of the tests now follows.

2.1.1.1 Lock range and capture range.

The normally encountered definitions for capture range and lock range are provided below.

- Capture range: Refers to the range of frequencies that the PLL can lock to when lock does not already exist.
- Lock range: The range of frequencies that the PLL can remain locked after lock has been achieved.

For certain applications these parameters are particularly important. For instance, lock range would need to be evaluated for frequency demodulation applications. When considering edge sensitive CP-PLLs the lock range is usually equal to the capture range.

For a CP-PLL synthesizer the lock range would be ascertained in the following manner for a single division ratio:
1) The CP-PLL would initially be allowed to lock to a reference frequency that is in the correct range for a particular divider setting.
2) The reference frequency would be slowly increased until the CP-PLL can no longer re- adjust its output to keep the PFDs inputs phase aligned.
3) When the CP-PLL fails to acquire constant lock the reference frequency is recorded.

This sequence is often aided by use of lock detect circuitry, that is used to provide a digital output signal when the PLL has lost lock.

2.1.1.2 Transient type response monitoring.

Frequency Lock Test (FLT):

An initial test that is carried out before more elaborate tests are employed is the frequency lock test (FLT). This test simply determines whether the PLL can achieve a stable locked condition for a given operational configuration. Stability criteria will be determined by the application and may consist of an allowable phase or frequency error at the time of measurement. Typically, this test is carried out in conjunction with a maximum specified time criteria, that is, if the PLL has failed to achieve lock after a specified time, then the PLL is faulty. The start of test initiation for the FLT is usually taken from system startup. It is common for this test to be carried out for various PLL settings, such as, maximum and minimum divider ratios, different loop filter settings etc. Due to its simplicity and the fact that it will uncover many hard faults and some soft faults in the PLL, this test is often used in many production test applications. A graphical description of the FLT is given below.

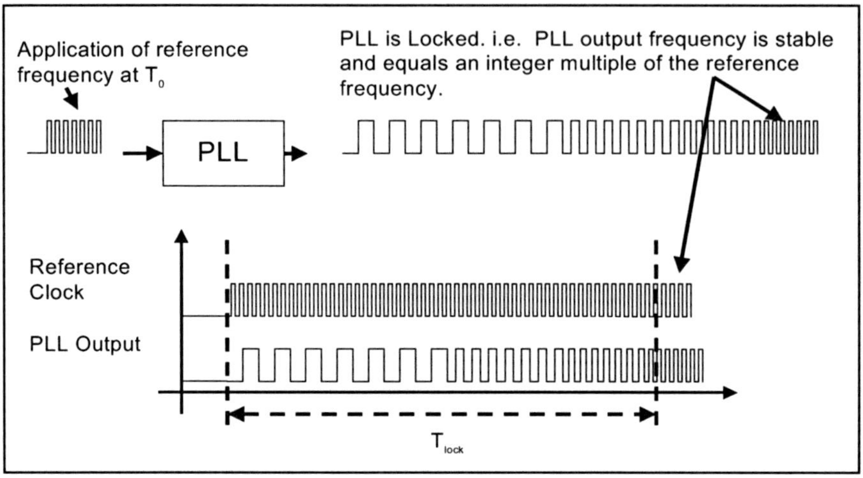

Figure 9 Graphical representation of a typical frequency lock test sequence

In the above diagram, T_0 represents the start of the test and T_{lock} indicates the time taken to achieve lock.

In many applications the output frequency is simply measured after a predetermined time, this is often the case in ATE (Automated test equipment) based test schemes, where the tester master clock would be used to determine the time duration. Alternatively, in some situations the PLL itself is fitted with LD (lock detect) circuitry that produces a logic signal when the PLL has attained lock [20]. In this situation, a digital counter is started at T_0 and stopped by the LD signal, thus enabling accurate lock time calculations to be made. Note that LD circuitry is not test specific, as it is often included in PLL circuits to inform other system components when a stable clock signal is available. However, sometimes a LD connection is fitted solely for DfT purposes. It must also be mentioned that in certain PLL applications it may be acceptable to access the loop filter node, if this is the case, the approximate settling time of the PLL can be monitored from this node. This technique is sometimes used for characterization of chipset PLLs, however, due to problems outlined in the previous sections it appears to be less commonly used for test of fully embedded PLLs.

Step Response Test:

The step response monitoring of PLL's is a commonly used bench characterization technique [2]; the basic hardware set-up is shown in Figure 10.

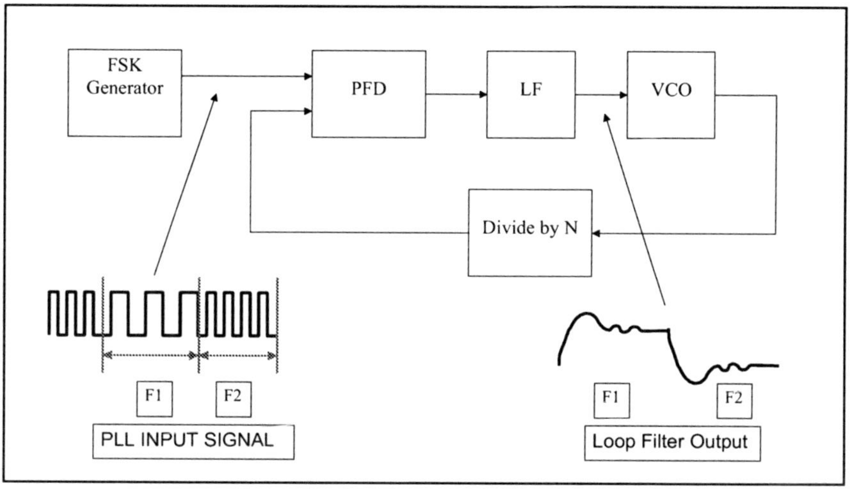

Figure 10 Basic equipment setup for PLL step response test

Further details relating to the above diagram are now given below:

- The input signal step is applied by using a signal generator set-up capable of producing an FSK signal. The signal is toggled periodically between F1 and F2. Note that a suitable toggling frequency will allow the system to reach the steady state condition after each step transition.
- If an external loop filter is used it is sometimes possible to measure the output response from the loop filter node. The signal measured at this node will be directly proportional to the variation in output frequency that would be observed at the PLL's output.

Also note that as the VCO output frequency is directly proportional to the loop filter voltage, the step response can also be measured at the VCO output. In fact, this is the technique that must be employed when loop filter access is prohibited. However, this technique can only be carried if test equipment with FT (frequency Trajectory) probing capabilities is available. This type of equipment allows a plot or oscilloscope trace of instantaneous frequency against time to be made, thus providing a correct indication of the transient step characteristics. Many high specification bench test equipment products incorporate FT functions, but it is often hard to incorporate the technique into a high volume production test plan.

An alternative method of introducing a frequency step to the system involves switching the feedback divider between N and N+1. This method will produce an output response from the PLL that is equivalent to the response that would be observed for application of an FSK input frequency step equal to the PLLs reference frequency. The technique can be easily verified with reference to equation 1.

The step response can be used to make estimates of the parameters outlined in section 1. To further illustrate the technique a graphical representation for a second order system step response is provided below.

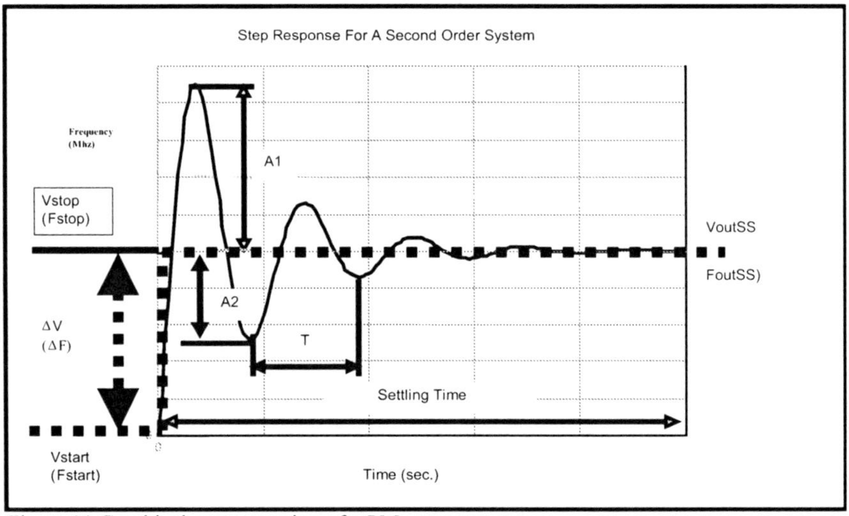

Figure 11 Graphical representation of a PLL step response.

In the above diagram the dashed line indicates the application of the input step parameter, and the solid line indicates the output response. Note that the parameters of interest are shown as Vparameters and (Fparameters) to indicate the similarity between a common second order system response and a second order PLL system response. An explanation of the parameters is now given.

- Vstart (Fstart): Is the voltage or frequency before the input step is applied.
- Vstop(Fstop): Is the final value of the input stimulus signal.

- $\Delta V(\Delta F)$: Is the amount by which the input signal is changed.
- VoutSS(FoutSS): Represents the final steady state output value of the system.
- Settling time: The amount of time it takes after the application of the input step, for the system to reach its steady state value.
- A1: Peak overshoot of the signal.
- A2: Peak undershoot of signal.
- T: Time difference between the consecutive peaks of the transient response.

Direct measurement of these parameters can be used to extract ω_n and ζ. Estimation of the parameters is carried out using the following formulas that are taken from [2] and are also found in many control texts [4]. The formulas are valid only for under damped system i.e. one in which A1,A2 and hence T can be calculated. However if this is the case other parameters can be used to assess the system performance such as delay time or rise time. This is true for many applications, when what is really desired is the overall knowledge of the transient shape of the step response.

The damping factor ζ can be found as follows:

$$\zeta = \frac{\ln\left(\frac{A1}{A2}\right)}{\left\langle \pi^2 + \left[\ln\left(\frac{A1}{A2}\right)^2\right]\right\rangle^{1/2}}$$

Equation 13

The natural frequency ω_n can be found as follows:

$$\omega_n = \frac{2\pi}{T\sqrt{1-\zeta^2}}$$

Equation 14

Further PLL system theory and control system theory texts [2][3][4] also contain normalized frequency and phase step response plots, where the amplitude and time axis are normalized to the natural frequency of the system. Design engineers commonly employ these types of plots in the initial system design phase.

2.1.1.3 Transfer function monitoring.

In many applications a PLL system is designed to produce a second order response. It must be noted that although second order systems are considered here, measurement of the transfer functions of higher order PLL loops can provide valuable information about system operation, and can be achieved using the methods explained here.

A bode plot of the transfer function of a general unity gain second order system is shown in Figure 12.

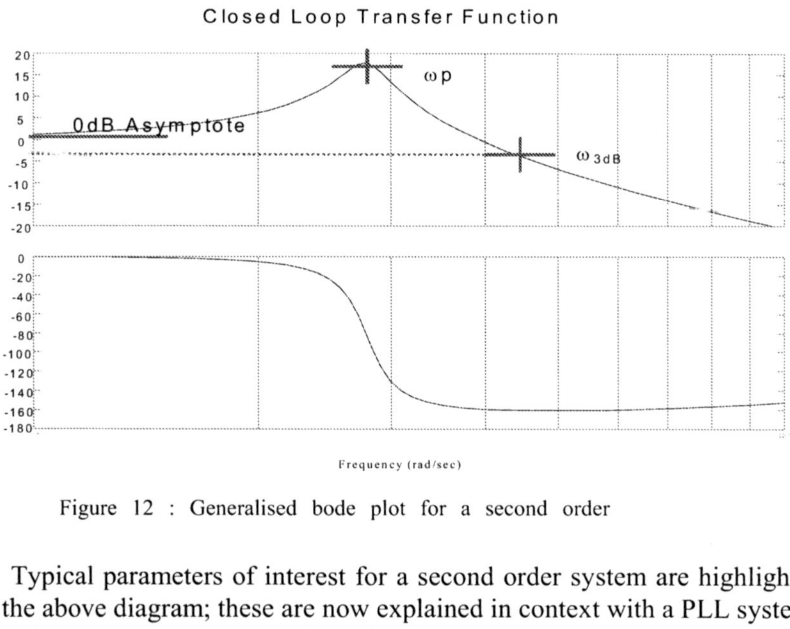

Figure 12 : Generalised bode plot for a second order

Typical parameters of interest for a second order system are highlighted in the above diagram; these are now explained in context with a PLL system:

<u>0dB Asymptote:</u> For a unity gain system, within the 3db frequency (see ω3dB below), the magnitude of the gain will tend to 1 (0dB) as the frequency of the excitation signal is reduced. The slope of this decrease will be determined by the damping of the system. In a similar manner the relative phase lag between the input and output of the system will tend to 0°. <u>Note</u> for a PLL the magnitude of the response within the loop bandwidth can be assumed to be unity [8]. As explained in later sections this is an important observation when considering PLL test.

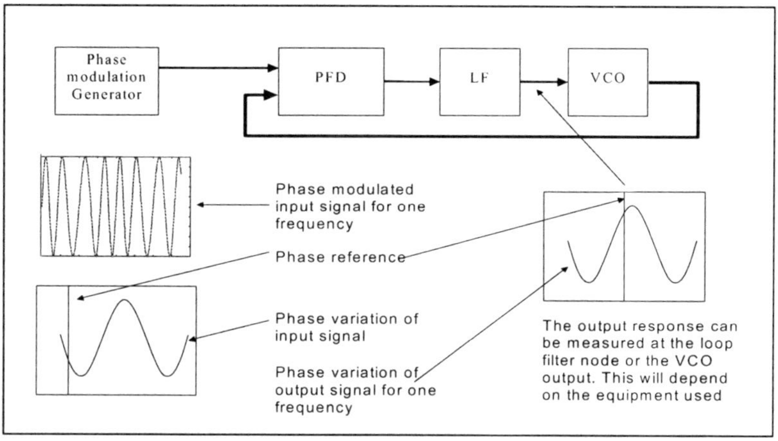

Figure 13 Illustrating application of a PLL transfer function measurement.

ωp: This is frequency where the magnitude of the system response is at its maximum. It is directly analogous to the natural frequency (ωn) of the system. In addition, the relative magnitude of the peak (above the unity gain value can be used to determine the damping factor (ζ) of the system. Relationships between ζ, the dB Magnitude and the normalised radian frequency are available in many texts concerning Control or PLL theory [2][3][4].

ω_{3dB}: Following [1] this defines the one-sided loop bandwidth of the PLL. The PLL will generally be able to track frequency variations of the input signal that are within this bandwidth, and reject variations that are above this bandwidth.

For normally encountered second order systems i.e. ones with a voltage, current or force inputs and corresponding voltage, current or displacement outputs, the bode plot is constructed by application of sinusoidally varying input signal at different frequencies, the output of the system is then compared to the input signal to produce magnitude and phase response information. However the situation is slightly different for the PLL systems we are considering. In this situation the normal input and output signals are considered to be continuous square wave signals. The PLL's function is to phase align the input and output signals of the PLL system.

It can be seen from equation (2) and figure (1) that to experimentally measure the PLL transfer function we need to apply a sinusoidal variation of phase about the nominal phase of the input signal $\theta_I(t)$, i.e. we sinusoidally phase modulate the normal input signal. The frequency of the phase change is then increased and the output response is measured. The block diagram for an experimental bench type test set-up is shown in Figure 13.

For the above test, the output response can be measured at the loop filter node or the VCO output. The output of the loop filter node will be a sinusoidal varying voltage. The output of the VCO will be a frequency (or phase) modulated signal. Magnitude measurements taken at a sufficiently low point below ωp can be approximated to unity gain i.e. the PLL exhibits 100% feedback within the loop bandwidth. Additionally, the phase lag can be approximated to 0°. This means that all measurements taken from the PLL output can be referenced to the first measurement. Figure 14 shows the method for phase measurement calculation between input and output.

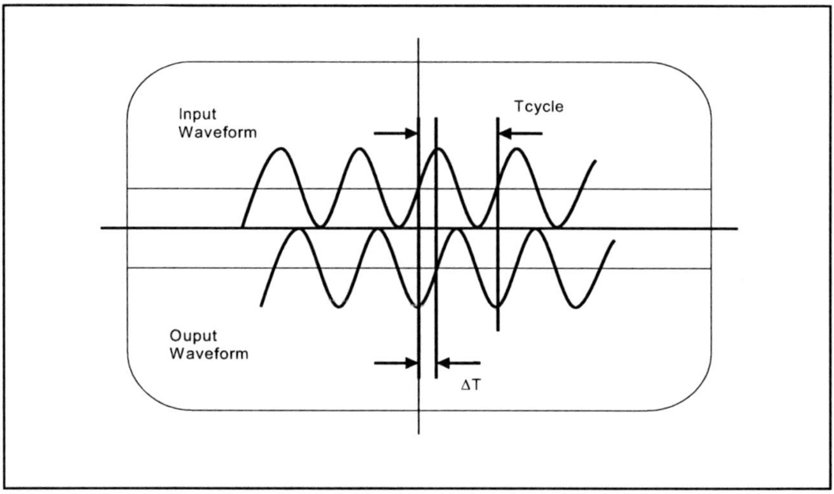

Figure 14 Measurement of phase difference between input and output.

Tcycle represents the time measured for one complete cycle of the input waveform. ΔT represents the measured time difference between the input and output waveform.
Using Tcycle and ΔT, the phase difference (Δϕ) between the input and output signal can be estimated using the following relationship.

$$\phi(jw) = \Delta\phi = \frac{\Delta T}{Tcycle} \cdot 360 \quad \text{(Degrees)}$$

Equation 15

For measurement of the magnitude response it must be recalled that well within the loop bandwidth the PLL response can be approximated as unity. It follows that an initial output signal resulting from an input signal, whose modulation frequency is sufficiently low, can be taken as a datum measurement. Thus, all subsequent measurements can be referenced an initial output measurement, and knowledge of the input signal is not required. For example, if an initial measurement was taken for a modulation of 100Hz, subsequent measurements could be carried out using the following relationship.

$$|H(j\omega)dB| = 20\log_{10}\frac{Vm_{100Hz}}{Vm_N} \qquad \text{(dB)}$$

Equation 16

Where Vm_{100Hz} is the peak-to-peak voltage measured at an input modulation frequency of 100Hz and Vm_N is the Nth peak to peak voltage output for a corresponding modulation frequency.

The technique described above for phase transfer monitoring is almost identical to a similar test technique known as jitter transfer function monitoring [21][22]. However, in this case a known jittery source signal is used as the phase modulation signal, as appose to the sine wave modulation mentioned in this text.

2.1.1.4 Structural decomposition.

This subsection will outline common structural decomposition tests that are often used to ease PLL characterization. In the interests of brevity emphasis towards the analogue sub circuits of the PLL will be given. With reference to section 1 and the associated equations it can be seen that the PLL system is broken down into three main analogue type blocks, consisting of the charge pump, the loop filter and the voltage controlled oscillator. These are considered to be critical parts of the PLL, hence much design effort is spent on these blocks. The blocks are often designed independently so that the combination of the associated transfer characteristics will yield the final desired PLL transfer function. In consequence, it seems logical to attempt to verify the design parameters independently. Typical parameters of interest that are often checked include:

- Absolute Charge Pump Current.
- Charge Pump mismatch.
- VCO gain.
- VCO linearity.

If direct access to the loop filter control node is permitted, all of these tests can be enabled using relatively simple methods. Also, to allow these tests to be carried out, extensive design effort goes into construction of access structures that will place minimal loading on the loop filter node. However, injection of noise into the loop is still a possibility and the technique seems to be less commonly used. A brief explanation of common test methods is now provided.

Charge Pump measurements:

A typical test setup for measuring the Charge pump current is shown in Figure 15.

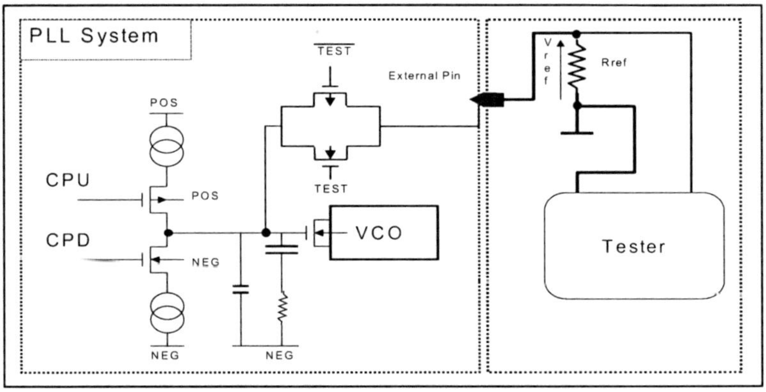

Figure 15 Direct access measurement of charge pump current.

Here, CPU is the up current control input, CPD is the down current control input, TEST is the test initiation signal that couples the loop filter node to the external pin via a transmission gate network, Rref is an external reference resistor, and Vref is the voltage generated across Rref due to the charge pump current. The tester senses Vref, and thus the charge pump current can be ascertained. A typical test sequence for the charge pump circuitry may contain the following steps.

1) Connect the loop filter node to the external reference network by enabling the TEST signal.
2) Activate the up current source by disabling CPD and enabling CPU.
3) Wait a sufficient time for the network to settle.
4) Measure the resultant UP charge pump current using the relationship.

$$Icpu = \frac{Vrefup}{Rref}$$

Equation 17

5) Activate the down current source by disabling CPU and enabling CPD.
6) Wait a sufficient time for the network to settle.
7) Measure the resultant DOWN charge pump current using the relationship.

$$Icpd = \frac{Vrefdn}{Rref}$$

Equation 18

An estimate of the charge pump current mismatch can be found by subtracting the results of equation 17 and equation 18. Also the CPU and CPD inputs can often indirectly controlled via the PFD inputs. Therefore, removing the necessity of direct access to these points and additionally providing some indication of correct PFD functionality.

Note that in the previous description, the test access point is connected to an inherently capacitive node consisting of the VCO input transistor, and the loop filter capacitors respectively. In consequence, if no faults are present in these components, there should be negligible current flow through their associated networks. It follows that this type of test will give some indication of the loop filter structure and interconnect faults.

VCO measurments:

A typical test set up to facilitate the measurement of the VCO gain and linearity is shown in Figure 16 where CPU is the charge pump up current control input, CPD is the charge pump down current control input, and TEST is the test initiation control input. A typical test sequence would be carried out as follows:

1) Initially both charge pump control inputs are set to open the associated charge pump switch transistors. This step is carried out to isolate the current sources from the external control pin.
2) A voltage V1 is forced onto the external pin.
3) The external pin is connected to the loop filter node by activation of the TEST signal.
4) After settling, the corresponding output frequency F1 of the VCO is measured.
5) A higher voltage V2 is then forced onto the external pin.

6) After settling, the corresponding output frequency F2 of the VCO is measured.

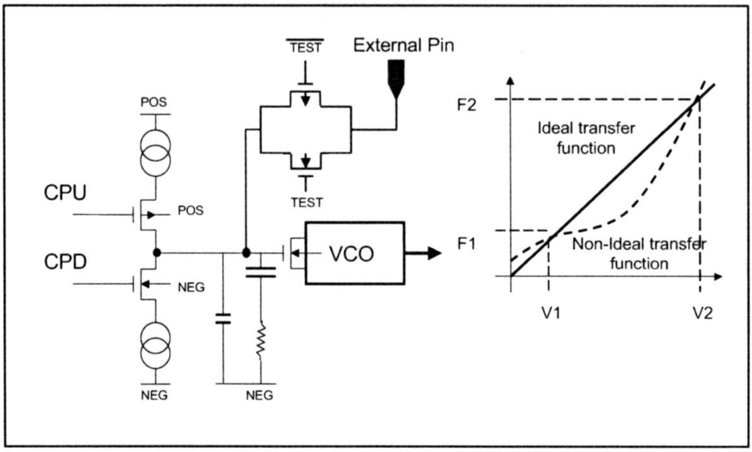

Figure 16 VCO test setup.

In the above sequence of events the values chosen for the forcing voltages will be dependant upon the application.

After taking the above measurements the VCO gain can be determined using the following relationship.

$$K_{VCO} = \frac{F2 - F1}{V2 - V1}(Hz/V)$$

Equation 19

In certain applications it may be necessary to determine VCO non-linearity, this can be determined by taking incremental measurements of the VCO gain between the points V1 and V2 in Figure 16.

2.2 Production Test Focused.

In many situations, due to the problems stated in previous sections the FLT may be the only test carried out on embedded phase locked loops. A particular test plan may therefore include the criteria that the PLL must lock within a certain time for a certain set of divider settings. Often to enhance the FLT results, structural decomposition and ad-hoc DfT techniques such as the ones outlined in the previous sections are used. The PLL is also generally provided with various DfT techniques incorporated into the digital structures, such as:

- Direct access to PFD control inputs and outputs.
- Direct control of feedback divider inputs.
- Direct clock propagation to feedback divider.
- Direct monitoring of feedback divider.

In addition, to these features, an embedded PLL will normally have a bypass mode that allows other on chip circuitry to receive clock signals from an external tester as appose to the PLL itself. This mode allows other on chip circuitry to be synchronized to the tester during system test. In this situation the PLL core is often placed in a power down mode. Particular examples of generic production test methodologies are provided in [18][19].

Ad-hoc DfT methods can be of use for PLL testing, however some of the associated problems, such as, noise injection, and analogue test pin access can introduce severe limitations, especially when considering test for multiple on-chip PLLs. In consequence, there has been recent interest into fully embedded BIST techniques for PLLs. An overview of BIST strategies is presented in section 3.

2.2.1 Jitter Measurements.

This section will provide an outline of typical jitter measurement techniques. Accurate jitter measurements generally require some form of accurate time based reference. Accuracy in this context refers to a reference signal that possesses good long-term stability and small short-term jitter as the reference signal jitter will add to the device under tests generated jitter and be indistinguishable in the final measurement. In consequence, the reference jitter should be at least an order of magnitude less than the expected output jitter of the device under test. For the following discussions it will be assumed that a good reference signal exists. It must be noted that much of the literature devoted to new jitter test techniques, appears to concentrate upon generation of accurate time based measurements, however the basic analysis principal often remain similar. Commonly used measurement and analysis techniques include period measurements and histogram measurements. These techniques are explained below.

Period based jitter measurments:

Period based measurements essentially consist of measuring the time difference between equally spaced cycles of a continuous periodic waveform. A graphical representation of the technique is shown in Figure 17.

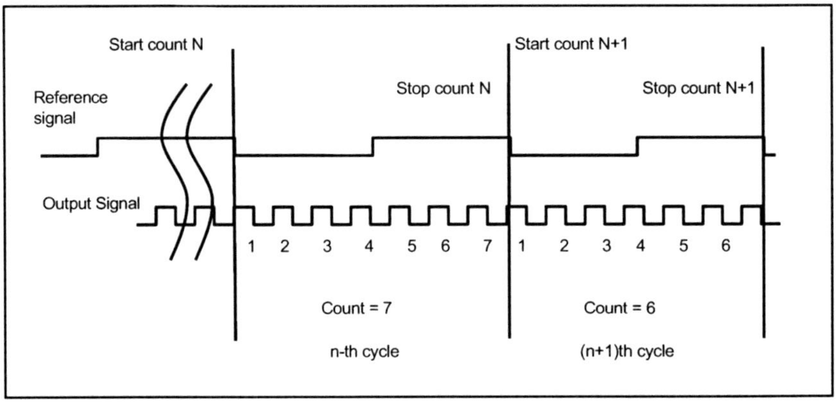

Figure 17 Representation of period based measurements.

This technique essentially carries out a frequency counting operation on the PLL output signal and will measure or count the number of PLL output transitions in a predetermined time interval (gate time) determined by the reference signal. The difference between successive counts will be related to the average period jitter of the PLL waveform. Obviously this method cannot be used to carry out the short-term cycle to cycle jitter measurements. Accuracy of this technique will require that the PLL output signal frequency is much higher than the gate signal.

The signals would be gated as shown in Figure 18 and would be used with fast measurement circuitry to initiate and end the measurements.

Histogram analysis

Histogram based analysis is often carried out using a strobe based comparison method. In this method the clean reference signal reference signal is offset by multiples of equally spaced time intervals. That is the reference signal edge can accurately be offset from the ideal position by.

$$N \cdot \Delta T$$

Equation 20

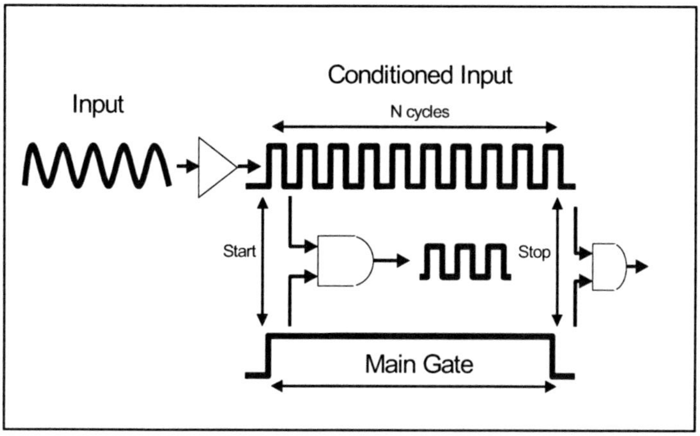

Figure 18 Illustrating gating of period measurement signals.

Where N represents the maximum number of time delays that the reference signal edge can be displaced by and ΔT is the minimum time resolution.

The measured signal is then compared to ascertain how many times its edge transition occurs after consecutive sets of the displaced edges. A Jitter histogram of the measured waveform is then constructed by incrementing N and counting the occurrences of the rising edge over a predetermined set of measurements. A value of N = 0, will correspond to a non-delayed version of the reference signal. The measurement accuracy will be primarily dependant upon ΔT, and ΔT should be an order of magnitude below the required resolution. For example, 100ps measurement accuracy would require a ΔT of 10ps.

An illustration of strobe edge placement for N=7 is shown in Figure 19.

As an example the count values could be collected from a given set of 100 measurements.

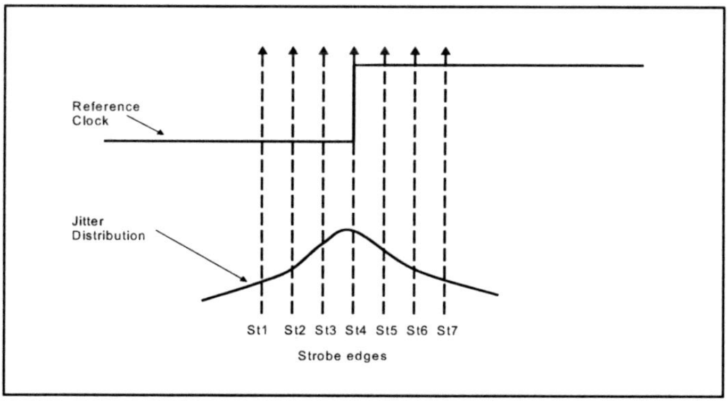

Figure 19 Example of strobe edge placement for 7 edges

Strobe position number	Failure Count	Pass Count
St1	4	96
St2	15	85
St3	35	65
St4	50	52
St5	65	35
St6	85	15
St7	90	10

Table 2 Example values for histogram-based measurement.

The values from the table would then be used to construct the appropriate jitter histogram as shown in Figure 20.

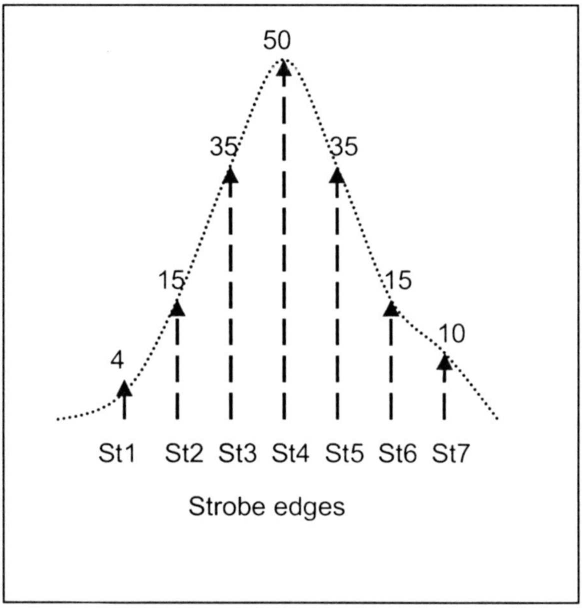

Figure 20 Jitter Histogram constructed from the values in table 2

It must be mentioned that various other techniques exist and are used to facilitate approximation of jitter such as, indirect measurements and Fourier based methods. For indirect measurements, a system function reliant on the

PLL clock signal is tested, typical examples may include SNR testing of ADC or DAC systems. For Fourier based methods the signal of interest is viewed in the frequency domain as appose to the time domain, and the resultant phase noise plot is examined. Proportionality exists between phase noise within a given bandwidth and the corresponding jitter measurement, thus allowing jitter estimation. [23][24].

3. BIST TECHNIQUES.

Although the primary function of a PLL is relatively simple, previous sections have shown that there are a wide range of specifications that are critical to the stability and performance of PLL functions that need to be verified during Engineering and Production Test. These specifications range from Lock time and Capture range to key parameters encoded in the phase transfer function such as damping and natural frequency. Parameters such as jitter are also becoming more critical as performance specifications become more aggressive.

The challenge therefore associated with self-testing PLL's is to find solutions that can;

- be added to the design with minimal impact of the primary PLL function
- have minimal impact of the power consumption
- involve minimal silicon overhead
- can be implemented by non-specialists
- guarantee's detection of faults through either direct or indirect measurement of key specifications and
- be tolerant to normal circuit noise, component offsets process spreads, temperature and supply variations.

The following section identifies several BIST strategies proposed for PLL structures. Only the basic techniques are described here. The reader should consult the publications referenced for more information on practical implementation issues, limitations and potential improvements.

A fully digital BIST solution was proposed by Sunter & Roy [10]. This solution is restricted to semi-digital types of PLL and based on the observation that the open loop gain is linearly dependent on the key parameters associated with the PLL , that is:

$$G_{OL} = \frac{K_p K_v G(s)}{N.s}$$

Where :

K_p - Gain of the phase detector which is a function of the charge pump current

G(s) - frequency dependent gain of the low - pass filter or integrator

K_v - gain of the VCO in radians/sec/volt

N - digital divider integer

s - Laplace variable

The BIST solution opens the feedback loop to ensure the output of the phase detector is independent of the VCO frequency. This is achieved by adding a multiplexor to the PLL input as shown in fig 21. A fully digital method is used to derive an input signal with a temporary phase shift. The method uses signals tapped from the divide by N block in the feedback loop to generate an output that is 180° out of phase relative to the input for 1-cycle of the reference clock only. This phase shift is activated on receipt of a logic "1" on the "Delay Next Cycle" signal.

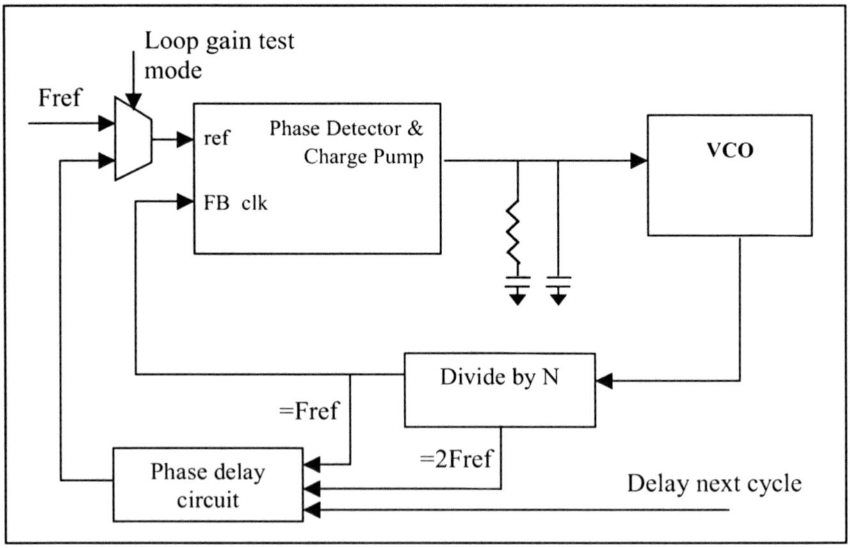

Fig. 21 Circuit for measuring loop gain

This phase shifted signal is applied to the phase detector via the input multiplexor. The strategy used eliminates problems in measuring very fast frequency changes on the VCO output that would result if a constant phase shift was applied. In this architecture, the VCO frequency will change only during the cycle where the phase is shifted. Following this cycle the VCO will lock again (zero phase shift on the input) hence it is relatively easy to measure the initial and final VCO frequencies and calculate the difference. The relationship between the change in VCO frequency as a function of the phase shift on the input, and the reference clock is:

$$\Delta \omega_{VCO} = \frac{K_v \delta . I_{cp}}{f_{ref} . C}$$

and the open loop gain is given by:

$$G_{OL} = \frac{\Delta \phi_{FB}}{\Delta \phi_{FB}} = \frac{K_v \delta . I_{cp}}{2\pi f^2{}_{ref} . NC}$$

So in summary, a digital circuit can be added to the divide by N block to generate a known phase shifted input for 1 cycle (stimulus generator) and a multiplexor added to the PLL input to allow the loop to be broken and the phase shifted signal from the stimulus generator to be applied. All that remains is the implementation of an on-chip solution to measure the output frequency change. This can be achieved digitally using a gated binary counter with a known reference frequency input.

Capture and Lock range measurements are also possible using this architecture by measuring the maximum and minimum frequencies of the lock / capture range. This is achieved by continuously applying a frequency error to force the output frequency to its maximum or minimum value at a controlled rate. The implementation involves connecting the PLL input to an output derived from the divide by N block with a frequency equal to, double or half the VCO output frequency. The VCO output frequency is continually monitored until the rate of change approaches a defined "small" value. This last frequency measurement is recorded. Lock time can also be measured in this testing phase by closing the PLL loop after this maximum or minimum frequency has been achieved and counting the number of reference clock cycles to the point at which the PLL locks.

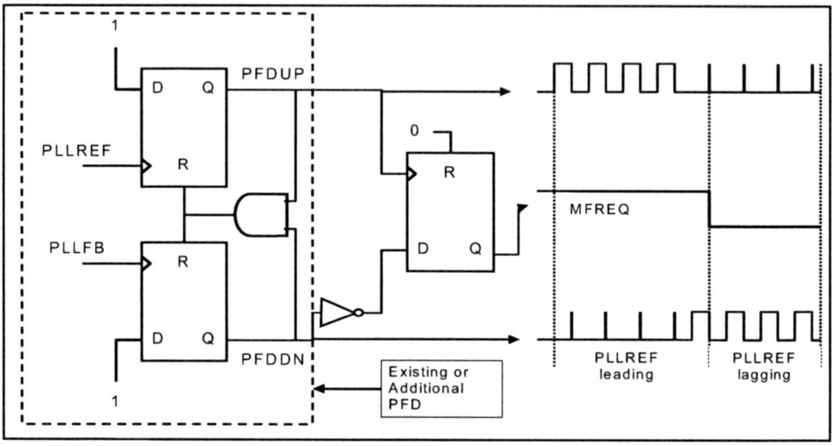

Figure 22 Sampling of output frequency

An alternative architecture for on-chip measurement of the phase transfer function is described Burbidge & Richardson [25] and utilises an input multiplexor as above, a digital control function and a phase frequency detector with a single D-type flip flop added to its output as shown in Figure 22.

The purpose of this modified phase detector is to detect the peak output which corresponds to the response of the PLL to the peak value of the input phase. If a "strobe" signal is generated by the input stimulus generator when the input phase is at its peak, measurement of the time delay between the input strobe and the output change on the Q output of the D-type can generate the phase response at a fixed input frequency. In addition, the point at where this D-type output changes corresponds to the PLL being locked hence measurement of the output frequency at this point will allow the magnitude response of the PLL to be calculated at the reference frequency of the input stimuli. Repeating this process for different values of input frequency will allow the phase transfer function to be constructed. This modified phase detector and the methodology described above is used within the overall BIST architecture shown in Figure 23. The input multiplexor M2 is used to connect or break the feedback loop and apply identical inputs to the PLL forward path to artificially lock the PLL.

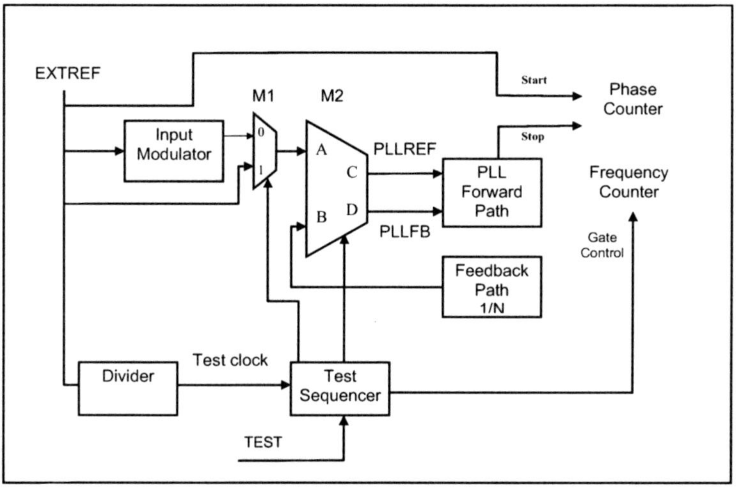

Figure 23 BIST architecture

The algorithm used to construct the phase transfer function is as follows:

Test Stage	M1	M2		Comments
(1) Ref set	0	A=C	B=D	Apply digital modulation with frequency **FN** Start Phase counter (counter referenced to EXTREF)
(2) Set Phase counter	0	A=C	B=D	Start phase counter at peak of input modulation
(2) Monitor Peak	0	A=C	B=D	Monitor for peak output signal frequency
(3) Peak occurred Lock PLL stop Phase counter	X	A=C	A=D	Holds the output frequency constant.
(4) Measure frequency and phase	X	A=C	A=D	Count output frequency and store. Store the result of the phase counter.
(5) Increase Modulation Frequency **FN** and repeat steps 1 to 4 until all frequencies of interest have been monitored.				

Table 3 - Test Sequence

Note that this technique requires an input stimulus generator that provides either a frequency or phase-modulated input with a strobe signal generated at the peaks. Either frequency modulation using a digitally controlled oscillator or phase modulation using multiplexed delay lines can be used.

A third method of achieving a structural self-test of a PLL structure was proposed by Kim and Soma [9] and involves injecting a constant current into the PLL forward path and monitoring the loop filter output which is usually a multiple of the input current injected and a function of the impedance of the forward path.

In this approach additional circuitry is placed between the PFD and CP with the primary objective of applying known control signals directly to the charge pump transistors. In the test the PLL feedback path is broken and control signals referenced to a common time base are applied to the charge pump control inputs. The oscillator frequency will be proportional to the voltage present at the loop filter node, which is in turn dependant upon the current applied from the charge pump. Thus if the output frequency can be determined information can be obtained about the forward path PLL blocks. The test proposal suggests that the loop divider is reconfigured as a frequency counter. The test basically comprises of three steps as follows. Initially closing both of the charge pump transistors performs a DC reference count, if the charge pump currents are matched the voltage of the loop filter node should be at approximately half the supply voltage. The measurement from this test phase is used as a datum for all subsequent measurements. In the second stage of the test the loop filter is discharged for a known time. Finally the loop filter is charged for a known time. For all of the test stages the output is measured using the oscillator and frequency counter and is stored in a digital format. In all of the tests the output response is compared against acceptable limits and pass or fail criteria are evaluated.

This BIST strategy covers most if not all analogue faults and can be extended to test the phase detector by increasing the complexity of the timing and control of the input vectors to the PLL.

Finally it should be noted that methods to measure jitter either directly or indirectly are currently being addressed. In [10] a method is proposed that has the additional advantage of utilizing minimal additional digital functions for implementation. To date however there are few other credible on-chip solutions. This is an important goal as frequencies continue to rise.

4. SUMMARY AND CONCLUSIONS.

This chapter has summarized the types of PLL used within electronic systems, the primary function of the core blocks and key specifications. Typical test strategies and test parameters have been described and a number of design-for-testability and built-in-self-test solutions described.

It is clear that as circuit speeds increase and electronic systems rely more heavily on accurate and stable clock control and synchronization, the integrity and stability requirements for of PLL functions will become more aggressive increasing test time and test complexity. Methods of designing PLL's to be more accurately and more easily tested will hence become more important as the system-on-chip industry grows.

REFERENCES

[1] F M Gardner, "Phase lock Techniques"; Second edition; Wiley Interscience; 1979.

[2] R Best, "Phase Locked Loops, Design Simulation and Applications", 4th Edition, Mc-Graw-Hill, ISBN 0071349030

[3] F M Gardner, "Charge-Pump Phase-Lock Loops", IEEE Trans Comm, 28th Nov 1980, pp 1849-1858.

[4] R Gayakwad, L Sokoloff; "Analog and Digital Control Systems"; Prentice Hall; 1998; ISBN: 0-13-032673-9.

[5] T H Lee, "The Design of CMOS Radio-Frequency Integrated Circuits", Cambridge Universtiy Press, 1998, pp 438-549, ISBN 0-521-63922-0

[6] D A Johns, K Martin, "Analog Integrated Circuit Design", John Wiley & Sons Inc, 1997, pp 648-695, ISBN 0-471-14448-7

[7] M Sachdev, " Defect Oriented Testing for CMOS Analog and Digital Circuits", Kluwer Academic Publishers, 1998, pp 37-38 and 79-81, ISBN 0-7923-8083-5

[8] S. Kim M. Soma, "Programmable Self-Checking BIST Scheme for Deep Sub micron PLL Applications.", Dept Electrical Engineering, Box 32500, University of Washington, Seattle, WA 98195-2500.

[9] S. Kim, M. Soma, D. Risbud "An Effective Defect-Oriented BIST Architecture for High-Speed Phase-Locked Loops", 18th IEEE VLSI Test Symposium, 30th April 1999, pp 231-237.

[10] S. Sunter, A. Roy, "BIST for Phase-Locked Loops in Digital Applications" Logic Vision, International Test Conference 1999. IEEE Proceedings 1999, pp.532-40.

[11] P. Goteti, G. Devarayanadurg, M. Soma, "DFT for Embedded Charge-Pump PLL Systems Incorporating IEEE 1149.1", IEEE Proc. Custom Integrated Circuits Conference, 1997.

[12] F. Azais, M. Renovell, Y.Bertrand, A. Ivanov, S. Tabatabaei, "A Unified Digital Test Technique For PLL's; Catastophic faults covered", 5th IEEE International Mixed Signal Testing Workshop.

[13] B. Vinnakota, "Analog and Mixed-Signal Test", Prentice Hall, 1998.

[14] Philips Semiconductors, Data Sheet PCK857 66–150MHz Phase Locked Loop Differential 1:10 SDRAM Clock Driver, Dec 1st 1998.

[15] Texas Instruments, Data Sheet TLC2933, HIGH-PERFORMANCE PHASE-LOCKED LOOP, APRIL 1996 – REVISED JUNE 1997

[16] MAXIM Semiconductors, Data Sheet *19-1537; Rev 2*, 622Mbps, 3.3V Clock-Recovery and Data-Retiming IC with Limiting Amplifier, *19-1537; Rev 2; 12/01*

[17] MAXIM Semiconductors, Application Note: HFAN-4.3.0Jitter Specifications Made Easy: A Heuristic Discussion of Fibre Channel and Gigabit Ethernet Methods, Rev 0; 02/01.

[18] Mark Burns, Gordern Roberts ,An Introduction to mixed signal IC Test and Measurement, Oxford University Press, 2002.

[19] O K I A S I C P R O D U C T S, Data Sheet, Phase-Locked Loop 0.35µm, 0.5 µm, and 0.8 µm Technology Macrofunction Family, September 1998

[20] Chip Express, Application Notes, APLL005 CX2001, CX2002, CX3001, CX3002 Application Notes, May, 1999

[21] Lecroy, Application Brief, PLL Loop Bandwidth: Measuring Jitter Transfer Function In Phase Locked Loops. , 2000.

[22] B. R. Veillette, G. W. Roberts, "On-Chip measurement ofthe jitter transfer function of charge pump phase locked loops" Proc. Int. Test Conf, 1997, pp 776-85

[23] Bar-Giora Goldberg; "Digital Frequency Synthesis Demystified"; 1999; LLH Technology Publishing; ISBN 1-878707-47-7.

[24] Bar-Giora Goldberg; "PLL Synthesizers: A Switching Speed Tutorial"; Microwave Journal; 2001.

[25] M. Burbidge, A Richardson, J. Tijou, "Techniques for Automatic On-Chip Closed Loop Transfer Function Monitoring for Embedded Charge Pump Phase Locked Loops", DATE (Design Automation and Test in Europe) 2003.

Chapter 5

BEHAVIORAL TESTING OF MIXED-SIGNAL CIRCUITS

Veikko Loukusa, Lic. Eng.
Nokia Mobile Phones

Abstract: Behavioral modeling of analogue and mixed-signal components is gaining a lot of attention because of the significant speedup that high-level simulations can provide. This chapter describes a behavioral model for a design-to-test methodology. The advantages of behavioral testing of mixed-signal circuits will be presented as well as the speedup for test model generation.

Key words: Mixed-signal test, Behavioral testing, Design-to-Test methodologies

1. INTRODUCTION

In the fast developing field of mobile communications companies are faced with many challenges, some of which are straightly related to the design-to-test of integrated circuits. Design-to-test describes successive process phases in developing integrated circuits, called ASICs. The diversity and multitude of future products and their features place strict requirements on the efficiency of the design-to-test process, especially as at the same time the design time, size, cost and power consumption need to be minimized.

While many electronic systems are almost entirely digital devices, there are most that include at least a small part that are analog in nature. Indeed, any systems, which interface to the real world, must contain some analogue components, such as transducers and analog/digital converters. Designers are faced to utilize the advances in IC integration, which leads to System-on-Chip creation. This has resulted in mixed-signal chips continuously growing in complexity with analog-digital interaction and it is predicted to an enormous transfer from digital to mixed-signal ICs [1].

J.L. Huertas (ed.), Test and Design-for-Testability in Mixed-Signal Integrated Circuits, 137–162.

Since the complexity of mixed-signal systems is growing, behavioral modeling of analogue and mixed-signal components, blocks gains more attraction. Behavioral models are descriptions of systems, which can be done in many levels. Using behavioral models a significant speedup of higher-level simulations can be achieved [3, 4]. Also, they allow early design verification, which is uncommon in traditional approaches. In addition, behavioral models may be used for software prototype exchange without giving away any design knowledge. An analog behavioral model is typically a description of a system with differential equations. This model can consist of many levels of accuracy. Analog hardware language description (AHDL) standardization [1] has helped to increase the acceptance of behavioral modeling within a design community. Behavioral models can be developed in a top-down manner, where ideal system characteristics are modeled or bottom-up, where higher-level models are extracted from a detailed circuit realization [5]. In this chapter a top-down methodology is chosen. Top-down models are suitable during initial design phases. For final design verification the models need to account for circuit realization effect. Behavioral models can be derived either empirically or physically.

Executable system descriptions of analog circuits at component level result for practical applications in extreme simulation time (weeks, months, years) and storage requirements (Gbytes). Therefore, more efficient methods than SPICE like simulation of structural system descriptions are required. Unlike in the digital domain, a widely accepted hierarchy of abstraction levels and transformations between different levels is coming into analog design. A possible abstraction hierarchy is presented in fig. 1.

The hierarchy is related to the structural view of a design entity. The component level uses transistors and passive components as basic elements. An entity in circuit level is described by means of opamps or comparators as components. The macro level means that the functionality of a basic element is replaced by an abstract description, for example an ideal voltage source. Models at the behavioral level are constructed by using mathematical relationships between input and output signal. A filter is a good example. The connection pins carry physical signals, which are subjected to conservation laws, in this case Kirchoff's current and voltages laws. Finally, the functional level uses complex basic elements. Typically these models have been designed during system design, which takes place before design phase. The main difference is that conservation laws are not valid at connection pins of functional level; signals are not physical anymore [2].

This chapter bears a behavioural modeling on design-to-test phase. The advantages of behavioral testing of mixed-signal circuits will be presented.

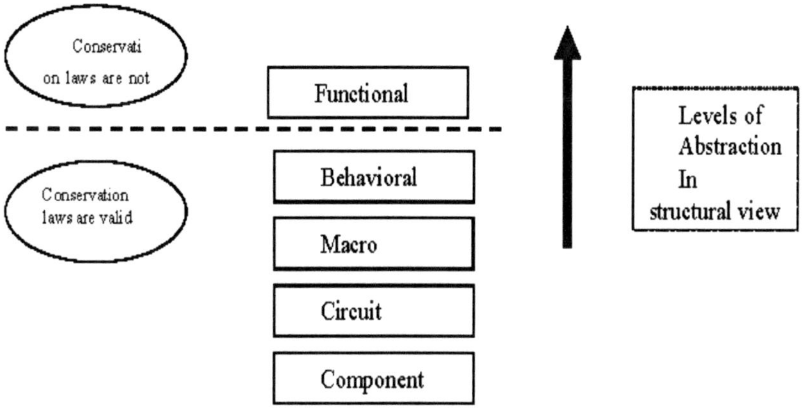

Figure 5-1. Abstraction hierarchy [2].

The speedup of model generation for tests is shown. A demo approach is also presented.

Major technical improvements have made it possible to reduce design cycle time in the IC design. But quite little development work has been made in making the IC test-development cycle more efficient concerning mixed-signal design. Developing IC test programs for ATE is a time-consuming task that takes place near the end of design cycle. In mixed-signal test development test engineers haven't been able to automatically translate design simulation data into test patterns and stimuli tailored for the target test system. There has been lack of top-level mixed-signal simulation data and transfer mechanism into the test system. As a result, manual mixed-signal test vector generation followed by time-consuming test debugging causes a big delay in mixed-signal IC verification [2, 3, 4, 5].

In this chapter, a new approach, utilizing top-level behavioral mixed-signal simulation followed by automated transfer into the test system, is presented. This method replaces manual mixed-signal test vector generation, generates the right mixed-signal stimuli (test patterns) in the target test system. Now, test engineers can fully debug their mixed-signal tests before first silicon in the test system, which is used in characterization or product verification. The motivation, importance and necessity of introducing top-down design-to-test methodology will be presented and a tool flow supporting this methodology will be proposed in this chapter. The importance of a widely accepted multi-level description and model language will be discussed. The chapter ends with a summary including the current status and the future developments in mixed-signal verification. An application of behavioral modeling for mixed-signal test generation and applied results are presented. It is shown that test debugging can be provided in the verification test system before silicon by utilizing simulated behavioral mixed-signal models. Due to the behavioral modeling technique, the computational performance was enhanced to a level allowing efficient test development and debugging. Influence on efficiency in design methods is described.

2. A PRINCIPLE OF TOP-DOWN DESIGN-TO-TEST METHODOLOGY FOR MIXED-SIGNAL TEST GENERATION [6]

The need for improved methods has grown out of the time-to-market requirement and a widespread cry in the mixed-signal community for better handling efficiency of complex mixed-signal ICs during test generation. Conventional methodologies for mixed-signal design and design-to-test do not support these requests well. Fig. 2 a) presents the conventional methodology for mixed-signal design and test development.

As a consequence, in OPTIMISTICS (EU project ESPRIT 24.887) it was decided to develop a new approach. The active players in devising this approach, in the following referred to as *behavioural test generation,* were chosen to represent in-depth knowledge of the problems from the user point of view as well as from the tool development point of view.

The basic objective of the top-down test generation approach was to generate and debug the test program for a mixed-signal chip design before the hardware is built. This would reduce the overall validation and debugging time associated with new mixed-signal prototype chips, hence improving the time-to-market in the fast evolving business segment of mobile telephony. Another important aim was to reduce the entire job of creating a test and validation suite for new complex mixed-signal components. This significantly improves program development productivity and allows the user company to handle even more complex solutions within a reasonable resource.

The plan was to show clear improvement over existing design verification/validation techniques for mixed-signal applications. Reductions in testing costs (e.g., improved productivity, lower total costs, etc.) were a clear objective. Part of the strategy was to obtain better technical performance in terms of accuracy and improved measurement techniques, such as multi-tone testing and other DSP-based testing methods. Improved transfer of results from the mixed-signal design environment was also seen as an important aim, and so was the ease-of-use and openness of the approach.

Over time, software has become a major element in improving the functionality and performance of test systems. A larger and larger part of the system owes its functionality to advanced system tools. This also applies to the approach taken in behavioural test generation.

The principle applied uses a simulation environment to emulate the given mixed-signal function. This emulated function is then used to drive electrical generators that generate the given electrical signals. Hence, complicated equivalent electrical signals can be generated that emulate the electrical signals the designed chip will eventually have when the first silicon is available. As a result, an electrical signal suite is available for debugging of test programs long before the actual silicon can be made available. Fig. 2 b) shows the method for behavioural test generation. The method utilizes top-down design methodology.

The rationale for behavioural test generation is that in the verification approach used by the major user hitherto, debugging of the test program used for chip validation (verification) cannot start until the first prototype hardware is delivered from the chip foundry. Since debugging of the test program today typically takes from 3-4 weeks up to 3 months, valuable time is lost in the design verification process.

The new approach, based on the top-down test generation principle, is to use the analogue and digital test hardware of the engineering test system as part of the emulation of more difficult electrical functions of the chip being prototyped, especially the analogue functions of this chip. This allows verification of such functions before the actual ASIC prototype is available. It further means that the test program itself can be verified before the silicon is available (Fig. 3).

The necessary function to drive the test hardware can be obtained from a mixed-signal simulator using a model of the given chip or a subset of the functionality of this chip. Actually, it is likely to be the case in many situations that only a subset of the functionality is simulated in detail. The performance of mixed-signal simulators is restricted in reaching the maximum level of simulation complexity. Depending on the level of details required for the electrical function of analogue functions, in particular, the

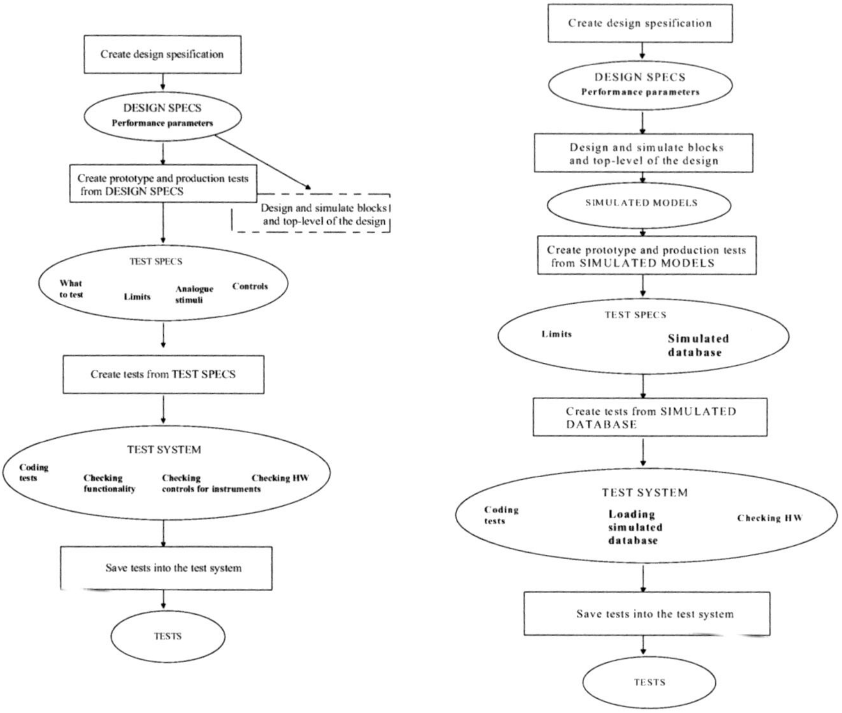

Figure 5-2. (a) Conventional methodology of mixed-signal design and test development, (b) Top-down design-to-test methodology of mixed-signal test generation.

simulator may have to deliver some simulation at the transistor level, while other functionality can be derived from higher-level simulations. On the other hand, the design can be portioned in simulation to different parts that the mixed-signal simulator handles at different levels. One level can be the transistor level; one level can be the behavioural model level, etc. The trade-off between simulation levels, provided that a hybrid approach is available, is an issue related to computational time, accuracy of the electrical signal, ability to subdivide the chip function, etc.

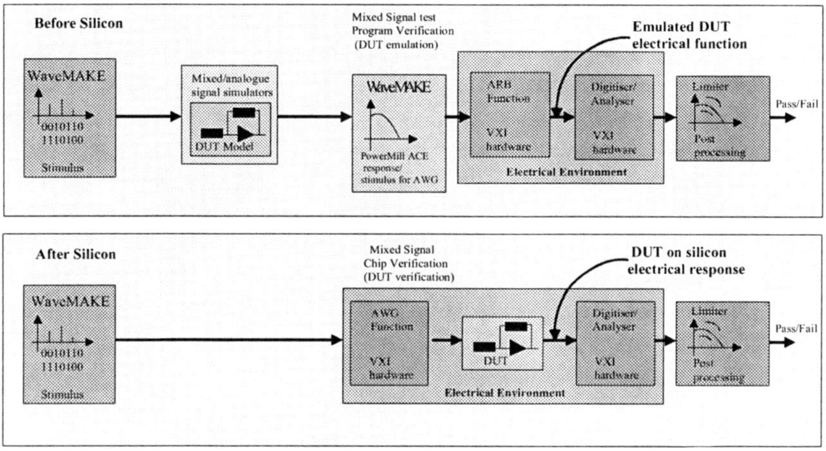

Figure 5-3. Illustration of the principle of top-down design-to-test for mixed-signal test generation. In the upper picture, simulation of a given DUT function takes place. This stimulus is used for the electrical hardware to emulate the function. The function can then be used for test program debugging. In the lower picture, the actual chip is being tested using the hardware modules that were earlier used for test program verification.

3. THE IMPLEMENTATION OF BEHAVIOURAL TEST GENERATION FOR MIXED-SIGNAL TEST

3.1 Modeling for mixed-signal test generation [6]

Modelling is based on design blocks and the top level of the mixed-signal ASIC. A Codec was selected for the demo case. The Codec RX path is presented in Fig. 4. The Codec includes a one-bit digital-to-analog converter, a one-bit analog-to-digital converter, analog filters and gain stages. A digital block controls these blocks. The Codec models are transistor-level ones, which have been designed during the basic bottom-up design procedure

Parallel to these models, behavioural models based on HDL-A and VHDL-AMS were designed. In this case, the design uses top-down methodology. The digital block model uses VHDL coding in top-level behavioural simulation and transistor-level models in top-level transistor-level simulations.

Evaluation of modelling is based on analogue performance estimation of the Codec blocks, mixed-signal functionality of the Codec and start-up functionality checking. Table 1 and Table 2 show analogue performance figures created in different simulations. The S/N ratio figures show that behavioural models form a necessary model base for good quality signals.

Just generating the signal - generating the emulated stimulus - is somewhat of a time consuming task. Tables 3 and 4 clarify this problem. Table 3 shows the reality in mixed-signal top-level simulation until now. Adding the digital block as a gate-level model schematic simulation in general is possible in PowerMILL ACE. Behavioural models can be simulated in a time that is realistic for test generation for hundreds of different measurements.

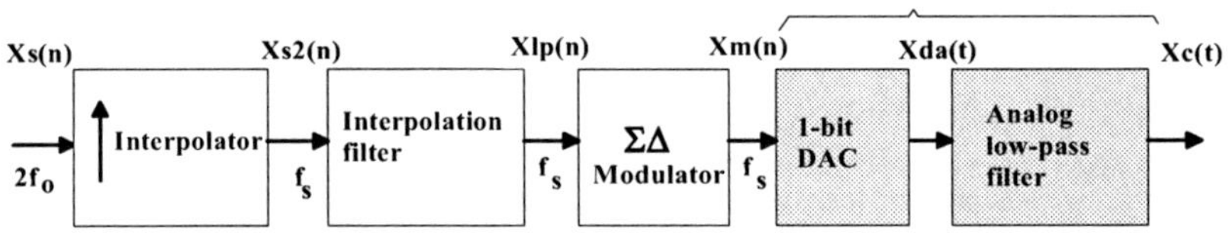

Xs(n)	digital input signal
Xs2(n)	upsampled signal
Xlp(n)	multibit digital signal
Xm(n)	***1-bit digital signal PDM data***
Xda(t)	DAC output signal
Xc(t)	output signal

(b) DAC

Figure 5-4. Codec RX path.

Table 5-1. Transient simulation times for a Codec without any digital block.

Block	Accusim II [days]	*PowerMILL ACE [hours]*
TX path	28	26.4
RX path	23	8.9

Table 4 shows a very important case where successive functions - start-up and use of a block - can be simulated. The start-up sequence is too slow to be simulated in a traditional way, even in an accelerated mode. And as

Table 5-2. Simulation times and performance (S/N [dB]) excluding the digital block

Accusim II		PowerMILL ACE	*VHDL-AMS Design Station (beta)*
Transistor-level	HDL-A model	Transistor-level	VHDL-AMS model
28 days	15 hours	26.4 hours	4.5 hours
Accusim II		*PowerMILL ACE*	*VHDL-AMS Design Station (beta)*
Transistor-level	HDL-A model	Transistor-level	VHDL-AMS model
79,52	88,7	51,5	86,5

another result, the behavioural simulation allows running of these functions in a single run.

As a result of this evaluation, the method allows the user to mix different models to create an accurate enough model for mixed-signal test generation.

All simulations have been run in Sun SPARCstation 450, which is driven by four CPUs at a 450 MHz clock frequency with 4 GB of cache memory.

Table 5-3. Top-level simulation.

Block	PowerMILL ACE *Transistor-level Model*	Continuum HDL-A – VHDL Model	VHDL-AMS Design Station *VHDL-AMS Model*
TX path	48.5 hours	6.5 hours	Not done
RX path	31.0 hours	5.5 hours	Not done

Table 5-4. Mixed-signal test generation

Function	PowerMILL ACE Transistor models	Continuum VHDL and HDL-A models
Startup *230 ms*	16 days	1.5 hours
Audio Codec TX path *15 ms*	48 hours	6.5 hours

3.2 An implementation of top-down methodology design- to-test for mixed-signal test generation

A top-down methodology and a test generation use of behavioral simulation have been presented as a level of principle earlier. Next these methods are made concrete in the following ways:

A new reallocation of a mixed-signal design and test is shown

An influence of introducing an A-HDL (Analog High level Description Language) mixed-signal test generation is presented by a mixed-signal ASIC design process

A demo for A-HDL test generation

A mixed-signal ASIC design process for test generation, described in section 2, shows how design and test specifications will be presented. An A-HDL test generation means that validation tests are generated from simulated models. As a result we get test specifications, which consist of simulated database and limits. Running an existing test is just a load operation and a test run. Running a new test consists of test coding, loading simulated database, test system HW checking and saving the test.

Next this methodology is described in detail by showing a mixed-signal design process for a 100 000 transistor mixed-signal design. Fig. 5 shows the ASIC design flow as a traditional process. A 100 000 transistor mixed-signal design means in this case minimum a one and half year design process with 15 design engineers and 5 test engineers.

The work of design engineers can be split into ASIC specification, block level design and simulation and top-level design and simulation. The top-level design is delivered to an ASIC vendor who processes the design onto silicon. A typical figure for an ASIC design can be fixed as 6000 transistors per one man-year design work. A quick look gives an impression that a design work is slow. A mixed-signal design work really is quite slow, because analog design is quite totally manual work.

ASIC specification produces ASIC specification documents, which define the top-level and block level designs of transistor-level ASIC implementation. During the design a design engineer produces block-level transistor sheets and ties them hierarchically into the top-level transistor design. Block-level and top-level simulations guarantee the process. The block-level simulations are SPICE simulations, mostly checking the performance, which is specified in the design specification. The top-level simulations are run in a table based event-driven simulator. The models are in both simulations transistor-level models. The used a table based event-driven simulator was PowerMILL ACE (trademark by Synopsys).

The conventional validation test system development consists of test plan design, test system & test program design and test system verification. After test system verification the ASIC prototype validation takes place.

The resource consumption is presented step by step presented in table 5.

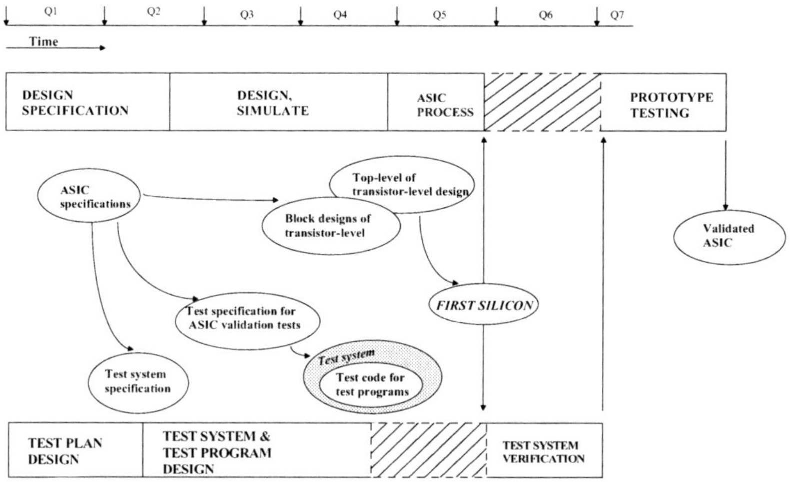

Figure 5-5. A conventional mixed-signal ASIC design-to-test process..

In addition the table shows the time gaps, which generate into both design and test flows.

Fig. 6 fixes the ASIC design flow into an A-HDL mixed-signal test generation for the same ASIC design process with the similar resources. There are two main differences compared to the conventional process. First, behavioral-level models are taken account for use for test system development. Second, there is no time-slot between design and test development.

ASIC specification produces ASIC specification documents, which define the top-level and block level designs of transistor-level ASIC implementation. During specification a design engineer guarantees his/her specification by modeling the design block and top-level. Models are

Table 5-5. Resources in a conventional ASIC mixed-signal design (my = man-year)

Task	Design Engineers	Test Engineers
Design Specifications	6 my	
Test Plan Design		1 my
Design & Simulate	8 my	
Test System&Test Program Design		2 my
Test system verification		1 my
Prototype Testing		1-2 my
Design development gap	*1 my*	
Test development gap		*1 my*
ALL TOGETHER	*15 my*	*5-6 my*

behavioral models, in the beginning they are theoretical, during the specification models will get practical features, for example fixed output impedance. Block level designs are tied to top-level design hierarchically. Later these models can be automatically replaced by transistor-level designs. The models are guaranteed by H-DLA (trademark by Mentor) or VHDL-AMS (standard behavioral modeling language) simulations. These models are used for test system programming and test system verification. At first the models include theoretical signals, later timing and delays are inserted into them. A WaveMAKE filter transfers the models. WaveMAKE is a signal waveform viewer, which has been developed by MicroLEX Systems, Denmark. The transistor-level design takes place after the models have been fixed.

The validation test system development consists of test plan design, test system & test program design and test system verification. After test system verification the ASIC prototype validation takes place. And this can start immediately after the silicon has arrived. Test design plan produces a test system specification and test specifications for ASIC validation tests. Now test specifications include the controls, which are received from behavioral simulations. There is no risk for a mistake in control programming. In test system verification phase simulated responses can be used for test debug. All these signals are transferred by an appropriate WaveMAKE filter.

Table 5-6. Resources in an A-HDL based ASIC mixed-signal design (my = man year).

Task	Design Engineer Now / Savings [my=man-year]	Test Engineer Now / Savings [my=man-year]
Design Specifications	6 my / - 1 my	0 my / 0 my
Test Plan Design	0 my / +0.5 my	1 my / -0.5 my
Design & Simulate	8 my / 0 my	0 my / 0 my
Test System&Test Program Design	0 my / +0.5 my	2 my / -1 my
Test system verification	0 my / +0.2 my	1 my / 0 my
Prototype Testing	0 my / 0 my	1-2 my / 0 my
Design development gap	*1 my / -1 my*	
Test development gap		*1 my / –1 my*
ALL TOGETHER	*15 my / -0.8 my*	*5-6 my / -1,5 my*

The resource consumption is presented step by step presented in table 6. In this case resource save is 0.8 man-years in design engineer work and 1.5 man-years in test engineer work.

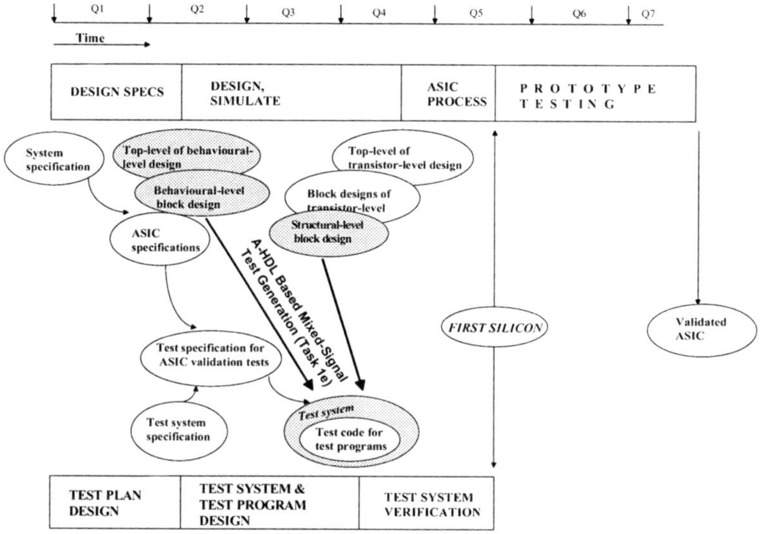

Figure 5-6. An A-HDL based mixed-signal ASIC design-to-test process.

This approach helps test engineers' test vector setup, test debug and verification. Test engineers can perform IC test development prior to first silicon. Test development time shortens, the overall product development time as well, test development and debugging can occur in parallel with design, layout and fabrication. Savings in man-years can be seen in table 6. In the research case total savings are 2.3 man-year resourcing. One big benefit is that simulation and verification cases can be merged. A second benefit is that this methodology creates accurate mixed-signal test vectors. The complex patterns have been verified in simulation. The signal coherency, which is necessary in mixed-signal test, is reached [12].

3.3 A demo test generation [6]

During the research, for a demo for behavioural testing of mixed-signal circuits, Codec RX path test generation methodology was implemented. In this case, top-level simulations generate top-level models of Codec RX. These mixed signals can be captured by wave viewer WaveMAKE. In WaveMAKE the test engineer/design engineer sets the vector formats of these signals and maps them into a digital module. Now the test has the right controls and test stimuli in the test system. Fig. 7 shows the principle of the demo case. SequenTEST is a test sequencer, which has been developed by MicroLEX Systems, Denmark.

As a typical mixed-signal test, the use of simulated signals as controls and stimuli assumes synchronisation of the analogue response and the digital

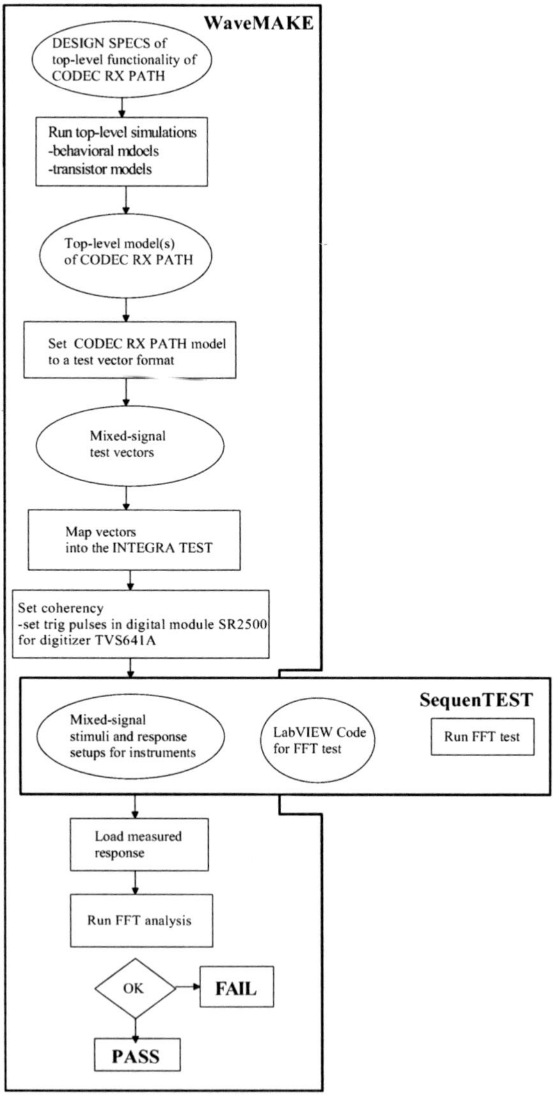

Figure 5-7. Codec RX path demo.

stimuli. Synchronisation has not been provided in today's instruments as a matter of course. In this case, one pin of digital module SR2500 was used as a trigger signal. This signal was cabled to a Tektronix digitizer TVS641A that captures the response of the Codec RX path.

The test outputs the result into WaveMAKE, which presents the simulated response signals v_earn_, v_earp_, v_hf_ and v_hfcm_, and the corresponding measured response signals m_EarN_, m_EarP_, m_HF_ and m_HFCM_ of the Codec RX path. The evaluation figure is the FFT analysis of the RX path output. Fig. 8 shows the WaveMAKE display at the end of the test. The measured and simulated responses can be compared and evaluated using FFT analysis.

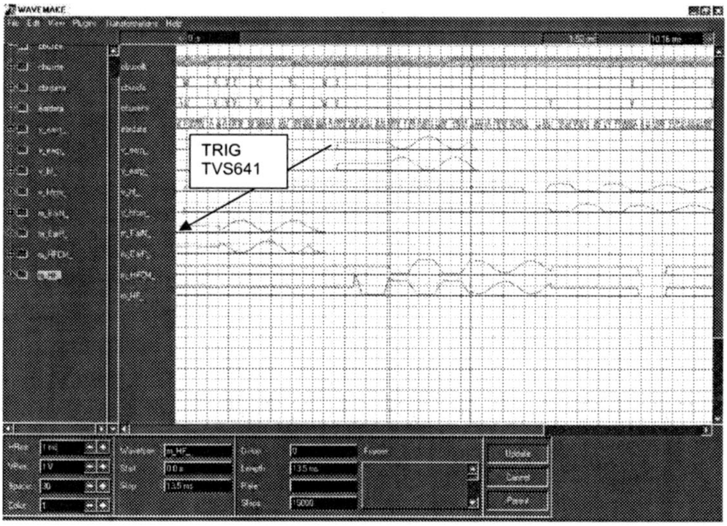

Figure 5-8. Demo design-to-test presented by WaveMAKE. Simulated and measured responses.

3.4 Test system approach

The test system concerned is a series of test systems focused on a variety of application sectors, but having a substantial part of the main software platform common for the systems of the different application domains. This prototype system meets the requirements of the user by offering openness, modularity and not least of all, performance, and it is planned to be integrated into the user's applications. The test system is a microLEX Systems IntegraTEST.

The system solution to be introduced yields a variety of test features like DSP-based test techniques for A/D and D/A converters and a wide range of mixed-signal functions. Test features include, for example: multi-tone testing, noise testing and a large number of mixed-signal testing capabilities, all operating in a coherent mixed-signal test environment.

An important feature of the system is the integration of the mixed-signal simulation software with that of the verification system. Fig. 9 shows the basic idea of test system IntegraTEST. The basic aim is to manipulate different types of signals in a true, common mixed-signal environment. Whether preparing for mixed-signal testing and verification or actually performing tests of mixed-signal components, the issue remains difficult. In the mixed-signal environment we need a tool that allows us to combine analogue waveforms and digital signals in a common visualisation window and to have access to support facilities for generating and manipulating signals, integrating signals generated through mixed-signal simulation, and adding constraints such as noise. On the other hand, the tool has to be able to combine results from several simulators used in testing. Simulation of complex mixed-signal designs is often based on partial simulation results with inputs from analogue simulators, like SPICE, and digital VHDL-based simulators or mixed-signal simulators, like PowerMILL ACE. In this case, the design can comprise behavioural models or transistor-level models. During the development of this method, transistor-level models were simulated in Accusim (SPICE) and PowerMILL ACE. Behavioural models have been simulated in Continuum (VHDL and HDLA models) and in AdvanceMS (VHDL-AMS models). The tool must be able to allow signals from various sources to be assembled and organised with respect to each other, including phase relationships. Using a common window avoids problems of incorrect phases between analogue and digital signal

The system is planned to be applied in the verification of prototypes, where the facilities offer quick programming and easy set-up as well as fast debugging and hence, faster time-to-market. The interface solution between mixed-signal simulation and mixed-signal testing is an important part of the solution. Such facilities will reduce the verification time even more.

4. GENERIC TEST GENERATION MODELING

Behavioral modeling and simulators are mature to enable the designer to generate a set of mixed-signal test vectors. The set can be specified in the simulation phase as one simulation case and comprehensively in testing

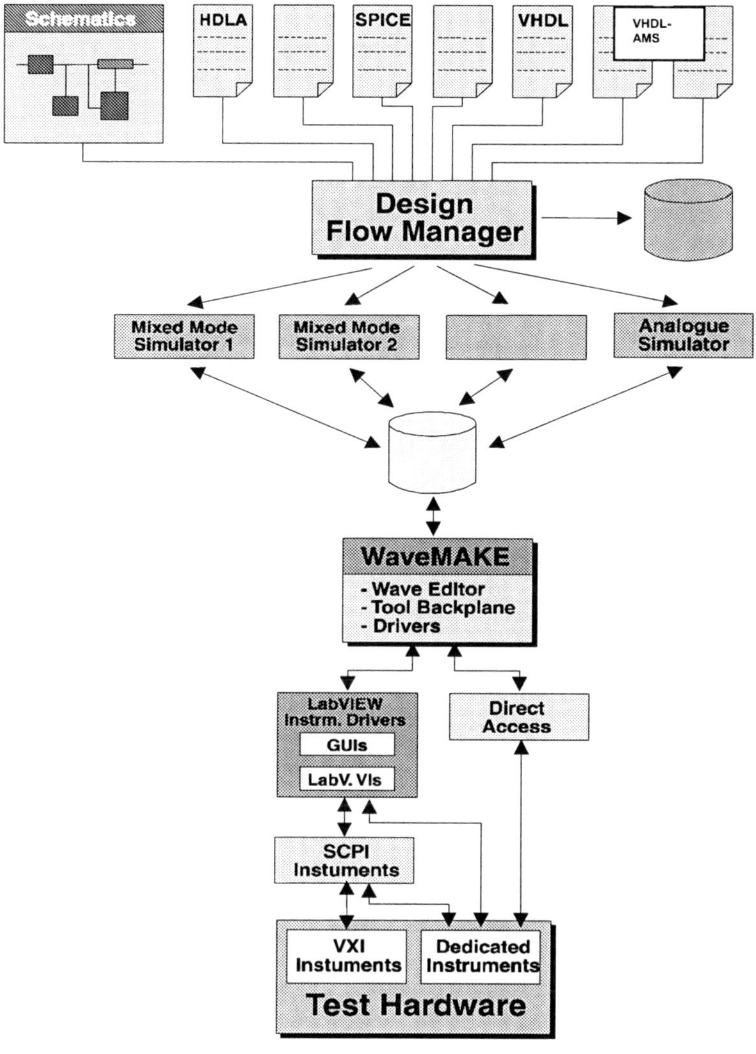

Figure 5-9. Interface between a design environment and the engineering test system

phase as one test case. Generic test generation modeling generates mixed-signal test vectors by utilizing the behavioral model simulation database, from which the method generates tester instrument format test vectors. The method can be compared to the corresponding digital test vector generation.

In this method, efficient top-level simulation requires the use of a mixed mode (analog and digital) simulator. In order to achieve considerable simulation times, behavioral models of key analog and digital blocks were developed. So in this project, ADVanceMS (Mentor trademark), a Modelsim like mixed-signal simulation environment was used. ADVanceMS supports several different description formats such as VHDL (behavioral and structural), VHDL-AMS (behavioral and structural), Verilog, SPICE/ELDO and C language.

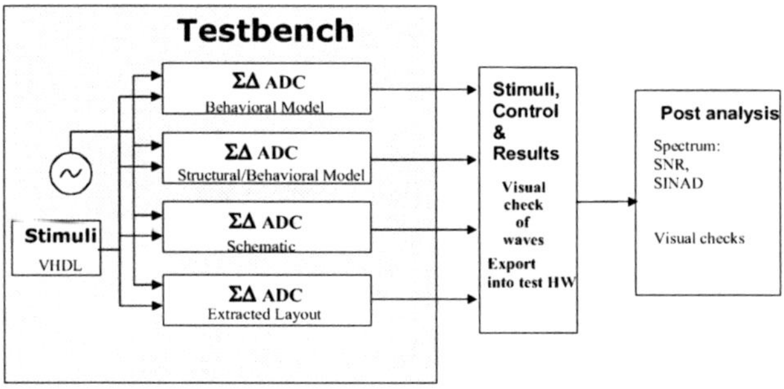

Figure 5-10. The principle of generic test generation model.

In order to accomplish generic test generation, analog blocks are modeled by VHDL-AMS and digital blocks modeled by VHDL. These models are available in the early design phase. As the design process goes on, the SPICE/ELDO models are available, too. ADVanceMS accepts a multi-modeling hierarchy.

The principle of generic test generation model is shown in fig. 10. At the beginning a behavioral model of a sigma-delta converter is created. It specifies the converter. The structural model is used after the behavioral simulations to find specifications for the circuit blocks and to verify the operation with the most important non-idealities. The behavioral model describes the high level behavior of the system, and the structural model describes and models one possible implementation of the system in more detail. The simulation results of the structural model should be compared with the results of the behavioral simulations, as shown in fig. 1, to verify the correctness of the model development.

After simulations the RTL level VHDL description of digital part can be used in the logic synthesis, though the code may require some modifications. The correctness of the operation of the structural model should be verified by simulating it in the same testbench with the behavioral model and by comparing the results. The structural model can be used as a reference in later circuit simulations.

The main features and advantages of the structural model are two-sided. Specifications for the circuit blocks such as amplifiers can be determined and optimized. Behavior of the system can be simulated taking into account implementation specific non-idealities. The structural code can be used as stand-alone with ADVanceMS simulator, or it can be imported to design frameworks and used as a part of a larger system.

After high-level simulations with VHDL-AMS descriptions and finding block specifications these blocks can be substituted one by one with transistor level descriptions (schematics). This way larger systems can be gradually and more easily developed to full transistor level designs.

The simulation case, concretely the mandatory testbench, can be described in two ways. It can be a simulation sheet or a VHDL-AMS code. The simulation sheet is formed from the top-level design sheet by adding necessary analog generators for stimuli, response line terminations and a test vector VHDL model, which generates the digital control and is labeled by a generic test number. The designer selects the test number in the simulation phase. The test number links the simulation to the corresponding test case. For ADVanceMS simulation environment the simulation case is a SPICE netlist, which is shown in fig.11 and which includes VHDL models for both the digital block and digital test vector model (one block in the sheet) and VHDL-AMS models for analog blocks. The description of the test vector model is shown in fig. 12.

Analog stimulus (current or voltage source):
```
V_I$189  miclp vss SIN( 1.35 0.5 3000 )
```
Mixed-signal test vectors (digital test pattern):
```
.MODEL   TestHarness_ams(Harness_ams)   MACRO
LANG=vhdlams LIB=work
Y_I$1240 TestHarness_ams(Harness_ams)
+ port : cbusclk cbusda cbusenx
```
Analog response (impedances):
```
C_I$9  vss earn 100p
C_I$8  vss earp 100p

          R_I$6  earp earn 33
```

Figure 5-11. A netlist view of a simulation sheet.

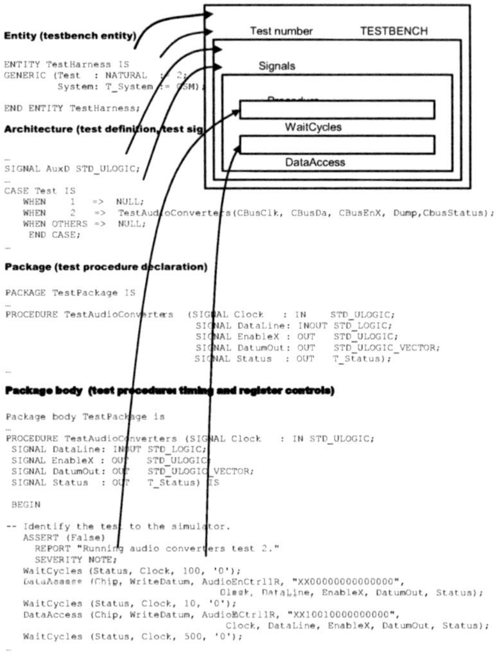

Figure 5-12. A test vector model.

The VHDL-AMS coded testbench, shown in fig. 13, describes the analog stimulus and simulated response as VHDL-AMS equations. The test case is, as earlier port-mapped test vector VHDL model in the testbench code. Similarly all simulated blocks, the digital block and analog blocks, are port-mapped into the testbench.

```
--Analog stimulus:
v_miclp == 0.5*sin(2.0*math_pi*3.0e3*NOW)+1.35;
--Mixed-signal test vectors (digital test pattern):
gen1: ENTITY TestHarness_ams(structural)
      PORT MAP(cbusclk cbusda cbusenx);
--Analog response (impedances):
v_earp== i_earp*R_LOAD;
```

Figure 5-13. A VHDL-AMS coded testbench.

During and after the design phases of a mixed-signal IC it is important to verify the correctness of the developed models and descriptions, i.e. those descriptions at various phases have the same functionality and performance. For generic test generation it is important that the various descriptions (behavioral model, behavioral/structural, schematic level, extracted layout) can be simulated in the same testbench using the same stimuli and test vectors (mixed-signal stimuli). The evolution of design is transferable into test debugging in real time.

5. EXTRACTED BEHAVIORAL TEST GENERATION MODELING

The test vectors, test stimuli and expected test response for test debugging are extracted from the design phase simulations, and the resulting signal formats are used for test vectors, test instrument setups of the analog generators and meters, measured responses.

Test vectors are set into tester acceptable format and mapped into the target pins of the digital module of the test system. The digital module generates mixed-signal test vectors in tests. The Mentor wdb-format database is handled by SimView list window. Selected time-equidistant signal are stored into a file, which the wave viewer WaveMAKE is able to import. WaveMAKE exports the database by utilizing its filters into the digital module in the test system. The digital module emulates the digital control of the mixed-signal chip. Extracting a behavioral testbench model, which is test case number 2 in the VHDL-AMS simulation, forms the digital control. In fig. 14 is shown the extraction of the digital control.

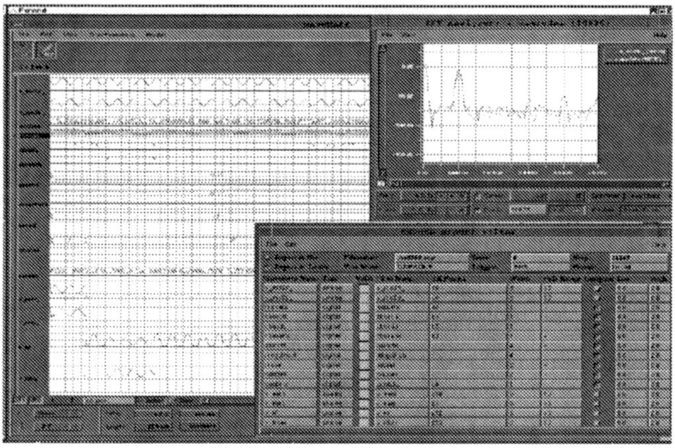

Figure 5-14. Extracted behavioral test generation modeling of digital or arbitrary signals.

Respectively, analog stimulus and response are used for generating needed SCIPI commands to drive the instruments during tests. So a mixed-signal test is run and debugged in the real test system by using extracted behavioral modeling. It is no need simulate test system itself, as it is in virtual testing [7]. In fig. 15 it is shown, how the SPICE netlist or VHDL-AMS testbench is extracted. The abbreviation SIN selects the LabVIEW driver of a function generator with the parameters (amplitude, offset, frequency) in corresponding expressions. Signal terminals cause a relay setup with appropriate terminations in the test system.

Test development phases include now extracted test generation modeling instead of manual test vector setup and instrument driver setups. Test specification consists of limits and extracted test generating models. Test system debugging can be concentrated on HW checking. So test structure is an assembly of necessary HW.

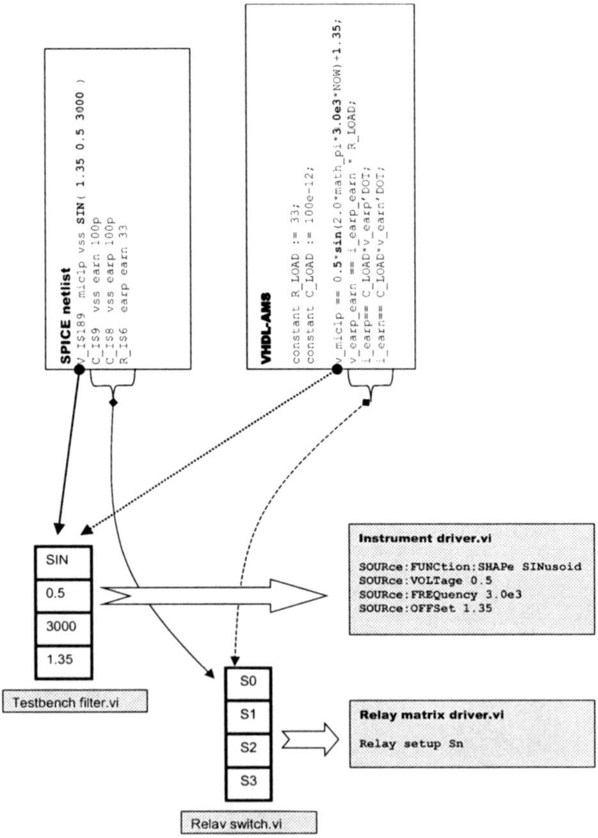

Figure 5-15. Extracted behavioral test generation modeling of analogue signals.

6. A CASE STUDY

For the evaluation, we used an IntegraTEST mixed signal test system as the tester and Mentor Graphics ADVanceMS simulator with Mentor framework as producing necessary behavioral simulation models.

The device under test was a mixed-signal design with 100 000 transistors. A 100 000 transistor mixed-signal design means in this case minimum one and a half-year design process with 15 design engineers and 3 test engineers. The device included the digital block and the analog codec block in addition to necessary analog power-up blocks. Analog blocks were modeled by VHDL-AMS language [8]. Respectively the common digital block is described in VHDL.

```
ENTITY sdm_mod IS
GENERIC (A: REAL := 1.0;
         B: REAL := 2.0;
         C: REAL := 2.0);
PORT        (SDMP        ELECTRICAL;
             SDMM        ELECTRICAL;
             Y           REAL := 0.0);
...
ARCHITECTURE behavior OF sdm_mod IS
QUANTITY X across SDMP to SDMM;

...
             VARIABLE Stage1     : REAL := 0.0;
             VARIABLE Stage2     : REAL := 0.0;
             VARIABLE Comparator : REAL := -1.0;
...

      Stage1 := Stage1 + X - A * C * Comparator;
      Stage2 := Stage2 + Stage1 - B * C * Comparator;

      IF (Stage2 <= 0.0) THEN
             Comparator := -1.0;
             Y <= V_MINUS;
           ELSE
             Comparator :=  1.0;
             Y <= V_PLUS;
      END IF;
...
END ARCHITECTURE;
```

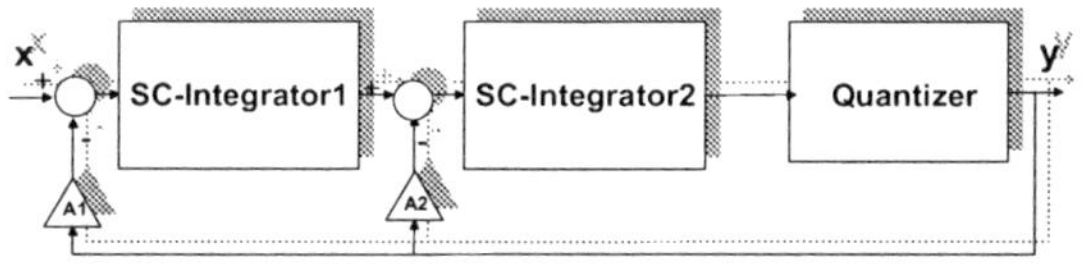

Figure 5-16. A behavioral model of a one-bit second-order sigma-delta modulator.

The modeled analog block is a one-bit second-order sigma-delta modulator. Fig. 16 shows the VHDL-AMS behavioral model. The behavioral model enables to tune feedback coefficients, verify stability and SNR figure. The structural modeling, in fig. 17, is refining the model towards the physical implementation. It models the actual structure. The structural model aids to find amplifier and quantizer specifications, to check clocking and timing and to find switch resistance and capacitance sizes. Table 7 shows the simulated S/N performance figures in dBs and computational times during the design. The simulation runs has been done in SUN SPARCstation with 450 MHz clock with 4 GB cache memory. The table shows respectively the speed of block-level and top-level simulations for behavioral and structural models. The Accusim by Mentor is a SPICE simulator. Respectively PowerMILL ACE is an event-driven circuit-simulator, which uses transistor-models of gates.

The testcase selection is based on VHDL test vector generation, shown earlier in fig. 13. The top-level parameters are the test number and the system selection (GSM, CDMA etc.). The designer has to select the test number and the system and recompile the testbench. In the case new test the designer has to add the new test into the VHDL case structure, and to code the needed controls for registers. Fig, 12 presents the test procedure calls and timing calls WaitCycles, which define the timing of a test vector pattern.

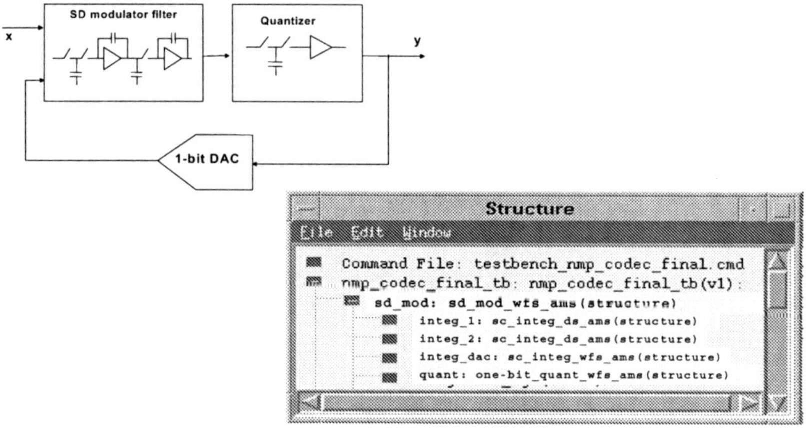

Figure 5-17. A structural model of a one-bit second-order sigma-delta modulator.

Using specific filters tested extraction from simulation into the IntegraTEST. The filters format simulation data into tester specific format. Digital test vectors and digital stimuli/response of codec were mapped into SR2500 digital module, which is the physical description of the VHDL test vector model. The codec analog response signals were mapped into digitizer TVS641A, which respectively is the physical description of analog response. For evaluation, it was used a DAC function of the codec. This method was presented in IMSTW01 [6].

Table 5-7. Simulation times for block level and top-level codec and block level performance (S/N [dB]) of the codec.

	Accusim	PowerMILL ACE	AdvanceMS Behavioral	AdvanceMS Structural
Block level	28 days – 79.52 dB	26.4 h – 51.5 dB	11 min – 86.5 dB	1.1 h – 81.5 dB
Top-level	NA	48 h	5.3 h	16.0 h

The generic behavioral test generation modeling means that verification tests are generated from simulated models. As a result we receive test specifications, which consist of extracted test generation models and test limits. Running an existing test is just a load operation and a test run. Running a new test consists of test coding, loading extracted database, test system HW checking and saving the test.

The testbench, used in modeling, is based on the top-level description of the device. The schematic based testbench is a natural and automatic output of design process. It is conveniently a SPICE netlist, which is made by a netlister. Before netlisting it just exists need to add stimulus and response elements, which further are the direct link to the instruments of the test system. Generic behavioral test generation can start immediately after the behavioral model of a sub-block is ready. AVHDL-AMS testbench has to be created manually, which supports to use the schematic based testbench as early as possible.

Mixed-signal behavioral modeling has been launched quite effectively during last years [9]. VHDL-AMS models are accurate; as well they are available early in a design process. The simulation times of top-level mixed-signal devices are now reasonable [10, 11]. Related to conventional high-level descriptions, behavioral models describe the silicon and real signal values in interfaces.

Reuse of design is today very important in IC design and verification. This methodology supports resuse of verification as follows. Testcases emulate top-level simulations. Tests themselves can be transferred into different test systems. The extraction filter library defines the supported hardware. Test generation modeling as well is transportable into different design environments.

7. CONCLUSIONS

It seems fair to conclude that behavioural testing of mixed-signal circuits for mixed-signal design-to-test generation is a promising approach for mixed-signal design-to-test. Especially for digitised analog applications, where a limited number of input and output pins are present, the emulation approach taken appears the most optimal solution, given the complexity of the mixed-signal functions.

One direct result up to now has been to introduce new solutions in the IC design-to-test flow. The main result is to demonstrate that the system approach is very applicable for verification of mixed-signal chips for mobile communications area. The gains in total development time are extremely encouraging and indicate savings of months in the total development time of such complex chips undertaken in the present activity.

Generic behavioral test generation modeling was shown to be technically suitable for generating and handling the complexity of mixed-signal device verification. Configuring mixed-signal instruments into a large amount of different testcases can be made in considerable time. New testcases, manually impossible, are feasible to take account.

8. ACKNOWLEDGEMENTS

The research results presented in this chapter has been partially supported by the European Union (EU) under ESPRIT programme, Best Practices in Electronic System design (ESD) domain, Project No. 24.877.

9. REFERENCES

[1] J. Oudinot, M. Robbe, Patrick Radja: Mixed-Signal ASIC Top-Down and Bottom-Up Design Methodologies using VHDL-AMS, 1999

[2] R. Rosenberger, S.A. Huss: A Systems Theoretic Approach to Behavioral Modeling and Simulation of Analog Functional Blocks,1998

[3] R. Saleh, S.-J. Jou and A.R. Newton: Mixed-Mode Simultion and Analog Multilevel Simulation, 1994

[4] G. Gasinovi, J.-M. Jang: Multilevel simulation of large analog systems containing behavioural models, IEEE Trans. Computer-Aided Design, vol. 13, pp. 1391-1399, 1994

[5] C. Borchers: Symbolic Behavioral Model Generation of Nonlinear Analog Circuits, IEEE Trans. On Circuits and Systems-II,vol. 45, no 10, 1998, pp. 1362-1371, 1998

[6] V. Loukusa et. al: Model based Test generation: A means of Improving Test Quality and Time-to-Market in Mixed-Signal Test, IEEE 7th Mixed-Signal Test Workshop, pp. 161-173 , 2001

[7] K. Einwich, G. Krampl, R. Hoppenstock, P. Koutsandreas, S Sattler: A Multi-level modeling Approach renfering Virtual Test Engineering (VTE) Economically Viable for Highly Complex Telecom Circuits, DATE 99, User Forum, pp. 227-231

[8] IEEE Standard VHDL Analog and Mixed-Signal Extensions, IEEE Std 1076.1-1999

[9] J. Oudinot, M. Robbe, Patrick Radja: Mixed-Signal ASIC Top-Down and Bottom-Up Design Methodologies using VHDL-AMS, DATE 2000, User Forum, p. 175

[10] B. Li, L. Zheng, H. Tenhunen: Hierarchical Modeling of Sigma Delta Modulators for Noise Coupling Analysis. SSMSD '99. 1999 Southwest Symposium on Mixed-Signal Design, 1999, pp. 58-62

[11] B. Li, L. Jia, H. Tenhunen: Optimization of analog modeling and simulation in Proceedings 5th International Conference on Solid-State and Integrated Circuit Technology, 1998, pp. 385 –388

[12] M. Mahoney: Tutorial DSP-BASED TESTING of Analog and Mixed-Signal Circuits. Computer Society of the IEEE, 1987, pp. 45-47

Chapter 6

BEHAVIORAL MODELING OF MULTISTAGE ADCS AND ITS USE FOR DESIGN, CALIBRATION AND TEST

Eduardo J. Peralías and Adoración Rueda
Instituto de Microelectrónica de Sevilla (CNM-CSIC/Univ. Sevilla)

Abstract: This chapter provides a general model for the static behavior description of multistage ADCs that intends to be independent of the topologies and design techniques used. Furthermore, under this general behavior, results for the correction of its main errors will be extracted by means of calibration in the digital domain. Finally, two well-consolidated state-of-the-art techniques will be explained by using the developed mathematical description.

Key words: Analog-to-Digital converters, Behavioral modeling of mixed-signal circuits, Digital calibration.

1. INTRODUCTION

Behavioral modeling of analog and mixed-signal circuits has become of increasing importance with the availability of complex systems in present advanced technologies. Behavioral models can not only facilitate and accelerate system simulation and verification, but also allow the designer the early exploration of new ideas as well as of the design space. Two fundamental properties should be inherent to a practical behavioral model: easy to generate and compatible with high-level description languages.

Analog-to-digital Converters (ADC) are the circuit blocks more required in many current applications. New design and test methodologies for high performance and embedded ADCs are claimed, and both of them benefit

J.L. Huertas (ed.), Test and Design-for-Testability in Mixed-Signal Integrated Circuits, 163–214.

from good behavioral models. In the last decade, two architectures of CMOS ADCs are monopolizing most of applications where a medium/high precision is required ([1]-[3]). One is the Sigma-Delta based on the oversampling and the noise conformation. Another is the Pipeline architecture, based on the subdivision in cascade of the conversion process. The first one is located as predominant for the low (<50kHz) and medium (<5MHz) signal frequencies. Pipeline architectures have a good present in the intermediate frequencies (<50MHz) and a good future is foreseen in the high frequencies (<500MHz) especially when operation in the *time-interleaved* architecture is employed [2]. This chapter is almost completely dedicated to the pipeline architecture (more concretely to the *multistage architecture*[1] to which it belongs). We will provide a general model for its static behavior description that intends to be independent of the topologies and design techniques used ([4]-[6]). Although it is possible to deduce the formation of the multistage structure starting from the mathematical theory of the numeric code [7], in this chapter we will assume a well-known structure, which will be sketched in Sect. 4.1. Under very general behavior hypothesis, we will extract results useful for the ADC design and for the correction of its main errors by means of calibration in the digital domain. Previously, in Sect. 2 we will introduce a general mathematical formalism, valid for the static description of any converter and it will provide its maximum utility in the modeling of converters with multistep operation (Sect. 6). The sections 4 and 5 have been reserved to explain what we understand for ADCs of multistage type and to indicate the conventions and notations that we will use in its description, as well as the introduction to the calibration concepts and digital self-calibration in any converter class. Section 6 is entirely devoted to the description, to the modeling and to the processes of self-calibration of multistage converters. In the part dedicated to calibration, we explain two techniques very consolidate in the state of the art, but we will make it using the mathematical description introduced along the chapter.

In the whole treatment that follows we will consider to the ADC as the subsystem that produces quantized versions Z (output or digital code) of the amplitude x of an analog signal (input). The maximum number M of possible quantizations of a signal that covers the whole input operation range fixes the resolution of the ADC. Usually, the quantized output is coded in binary, for that reason, the resolution is defined as the smallest number that verifies: $N \geq \log_2(M)$. The conversion range of the ADC is the interval of amplitudes in which it can quantize without ambiguity, and therefore it can

[1] Also denominated iterative or multistep architecture.

be defined by a function $Z = f_{AD}(x)$. For amplitudes outside of this interval the ADC responds always with the same unique value; for $x \geq R_u, Z = zu$, and for $x \leq R_u, Z = zd$. These output values are denominated *saturant codes* of the ADC. The longitude of the conversion range is denominated *Full Scale*: $FS = R_u - R_d$, and its limits are known as *Characteristic References* of the ADC.

2. DESCRIPTION OF THE STATIC BEHAVIOR OF AN ADC

To describe the static input-output relationship of an ADC it is necessary to characterize its transfer curve, which is always a staircase function. The mathematical definition of a staircase function needs only of its discontinuity points (habitually denominated *code transitions*) and the values between what the transition happens (output codes of the ADC)[2]. Based on this, the most widespread nominal behavior in an ADC is that of a growing staircase function with a unique transition between each couple of continuous codes. In this way, the number of transitions is *M*-1, where *M* is the number of possible output codes of the ADC, and the value of the discontinuity in each transition is +1. This behavior is defined by the functional form:

$$Z = Z(x) = f_{AD}(x) = zd + \sum_{j=1}^{M-1} u(x - t_j) = zd + \sum_{j=1}^{M-1} (x \geq t_j) \tag{1}$$

where x is the value of the ADC input, Z the value of the output code, $\{t_j\}$ the set of transitions and *zd* the inferior saturant code. In that follows, the unit-step function $u(x-a)$ will be considered as equivalent to the logical evaluation of the following parenthesis $(x \geq a)$.
With this behavior it is always possible to define a unique interval for each code and therefore its code-width (width of the step), $w_k = t_{k+1} - t_k$.
Any behavior that cannot be described by means of (1) is considered anomalous. Two important anomalies are: a) the loss of monotonicity growing of the transfer curve (this implicates that there are discontinuities of negative value), and, b) the loss of codes (this implicates that there are

[2] If $\{t_k\} \in D_x$ is the set of transitions that there is in the function domain D_x and $\{\Delta Z_k\}$ the values of the discontinuities in each one of them, the staircase function is $f(x) = \sum_k (\Delta Z_k \cdot u(x - t_k)), (x \in D_x)$ where $u(.)$ is the Heaviside unit-step function.

transitions with discontinuities bigger than the unit). This last anomaly could be included in (1) if we assign to each nonexistent transition a value equal to the transition of the closest smaller code. If for example t_{kp} is a transition point among the codes k and $k+p$ (with $p>1$), then the codes $\{k+1, k+2, \ldots, k+p-1\}$ never appear at the output of the ADC. There is loss of codes in this ADC. To describe the behavior by means of (1) we will assign to each code that does not appear, a transition value equal to t_{kp}. This means all the code transitions $\{k, k+1, k+2, \ldots, k+p-1\}$ are in the same point.

Now, let us introduce in the next subsection the solution we propose to include anomaly a) in the ADC input-output description.

2.1 Prelatio. Types of ADC transfer curves.

An ADC with the anomaly of loss of monotonicity can be described with the form (1) whenever we find a form of assigning to it a new set of transitions of unit jump (+1). This can produce a description if not equivalent to the original at least with the minimum loss of information[3]. To do that, we introduce the concept of *prelatio*.

– **Prelatio**

We will denominate *prelatio* (or preference) to any functional operation or procedure that provides a representative set of transitions for the description (1) of the ADC. The prelatio gives a value for each transition and this is of unit jump (+1). In any real converter, the transition among codes is more or less a wide interval (*transition zone*) due to the existence of noise, and, the physical impossibility of eliminating it completely. The prelatio assigns a single and representative value for the transition zone. The assigned value does not have to belong necessarily to the transition zone.

A prelatio that is broadly used for the experimental characterization of converters is based on histogram techniques ([8]-[11]) (model of uniform probability). Now we indicate the basic relationship of this prelatio, which has been used in the example of the Fig. *6-1* for illustration purpose,

$$t_z = t_1 + \sum_{k=z_d+1}^{z-1} D(k) \quad , \forall z \in (zd, zu] \tag{2}$$

[3] It can measure the loss of information using probability techniques, evaluating the power of the error that exclusively is due to the prelatio process. It will depend on the application type that the ADC carries out and therefore of the typical probability distribution of the input signals to the converter.

where $D(k)$ is the result of adding the longitude of all the associated intervals to the code k, and t_1 is the associated value to the first transition of zd to $zd+1$ and it is usually made coincide with a value of the first measured transition zone.

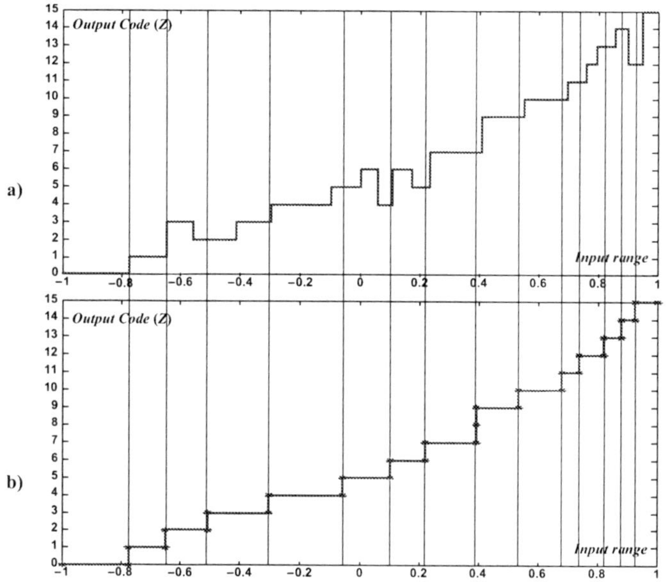

Figure 6-1. **a)** Real transfer curve of an ADC with range [-1,1] and resolution 4
b) Assigned transfer characteristic after applying a prelatio

It is evident that the behavior of the ADC with the transfer curve in the Fig. 6-1a is non-monotonic and also is noticed the loss of codes. The Fig. 6-1b shows the new transfer curve described by (1) after applying the prelatio defined in (2).

– **Types of ADC transfer curves**

If after carrying out a prelatio in an ADC, the set of assigned transitions follows a certain mathematical law, the ADC is classified according to the type of this law. Among the most important for its technological applications, we find the linear and the logarithmic ADC. In a *linear or perfect ADC* the transitions law is linear, that is,

$$\tilde{t}_z = a\cdot(z-z_d)+b \quad ,\forall z\in(zd,zu]\subset\mathbb{Z} \qquad (3)$$

where $\tilde{t}_z$ is the transition assigned to the jump of the code z to the $z+1$, a and b are the coefficients of the linear law and *zd*, *zu* are the extreme output codes of the converter or saturant codes. Notice that the first transition happens for the code *zd*+1. The tilde on the t will remind us that the transition follows a linear law.

The linear ADC of reference for any design is the *ideal ADC*. In this type, all the possible codes for a given resolution *N*, $[zd = 0, zu = 2^N - 1]$, are uniformly distributed in the input range of the ADC. In this way, all the codes have intervals or uniform steps in the staircase transfer. The longitude of each step is $q = FS/2^N$, the *quantum*, *step* or **LSB** (**L**east **S**ignificative **B**it) of the ideal ADC.

For calculation convenience we will consider the input range bipolar and symmetrical [-*R*, *R*] and therefore determined by a single value, the converter *Reference* (*R*). For this, Full Scale is $FS = 2R$ and the law (3) for the ideal ADC becomes:

$$l_z = l(z) = -R + q \cdot z \quad ,\forall z \in (0, 2^N - 1] \subset \square \tag{4}$$

Hereinafter, we will use the notation l_z for the transitions that follow the law (4).

For comparison among (3) and (4) we will denominate *effective quantum* of the perfect ADC to the coefficient a of (3), it will be denoted by q_e. The independent term b of (3) will be called *effective negative reference* of the ADC and denoted by R_{ne}. In this way, the law (3) of transitions of a perfect ADC can be written as:

$$\tilde{t}_z = R_{ne} + q_e \cdot (z - z_d) \quad ,\forall z \in (zd, zu] \subset \square \tag{5}$$

The *effective positive reference* (R_{pe}) of the ADC is obtained considering that the existent codes (*zu*-*zd*+1) are uniformly distributed in the input range defined by the effective references [R_{ne}, R_{pe}]: $q_e = (R_{pe} - R_{ne})/(zu - zd + 1)$.[4]

Although the use of (1) is now possible to describes any ADC on which a prelatio has been carried out, for the linear-ADC type defined by means of its references and its quantum, a simpler and more efficient functional form exists given as,

– for an ideal ADC:

[4] The effective references of a perfect ADC do not have to coincide with its Characteristic References.

$$Z = Z_{id}(x) = \left\lfloor \frac{x+R}{q} \right\rfloor \quad ,\forall x \in [-R, R)$$
$$zd = Z_{id}(x < -R) = 0 \tag{6}$$
$$zu = Z_{id}(x \geq R) = 2^N - 1$$

– or in a general way for a perfect ADC:

$$Z = Z_{lin}(x) = \left(\left\lfloor \frac{x - R_{ne}}{q_e} \right\rfloor + zd \right) \cdot (R_{ne} \leq x < R_{pe}) + zd \cdot (x < R_{ne}) + zu \cdot (x \geq R_{pe}) \tag{7}$$

where $\lfloor . \rfloor$ represents the *floor function*. The expression (7) is valid for any input and it contains to (6) particularizing for the parameters of the ideal ADC.

3. CLOSED FORMS FOR THE STATIC BEHAVIOR DESCRIPTION OF ADCS

Although expressions (1) and (7) serve to define and to calculate numerically the static behavior of linear and non-linear ADC, they are very complex and not clear for a symbolic description due to the existence of the integer-part and unit-step functions. In this section, different functional forms for the static behavior description of an ADC will be proposed. They share in common that are closed equations including explicitly some ADC non-idealities considering that a previous prelatio operation has been carried out. It will be shown that they result very efficient not only to describe the static ADC behavior but also to explain and understand the operation of calibration and correction techniques. The distinction between them relies in the non-ideality terms they include.

3.1.1 Primitive Forms

Any input x belonging to the interval of the code k can be expressed mathematically as:

$$x = x(k) = t_k + \varepsilon_k \quad ,\forall k \in (zd, zu] \tag{8}$$

where $\varepsilon_k \in [0, w_k]$ is a no-negative real number that measures the distance between x and t_k; it is not bigger than the width of the code k. Notice that (8) is an inverse functional form of the ADC transfer curve since the independent variable is now the code of the ADC. We can interpret it as a partition of the information of the input signal in two terms. The first one is the part of information that is not missed because it is integrated in the law of transitions of the ADC (which we assume known with certain degree of perfection when having characterized the static behavior of the converter with a certain prelatio). The second term is the part of information of the input signal that is missed in the conversion process and that we will denominate *residue of the conversion*. Notice that this is so because any input corresponding to the interval $[t_k, t_{k+1}]$ presents the same portion of signal: t_k. In the ideal converter, the residue is denominated *quantization error* and it is denoted by $\varepsilon_q(k)$.
The form (8) is not defined for the saturant inferior code since this does not have transition associated in the prelatio. To avoid this ambiguity, the prelatio should also assign a transition value for the lower code extrapolating the behavior of the law of transitions. So in the linear converters the transition associated to the saturant inferior code is the effective negative reference R_{ne} as it can be verified in (5) (equal to -R in ideal ADCs).

In a general way any ADC that is designed to have a certain law of transitions $\hat{t}(k)$ will present in its implementation variations that are necessary to evaluate and to control. The primitive form (8) can make explicit these variations in the form:

$$x = \hat{t}_k + fni(k) + \varepsilon_k \quad , \forall k \in (zd, zu] \tag{9}$$

where $fni(\cdot)$ is the difference function between the law of transitions obtained by a prelatio on the implemented ADC and the expected or proposed in the design $\{\hat{t}_k\} : fni(k) = t_k - \hat{t}_k$.
For linear converters this difference function is denominated *non-linearity function* and it can be of two types depending on if the expected law of transitions is the corresponding to an ideal ADC or to a perfect ADC. In the first case, the difference function is denominated ***Integral Non-Linearity*** (INL) of the ADC, and (9) is written in this way,

$$x = l_k + q \cdot INL(k) + \varepsilon_k \quad , \forall k \in (zd, zu] \tag{10}$$

In the second case, the difference function is denominated ***error of non-linearity*** (enl) of the ADC, and (9) is now written as,[5]

$$x = \tilde{t}_k + q_e \cdot enl(k) + \varepsilon_k \quad , \forall k \in (zd, zu] \tag{11}$$

For linear converters, the most usual case is the first one. To use the form (11), the function that is obtained by linear regression of the "real" set of transitions $\{t_k\}$ is chosen as reference law. Between the expression obtained by linear regression:

$$\tilde{t}(k) = R_{ne} + q_e \cdot (k - zd) = fit\_poly1(\{t_k\}, \{k\}\big|_{zd+1}^{zu}) \tag{12}$$

and the corresponding to an ideal ADC, $l(k) = -R + q \cdot k$ (both linear laws corresponding to the same converter) there are differences only in the slope and in a vertical offset. The ratio $g = q/q_e$ between the slopes is denominated ADC *gain*, and its difference from the nominal value (the unit), will be denominated ADC *gain error*, expressed as,

$$\varepsilon_g = g - 1 = \frac{q}{q_e} - 1 \tag{13}$$

The difference between the transitions associated to the central code $k_m = 2^{N-1}$ in the ideal output range will be denominated ADC *offset error*, $\tau_{os} = l(k_m) - \tilde{t}(k_m)$.
These definitions allow us to relate both expression (12) and the ideal (4) by means of a linear transformation,

$$l(k) = g \cdot \tilde{t}_k + x_{os} \tag{14}$$

where $x_{os} = g\tau_{os}$.[6]
The relationships between the parameters herein introduced and those that traditionally are defined in the literature [11] are the followings,

5 The scale or normalization factors q and q_e in (10) and (11) are introduced by convenience. Both the INL and the *enl* are expressed in LSB units.

6 The values of the gain and offset here defined should be interpreted in a logical way on the transfer function. So, a greater gain than the unit means that the staircase of the ADC has a greater inclination than the ideal ADC of reference. A positive offset indicates that the transfer curve presents a vertical displacement above of the ideal.

$$\begin{cases} e_g = \dfrac{I\tilde{N}L(zu) - I\tilde{N}L(zd+1)}{(zu - zd + 1)} = \dfrac{-\varepsilon_g}{1+\varepsilon_g} & \text{Gain error} \\ e_{os} = I\tilde{N}L(zd+1) = -e_g \cdot q \cdot (k_m - (zd+1)) - \tau_{os} & \text{Offset error} \end{cases} \tag{15}$$

where $I\tilde{N}L(k) = fit\_poly1(\{INL_k\},\{k\}|_{zd+1}^{zu}) = (\tilde{t}_k - l_k)/q$ represents exclusively the linear part of the *INL*.
From all the above development, we can conclude that the function $INL(k)$ includes all the information regarding degradations respect to the ideal behavior in gain, offset, and non-linearity. However, the function $enl(k)$ only incorporates degradation in the linearity. Therefore in the *INL* should be contained the information of the *enl*. It can be easily demonstrated that the relationship between them is

$$enl(k) = g \cdot (INL(k) - I\tilde{N}L(k)) \tag{16}$$

Equations (8), (9), (10) and (11) will be denominated *primitive forms* of the mathematical description of ADCs.

3.1.2 Canonical Form

In the primitive forms, the contributions of the gain factors, offset and non-linearity are not explicitly separated. We will consider the *canonical form* for the mathematical description of ADCs to an expression that achieves such partition. Using (11) and (14), it is easy to deduce that the canonical form is

$$g \cdot x + x_{os} = l(k) + q \cdot enl(k) + \nu(k) \quad ,\forall k \in [zd, zu] \tag{17}$$

From this expression it is easy to realize that the residue, due exclusively to the conversion process, is inferior to $g \cdot w_k = g \cdot (t_{k+1} - t_k)$. Traditionally the code-width w_k has been related with a characteristic denominated **D**ifferential **N**on-**L**inearity (*DNL*). The DNL is defined as the relative error of the width that each step of the transfer curve has with respect to the ideal, $w_k = (t_{k+1} - t_k) = q \cdot (1 + DNL(k))$. From this definition, it is deduced that due to the prelatio, the minimum value of DNL is (-1), since the minimum difference between two transitions is zero. Therefore, those codes that have its DNL equal to zero are codes that do not appear at the output of the ADC.
From the above definition and (10) it is deduced that the DNL is the first-difference of the INL, $DNL(k) = \Delta_1 INL(k) = INL(k+1) - INL(k)$. By

analogy, we can define the **d**ifferential **n**on-**l**inearity (*dnl*) of the ADC as the first-difference of the *enl*:

$$dnl(k) = \Delta_1 enl(k) = enl(k+1) - enl(k) \tag{18}$$

From this definition, it is easy to demonstrate that $1 + dnl(k) = g \cdot (1 + DNL(k))$, and that the residue in the canonical form is smaller or equal to the ideal LSB modulated by the *dnl*, $0 \le \nu(k) \le q \cdot (1 + dnl(k))$.

Whenever the $|enl|$ of the ADC stays below 1/2LSB, the *dnl* will be higher than (-1), this meaning that there is not loss of codes in the behavior of the ADC, that is always a desirable characteristic for the converter. We define the *precision* of the converter as the maximum resolution in which there is not code loss, and the *accuracy* as the maximum resolution in that the non-linearity error ($\max(|enl_k|)$) is maintained bounded to 1/2LSB. With this, the designs can be classified as *relaxed* if they require a precision equal to the resolution and *exigent*[7] if is the accuracy what must be equal to the resolution.

The incorporation of the effect of the electronic noise in the ADC can be made adding to the input a term n_x which models the total input referred noise:

$$g \cdot (x + n_x) + x_{os} = l(k) + q \cdot enl(k) + \nu(k) \quad , \forall k \in [zd, zu] \tag{19}$$

3.1.3 Pseudocanonical Form

Any functional form similar to (17) but whose coefficients and terms do not coincide numerically with those of (17), is a *pseudocanonical form* for the mathematical description of the converter. For example,

$$\alpha \cdot x + \beta_{os} = l(k) + q \cdot InL(k) + \upsilon(k) \quad , \forall k \in [zd, zu] \tag{20}$$

is a pseudocanonical form if $\alpha \ne g$. In this case, the equality $InL(k) = enl(k)$ does not apply. If the function $InL(k)$ has certain linear content which can be extracted carrying out a linear regression, the slope σ of that linear part is the factor that is needed to get the equality,

[7] In very strict designs, it can be required that the accuracy is evaluated based on the bounding of the INL.

$g = \alpha/(1+\sigma)$. Only in case the regression on the *InL* produces an identically null function, we could say that the pseudocanonical form is in fact the canonical form.
The pseudocanonical form will be of great importance in the development that we will do of the description of multistage ADCs. Its parameters will be denominated: α quasi-gain, β_{os} quasi-offset, InL_k quasi-non-linearity, and υ_k quasi-residue.

3.1.4 Potency-Residue Form

It is advisable in certain situations not to make explicit the degradations of gain, offset or non-linearity. Instead, it is better to include the whole behavior degradation inside the conversion residue. Then, it results a very simple primitive form similar to (9) and with a very concrete meaning,

$$x = \hat{t}_k + r_k \quad ,\forall k \in [zd, zu] \tag{21}$$

For a fixed code, *k*, the value $\hat{t}_k$ reduces to a single point all the information that there is in the input interval $I_k = [t_k, t_{k+1}]$ whose output code is *k*. The association of a dense interval to a single point is the reason for denominate to $\hat{t}_k$ as *potency* of the interval I_k. The residue r_k represents the information lost when an interval is reduced to its potency. The residues are the distances between the points belonging to a code interval and their associated potency. An expression as the (21) will be denominated *potency-residue form* and it will be of extended usage in the description of multistage ADCs, as we will see later.

From now on, as we will only work with linear ADCs, the potencies will fulfill a linear law $\tilde{p}(k) = p_0 + q \cdot k$ with a slope equal to the LSB of the corresponding ideal ADC. The independent term p_0 will depend on each particular description, but in general it will be chosen in such a way that in the nominal design, the distribution of probability of all the possible residues of the ADC will have even symmetry around zero whenever the input *x* is considered with uniform distribution in the whole input range (*criterion of symmetry of residues*). For example, for an ideal ADC, $p_0 = -R + q/2$, because in this way all the potencies are in the center of their associate interval, and consequently, the residues are distributed uniformly around zero and with an extension of 1/2LSB. Independently of the chosen value for p_0, an exigent design must propose and foresee the existence of a code transition among each couple of potencies, since in this way there is not loss of accuracy.

3.2 Block model for the computational evaluation of the static behavior of linear ADCs

As we have seen along the previous section, the complete static behavior of an ADC is given by the set of transitions. This set can be determined if the gain (or the gain error), the offset and the *enl*(*k*) function are known. So, in the first model of blocks we show in Fig. *6-2* (Flash-model), we evaluate, previously to the simulation, the complete set of transitions $\underset{\sim}{t} = \{t_k\}$, using the gain, offset and non-linearity specifications in the expression,

$$g \cdot t_k + x_{os} = l(k) + q \cdot enl(k) \quad ,\forall k \in [0, 2^N - 1] \tag{22}$$

which comes from the canonical form (17).
Having the set of transitions, the expression (1) is implemented in software according to the diagram of the Fig. 6-2, in which the corresponding p-code description also appears,

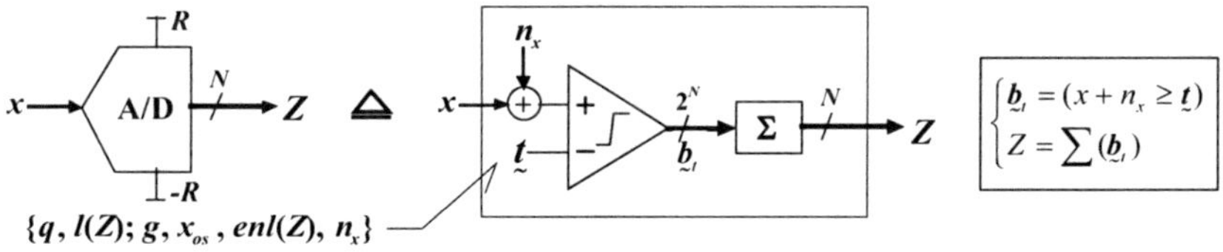

Figure 6-2. Flash-model for the numerical evaluation of the behavior of ADCs

If $\boldsymbol{x}$ is a vector that represents the record of a sampled signal, there is an alternative model to the Flash-model. This alternative, denominated SA-model, has a more efficient evaluation as N grows if it makes use of the vectorial algebra[8] In the Flash-model, it would be necessary to evaluate approximately $length(\boldsymbol{x}) \cdot 2^N$ comparisons while in the SA-model it is about $length(\boldsymbol{x}) \cdot N$. There are more vectorial arithmetic operations but these are very fast if the simulation motor is optimized for the vectorial evaluation. Next in (23) their p-code is presented:

– SA-model (Successive Approximations model)[9]

[8] The comparison of vectors $(\vec{a} \geq \vec{b})$ is a tensorial operation whose result is a matrix $(\vec{a} \geq \vec{b})_{ij} = a_i \geq b_j$ with dimension $length(\vec{a}) \cdot length(\vec{b})$.

[9] $\lceil \cdot \rceil$ represents the *ceiling* function.

$$\left|\begin{array}{l} ku = 2^N - 1,\, kd = 0,\, km = \lceil (ku + kd)/2 \rceil \\ \text{LOOP} \quad N \quad \text{times} \\ \underset{\sim}{\boldsymbol{b}}_t = (\underset{\sim}{\boldsymbol{x}} + \underset{\sim}{\boldsymbol{n}}_x \geq t(km)) \\ ku = \underset{\sim}{\boldsymbol{b}}_t \cdot ku + \overline{\underset{\sim}{\boldsymbol{b}}_t} \cdot km \\ kd = \overline{\underset{\sim}{\boldsymbol{b}}_t} \cdot km + \underset{\sim}{\boldsymbol{b}}_t \cdot kd \\ km = \lceil (ku + kd)/2 \rceil \\ \text{END LOOP} \\ Z = kd \end{array}\right. \tag{23}$$

A model of easy implementation, although strictly it does not verify the canonical form, is presented in the Fig. 6-3. It is specially recommended when the non-linearity error *enl* does not take very high values (<10LSB) and it is defined by a closed expression (for example, a polynomial).

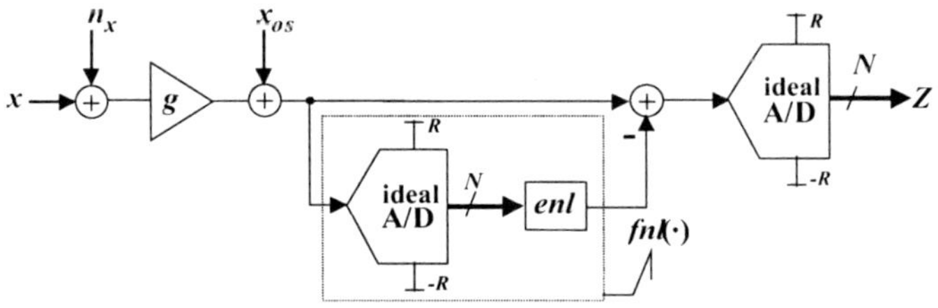

Figure 6-3. Block Model with a behavior approximated to the canonical model

The denominated "ideal A/D" block implements the expression (7). The inferior loop that provides the effect of the *enl*, can be reduced if we make a linear transformation of the domain of that function. Mapping the output range $[0,2^N]$ into $[-R,R]$, a single block remains corresponding to the function $fnl(\cdot)$. For example, a very simple polynomial expression for the non-linearity is,

$$fnl\left(\frac{x}{A}\right) = \mp\left(q \cdot enl_{\max} \cdot \sqrt{\frac{p^p}{(p-1)^{(p-1)}}}\right) \cdot \frac{x}{A} \cdot \left(1 - \left(\frac{x}{A}\right)^2\right)^{\frac{p-1}{2}} \quad ,|x| \leq A \tag{24}$$

where p is the degree of the polynomial (assumed odd), $A = R - q$ is the reduced reference of the input range, enl_{max} the extreme value for the non-linearity in LSBs. This polynomial has odd symmetry and it always reaches a maximum ($+enl_{max}$) and a minimum ($-enl_{max}$) inside the input range. It is canceled in the center $x = 0$ and in the $x = \pm A$ edges. If the negative sign is used in the expression, the maximum appears before the minimum, $x_{\max} < 0 < (x_{\min} = -x_{\max})$.

4. MULTISTAGE CONVERSION

4.1 The Structure of Multistage Converters

A multistage A/D converter is formed by the interconnection in cascade of several blocks $\mathbf{STG_i}$ denominated *stages* (see Fig. *6-4*). Each stage has the function of obtaining a digital subcode, c_i, that is only a reduced part of the information contained in the final output code of the ADC, Z. This output code is obtained by means of the appropriate combination of all the subcodes, being the final resolution less than or equal to the sum of all the resolutions. If two adjacent stages can process (although in a different way) a common part of information, it is said that the conversion presents *redundancy* (Sect. 6.1). The form in that the total process of conversion is performed in the time gives a subdivision of multistage converters into three types: *subranging*, *pipeline* and *cyclic*. In subranging converters, the input is maintained constant while all the conversion process takes place. All the stages evaluate, jointly in that time, parts of information corresponding to a single sample of the input. In the pipeline ADC each stage processes, in a same interval of time, parts of information corresponding to distinct input samples. The subcodes produced in each instant should be memorized, grouped together and combined those corresponding to the same input sample. The practical subranging ADC is limited to very low number of stages, in general two, while the pipeline ADC presents a great diversity being possible to find in the literature implementations using from two to twenty stages. The cyclic ADC changes the spatial structure of L stages for a structure of one stage whose process is carried out L times. The processing in the cyclic ADC requires that, like in the subranging, the input remains constant during all the conversion.

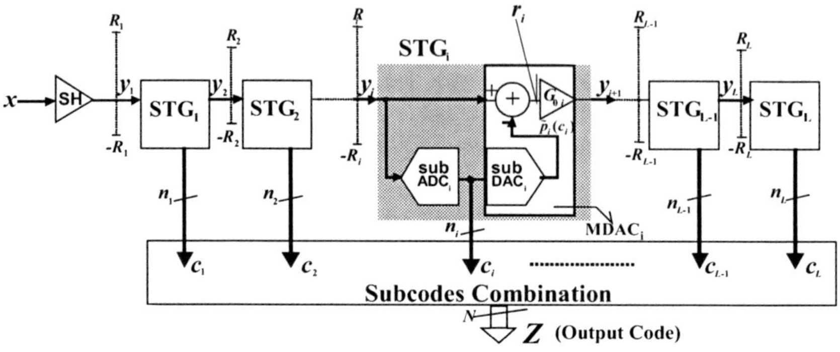

Figure 6-4. Block diagram of a multistage ADC

In each stage of a multistage ADC a local A/D conversion takes place by means of a subADC with smaller resolution than the total one[10]. The signal transmitted from a stage to other is analog and it is the local conversion residue r_i just as was defined in (21). Each stage **STG**$_{\mathbf{i}}$ (except the first one which evaluates the input) produces a low-resolution quantized version c_i of the previous residue r_{i-1}. For the process in cascade to become effective, the residues produced by a stage should be quantized for the following subADC. Consequently, or the quantum Q_i goes decreasing progressively in the cascade[11] or the residue value adapts to the following conversion range. The adaptation is obtained applying a linear transformation that in the practice is reduced to an amplification by means of a gain element $|G_{0i}| \geq 1$. The functional block that evaluates and adapts the residue for the following process is denominated MDAC (**M**ultiplying **DAC**). Therefore, each stage of a multistage ADC consists of two basic blocks: a subADC and a MDAC. In the last stage **STG**$_{L}$, it is only necessary the subADC, so it is denominated *Last Quantizer*. The functional structure of the MDAC is also represented in the Fig. *6-4,* it consists of: a Digital-to-Analog converter (subDAC) that evaluates the value of the potency in (21), and an adder-amplifier that calculates and amplifies the residue. The gain of the MDAC, G_{0i}, will be denominated: *interstage gain*. Here, we will suppose in a general way that each stage **STG**$_{\mathbf{i}}$ has an analog input range D_{xi} less or equal to the input range of its subADC (symmetrical and bipolar $D_{xi} \subseteq [-R_i, R_i]$, being R_i the reference of the subADC), and an analog output range D_{oi} less or equal to the

[10] If the local resolution n_i is very low compared with the global N, in general, there will be a great number L of stages in the structure so $N \leq \sum_{i=1}^{L} n_i$.

[11] In general, this implicates a waste of the available dynamic range.

input range of the following stage, $D_{oi} \subseteq D_{i+1}$. It is evident that the input range of the first stage must be that of the complete ADC $D_{x1} = [-R_1, R_1]$.

We will deal with linear multistage converters. The subADCs that form them will be consider also linear, for that reason, its potencies follow a linear law that we will write, for the i-th stage, as,

$$\tilde{p}_{i,k} = \tilde{p}_i(k) = p_{0i} + Q_i \cdot k = \delta_i Q_i + (-R_i + Q_i \cdot k) = Q_i \delta_i + l_{i,k} \quad , \forall k \in [zd_i, zu_i] \tag{25}$$

We have expressed $p_{0i} = -R_i + Q_i \delta_i$ to make explicit that the position of the stage potencies with regard to the ideal transitions $l_{i,k}$ of the subADC is determined by δ_i (*offset of potencies*). In general, its value is chosen following the *criterion of symmetry for residues* enunciated in the Sect. 3.1.4.

An algorithm compatible with the operation explained for the multistage ADC is the following:

$$\left| \begin{array}{l} y_1 = x \\ \text{FOR} \quad i = 1 \quad \text{to} \quad L-1 \\ c_i = f_{AD}^{(i)}(y_i) \\ y_{i+1} = f_{MDAC}^{(i)}(y_i, c_i) \\ \text{END FOR} \\ c_L = f_{AD}^{(L)}(y_L) \\ Z = Z(c_1, c_2, \ldots, c_L) \end{array} \right. \tag{26}$$

The functions of the subADC, $f_{AD}^{(i)}$, can be derived as it was exposed in Sect. 3.2. Those of the MDAC, $f_{MDAC}^{(i)}$, will be object of detailed study in later sections, but for the ideal situation its form should be of amplification of the stage residue, $f_{MDAC}^{(i)}(y,k) = G_{0i} \cdot (y - \tilde{p}_i(k))$. Also in a posterior section, the form in what the subcodes should be combined to obtain the final code Z will be explained.

5. CALIBRATION

In this section we introduce the concepts to understand the way the non-idealities of linear ADCs can be corrected acting on its own output code.

They are known as digital calibration procedures. First of all, we will consider a general procedure independent of the internal structure of the converter. In a posterior section (Sect. 6.3), we will explain digital calibration procedures specific for multistage converters.

5.1 Digital Calibration

Let us suppose that we develop the l_Z value in the primitive form (10), and then we group the value of the output code with the INL, obtaining

$$x = l_Z + q \cdot INL(Z) + \varepsilon_Z = -R + q(Z + INL(Z)) + \varepsilon_Z \tag{27}$$

From this expression we can infer that if we take as total output of the converter a value, not equal to but at least approximate to $Z + INL(Z) \approx Z^\dagger$, then the form (27) will result very similar to that of an ideal ADC. This means that whenever the natural output Z of the converter is corrected with a value close to its INL, the converter static behavior is improved (idealized) or in other words, it is calibrated.

If the own converter obtains an estimate of its INL and uses it to modify the output, it is said that a digital self-calibration procedure is implemented. Similarly, we could have considered the canonical form (17) and have taken as global output $Z^\dagger$ an approach of $Z + enl(Z) \approx Z^\dagger$, then the ADC will exhibit an improved linear response but we could not affirm anything about the degradation or not of the gain and offset[12]. To distinguish between both calibration procedures, we will denominate *Complete digital Calibration* or *Full Calibration* to that carrying out the approach of the *INL*.

If the sum $Z^\dagger = Z + \langle enl(Z) \rangle$,[13] is carried out with integers (with binary logic in the most general case of the A/D converters) it is immediate to observe that it should be made to a resolution $N' = N + p$ bigger than the nominal of the ADC. The reason is that the truncation of the value $\langle enl(Z) \rangle$ to N bits before the summation will produce, in general, an additional error bigger than if the truncation is carried out after the operation. In general, two additional bits $(p = 2)$ are usually enough. There are also

[12] The approach of the non-linearity term takes implicit the appearance of a remanent non-linearity. This can present a certain linear tendency with the code of the global output and hence, an additional gain factor and offset could appear.

[13] The brackets $\langle a \rangle$ indicate a certain approach carried out on the magnitude a. The approach type is stated explicitly when the design, architecture and technical application of the ADC are known.

procedures of digital calibration that pretend to correct the gain and/or the offset effects taking as basis the canonical form. In the offset digital calibration, the global output of the ADC is $Z^{\dagger} = Z - \langle x_{os}/q \rangle$. If we only calibrate the effect of the gain of the whole ADC, the global output is $Z^{\dagger} = Z \cdot \langle 1/g \rangle$. The global output including all these separate effects has the following form,

$$Z^{\dagger} = 2^{N'-1} + \left((Z - 2^{N'-1}) + \langle enl(Z) \rangle - \langle x_{os}/q \rangle \right) \cdot \langle 1/g \rangle \qquad (28)$$

The design of a good procedure for digital calibration requires a balance among the incorporation of new resources to make the operations in (28) and the remanent error that is accepted. It is also part of the design of the digital calibration procedure, the decision about what approaches are implementing, and, how they must estimate the values. For example, by means of direct measures on certain internal parts of the ADC that produce the most significant errors. If the estimates are carried out without interrupting the A/D conversion process, a *Background Calibration* procedure is implemented ([12],[13],[14]).[14]

The block diagram in Fig. *6-5*a shows an application of the complete calibration procedure in an ADC with resolution N'=12bits of which at least N=10bits of accuracy is required. The INL has been measured experimentally and it is shown in the Fig. *6-5*b where we observe that it is very superior to 1/2LSB. An estimation, rounding to the closest integer[15] at the level of 12bits, of the INL is carried out (shown in the Fig. *6-5*c) and it is memorized in a block of 2^{12}=4096 registers of 6bits each one (since the range of the INL is of the order of 50LSB). To make the correction of the code, in each conversion we added the natural code Z with the content of the register addressed by Z, $[INL(Z)]$.[16] Then, the result $Z^{\dagger}$ is considered with only the 10 most significative bits. In the Fig. *6-5*d, it is observed the INL that remains in the total converter with output $Z^{\dagger}$.

14 Usually it is said that we have a technique or calibration (or self-calibration) procedure, without more adjectives, when the estimations of the errors are carried out with the conversion process halted. To distinguish them of the Background calibration, we can denominate it *Offline calibration.*

15 The rounding to the closest integer [.] also denominated *round* function.

16 For the code $zd = 0$, the extrapolated value of the $\tilde{INL}(k)$ is chosen.

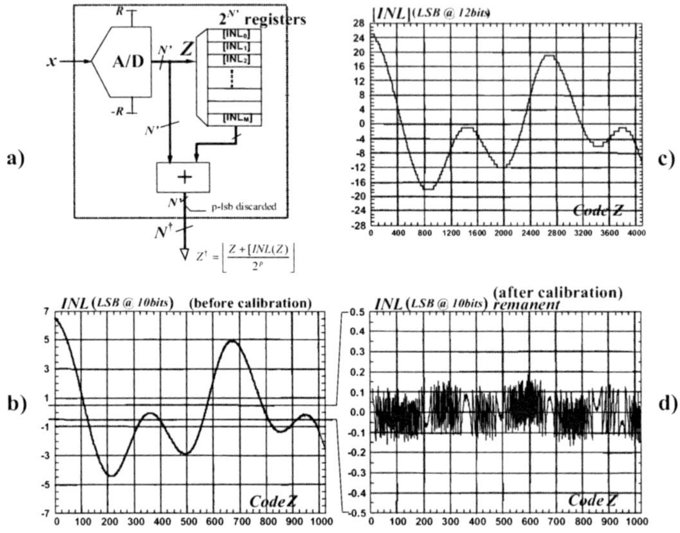

Figure 6-5. **a)** Block diagram of the complete calibration procedure
b) INL of an ADC example, measured at the level of 10bits
c) Estimation of the INL in **b)** by means of the closest integer
d) Remanent INL after applying a full calibration procedure with the estimation shown in **c)**

5.1.1 Offset Self-calibration

If the converter is considered linear with the canonical form,

$$g \cdot x + x_{os} = l(Z) + \zeta_Z \tag{29}$$

the offset x_{os} can be estimated and corrected by the ADC in an autonomous way and independently of its architecture. For that, it is necessary to apply a zero signal $x = 0$ to the primary input during a reserved timeframe to carry out the estimate. This time can be a single clock cycle, in which a unique sample of this signal is taken, or thousands of cycles to average the estimates. For each sample of the zero input we will obtain at the output a code Z_0 that will verify (29):

$$g \cdot 0 + x_{os} = l(Z_0) + \zeta_{Z_0} \tag{30}$$

After having calibrated the offset, the canonical form will be $g \cdot x = l(Z^{\dagger}) + \xi_{Z^{\dagger}}$, and the output $Z^{\dagger}$ associated to the zero input will be

$Z_0^\dagger = 2^{N'-1} = k_m$, if the converter resolution is *N*'. This value will verify the new canonical form:

$$g \cdot 0 = l(k_m) + \xi_{k_m} \tag{31}$$

Subtracting (30) and (31) we have

$$\frac{x_{os}}{q} = (Z_0 - k_m) + \frac{(\zeta_{Z_0} - \xi_{k_m})}{q} \tag{32}$$

Therefore, if the difference of residues is small and it becomes even smaller when we normalize it to the LSB of the calibrated ADC with the code truncated in *p*-bits, $\left|\zeta_{Z_0} - \xi_{k_m}\right| \ll q2^p$, then the first term of (32) $\left(\hat{Z}_0 - k_m\right)$ is a good estimation for the offset, where $\hat{Z}_0$ is the average value of all the obtained samples.

After this, the ADC will be connected again to its normal input and the estimation is applied to each resulting code, giving the definitive code:

$$Z^\dagger = Z - \left\langle \frac{x_{os}}{q} \right\rangle = Z - (Z_0 - k_m) \tag{33}$$

5.1.2 Gain Error Self-Calibration

The simplest and low-cost method to make the calibration of the gain error in the digital domain is equivalent to that we have shown in the example of the Fig. 6-5. Evidently, if the ADC is already supposed linear or linearized for other methods and it still shows offset and gain errors, then the values that are necessary to memorize in the registers are only those of the linear part of the INL, $(I\tilde{N}L(Z))$. The advantage of this procedure resides in that it is enough to carry out two independent measures and to use an integer-arithmetic algorithm to estimate all the corrections. Besides, both the gain and the offset are corrected simultaneously. The storage of the corrections has a practical purpose of facilitate their access and to use them in a short time. This is almost mandatory in medium and high-speed ADCs. However, if the operation speed is such that the algorithm can be executed in each conversion, the storage of values is reduced to several registers.

As for the offset self-calibration procedure, the ADC should be capable of providing the zero signal $(x = 0)$ for which the ideal output $Z_0^\dagger = k_m$ is known, in this gain self-calibration (and offset) procedure, the ADC should be capable of providing at the primary input two DC values (*self-calibration stimuli*) of the conversion range[17], $x = x_u$ and $x = x_d$, for which its ideal outputs are known[18], $Z^\dagger(x_u) = k_u$, $Z^\dagger(x_d) = k_d$.

Let us consider that the description of the ADC is given by the primitive form:

$$x = l(Z) + q \cdot I\tilde{N}L(Z) + \zeta_Z \tag{34}$$

If the ADC takes samples from x_u and x_d, the corresponding output codes Z_u and Z_d, should verify (34):

$$\begin{cases} x_u = l(Z_u) + q \cdot I\tilde{N}L(Z_u) + \zeta_{Z_u} \\ x_d = l(Z_d) + q \cdot I\tilde{N}L(Z_d) + \zeta_{Z_d} \end{cases} \tag{35}$$

After applying the calibration $Z^\dagger = Z + \langle I\tilde{N}L(Z) \rangle$, the description of the ADC is approximated to the ideal ADC $x = l(Z^\dagger) + \xi_{Z^\dagger}$, and we have for the known values:

$$\begin{cases} x_u = l(k_u) + \xi_{k_u} \\ x_d = l(k_d) + \xi_{k_d} \end{cases} \tag{36}$$

Subtracting the expressions (35) from the respective ones in (36), it results:

$$\begin{aligned} I\tilde{N}L(Z_u) &= (k_u - Z_u) + \frac{(\xi_{k_u} - \zeta_{Z_u})}{q} \\ I\tilde{N}L(Z_d) &= (k_d - Z_d) + \frac{(\xi_{k_d} - \zeta_{Z_d})}{q} \end{aligned} \tag{37}$$

Therefore, if the difference of residues is small and it becomes even smaller when we normalize it to the LSB of the calibrated ADC with the code truncated in p-bits, $\left|\xi_{k_x} - \zeta_{Z_x}\right| \ll q2^p$, then, we have good estimations for $\langle I\tilde{N}L(Z_x) \rangle = (k_x - \hat{Z}_x)$, where $\hat{Z}_x$ is the average value of the samples obtained with the calibration stimulus x.

17 One value away from the other to obtain a good estimate of the slope of the $I\tilde{N}L(Z)$

18 For example, for $x_u = R/2$, $x_d = -R/2$ we have: $Z^\dagger(x_u) = 3 \cdot 2^{N'-2}$ and $Z^\dagger(x_d) = 2^{N'-2}$.

With these estimations, any code correction can be obtained performing the algorithm (38).
Notice that the offset value estimated in the previous section is such that: $\langle x_{os}/q\rangle = -I\tilde{N}L(Z_0)$.

It is possible to find convenient cost reductions using symmetry arguments. For example, if we take $x_d = 0$, the number of registers to use decreases to the half. If the ADC needs a correction value that it is not stored we can obtain it of its symmetrical value regarding $x_d = 0$:
$\langle I\tilde{N}L(2Z_0 - Z)\rangle = 2\langle I\tilde{N}L(Z_0)\rangle - \langle I\tilde{N}L(Z)\rangle$.

$$
\begin{array}{l}
\left.\begin{array}{l}
I_\Delta = \langle I\tilde{N}L(Z_u)\rangle - \langle I\tilde{N}L(Z_d)\rangle \\
Z_\Delta = Z_u - Z_d \\
P_0 = 2|I_\Delta| - Z_\Delta \\
sI = \mathrm{sgn}(I_\Delta)
\end{array}\right] \text{Common to all values} \\
\textbf{if} \quad (Z < Z_d) \quad sI = -sI \\
I_z = \langle I\tilde{N}L(Z_d)\rangle, \quad P_z = P_0 \\
\textbf{LOOP} \quad |Z - Z_d| \quad \text{times} \\
\quad \textbf{if} \quad (P_z > 0) \quad I_z = I_z + sI, \quad P_z = P_z + P_0 - Z_\Delta \\
\quad \textbf{else} \quad P_z = P_z + P_0 + Z_\Delta \\
\textbf{END LOOP} \\
\langle I\tilde{N}L(Z)\rangle = I_z
\end{array}
\tag{38}
$$

6. MODELING MULTISTAGE ADCS: STATIC BEHAVIOR AND ERROR CORRECTION

The goal of this section is twofold. Firstly, a detailed behavioral model is proposed for multistage ADC where different error terms are explicitly included. Secondly, we will show how functionality errors can be corrected using digital calibration procedures. The model is developed for the general ADC structure specified in Sect. 4.1. It has been proven to be very appropriated for high-level simulations [15], for IP reuse [16], and for validation of *Design-for-Test* techniques [17].

6.1 Redundancy

In general, the main objective of the cascade formation of a multistage ADC is to make the residue of a stage, which is lower than the stage LSB, to be transmitted and quantized in the following stage (amplified or not). Serious functional problems appear in the multistage ADC when the non-ideality errors of the subADC produce greater residues than its LSB and there is not anything in the design that avoids them or take them into account. In this case, the incorporation of redundancy ([18]-[25]) in the design can correct all the problems. An example of this method is to reduce the interstage gain of the stage that produces excessive residues or either, to increase the conversion range of the stages that receive high residues. The objective of such modification in the design is to foresee that any residue produced by a stage is inside of the range of the following stage, and therefore, completely quantizable. In addition, these modifications implicate that the final code of the ADC is obtained by means of a concrete combination of the subcodes that is dependent on the way in what the residues are transmitted. Consider for example the situation illustrated in Fig. *6-6*, according to the diagram, we can express the value of the signal amplitude x in the following ways:
$x \approx 11Q_2 = c_1Q_1 + c_2Q_2 = 2Q_1 + 3Q_2 = 1Q_1 + 7Q_2$.

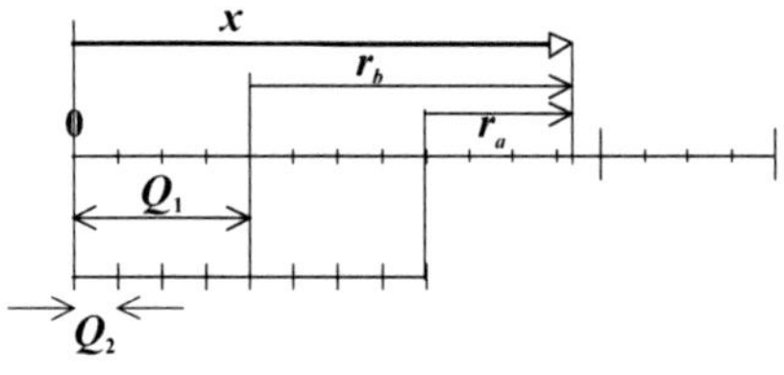

Figure 6-6. Redundancy: ambiguity in the partial quantizations,
$x \approx 2Q_1 + 3Q_2 = 1Q_1 + 7Q_2$.

This means that we make a coarse quantization of x with units of Q_1 and a fine quantization with units of Q_2. If the fine quantization admits residues r_x until 2 units of Q_1 (redundancy for range expansion), we can generate the two previous representations, one with $c_1 = 2$ and another with $c_1 = 1$.[19] In the first case the residue r_a is lower than Q_1 but quantizable with Q_2 in $c_2 = 3$ units. In the second case, the residue r_b exceeds of Q_1 but still inside the possibilities of the fine quantizer $c_2 = 7$. Both couples of values

[19] We discard the possibility of negative residues ($r_x < 0$).

$(c_1,c_2)=\{(2,3),(1,7)\}$ represent the same signal value $Z_x=11$. The way in what the elements of one of the possible couples have to be combined is as simple as $Z_x=4c_1+c_2=2^{(3-1)}c_1+c_2$. In this example, it is evident that due to the redundancy, the signal part that is not quantized in a level (the residue) will be in the following, but not in both levels.

Consequently, specifying the property of redundancy in the multistage structure will implicate the combined application of two design conditions. The first one must determine the range of residues that is tolerated in each stage; this condition will be designated as *redundancy level*:

- The *i*-th stage presents a redundancy level of s_i bits if all the residues are inside an interval with radio $2^{(s_i-1)}$ LSB of the stage. There is not redundancy when the level is of $s_i=0$ bits. That is:

$$|r_i|\le 2^{s_i}\frac{Q_i}{2} \tag{39}$$

The second consideration is a contour condition for which all the levels of residues received in a stage must be inside its conversion range; this condition will be designated as *full calibration among stages*:

- There is full calibration between the stages $\mathbf{STG_i}$ and $\mathbf{STG_{i+1}}$ when $|y_{i+1}|\le R_{i+1}$.

The combination of both conditions is satisfied if we impose in the design[20] the following relationship between the interstage gain, the redundancy level and the references:

$$|G_{0i}|=2^{n_i-s_i}\frac{R_{i+1}}{R_i} \tag{40}$$

If the stage $\mathbf{STG_i}$ has a resolution of n_i bits, the number of possible values for its subcode c_i is 2^{n_i}. The subDAC of the stage must generate a potency $\tilde{p}_{i,k}$ for each possible value $c_i=k$. Since, by definition, a distance of a LSB exists (Q_i) between two consecutive potencies, the combination of the nominal design of the subADC and of the subDAC must be such that a subcode transition appears between two potencies, for example in the middle point: $t_{i,k}=(\tilde{p}_{i,k-1}+\tilde{p}_{i,k})/2$. When we do this, the nominal residues are not bigger than 1/2LSB of the stage. The incorporation of a redundancy level of

[20] Although these conditions will be valid in all that follows, now they are only applied in case the MDAC is completely ideal and the residue of the stage is $r_i=y_i-\tilde{p}_i(c_i)$.

s_i bits is carried out because displacements of the transitions are expected that produce increases 2^{s_i} times the nominal residue value (this is equivalent to that the INL of the subADC reaches the value of $(2^{s_i}-1)/2$ LSBs).

The placement of the potencies in relation to the positions of the transitions of an ideal ADC of same resolution and range was specified by the *offset of potencies* δ_i, as we saw in (25) at the end of the Sect. 4.1. It is easy to show using (39) that for a redundancy level of s_i bits the possible values of the offset of potencies are limited to an interval:

$$\delta_i \in [1-2^{s_i-1}, 2^{s_i-1}] \subset \square \qquad (41)$$

In the case of not having redundancy, the interval is compressed to a single point $\delta_i = 1/2$. The cases of unit-redundancy $(s_i = 1)$ have possible offsets in the interval [0,1].
On the other hand, once fixed the offset of potencies, the set of subcodes that must be reproduced obligatorily by the subADC is that whose associated set of potencies $\{\tilde{p}_{i,k}\}$ is inside the conversion range of the stage $(\{\tilde{p}_{i,k}\} \subset D_{xi})$. This set is determined as:[21]

$$\{c_{i,k}\} = (-\delta_i, 2^{n_i} - \delta_i) \cap [0, 2^{n_i} - 1] \qquad (42)$$

For example, if there is not redundancy, the subADC must produce all the subcodes obligatorily. If there is unit-redundancy and an offset is chosen according to (41) equal to $\delta_i = 0$, the subADC can saturate inferiorly in $c_i = 1$. In this way, a reduction of the digital output range is produced and therefore it has the advantage of reducing the final cost of the design.[22]

In the Fig. *6-7* the transfer characteristic of a stage of 4bits is shown. It exhibits variations in the transitions of the subADC such that the residues can be up to 3/2LSB of the stage. From (39), the necessary redundancy level in the stage is 2bits. The chosen offset of potencies is $\delta_i = -1$. The implemented potencies correspond to the subcodes that verify (42) and are inside the input range of the stage, in this case from the $\tilde{p}_i(5)$ to the $\tilde{p}_i(14)$. Notice in this example that the references of the two involved stages are outside of their input ranges.

[21] Hereinafter we will denote $c_{i0} = \min\{c_{i,k}\}$, $c_{iM} = \max\{c_{i,k}\}$.

[22] This advantage and others, that we will indicate later, are missing, if when existing a redundancy level, values for the offset different from the ends of (41) are chosen. For example the choice $\delta_i = 1/2$, being $s_i = 1$.

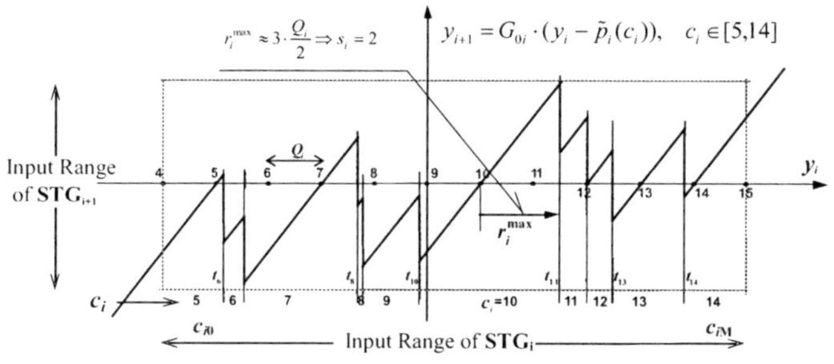

Figure 6-7. Transfer characteristic of a 4bit stage with redundancy level of 2bits

6.1.1 The multistage structure as implementation of a full calibration

We want to show here that the structure of a multistage ADC with the property of redundancy among the stages is a system implementing a digital full calibration. In particular, the INL error of the subADC in a stage is fully calibrated by the multistage structure following to that stage. To show this, we will consider the case of a two-stage subranging ADC (Fig. *6-8*). Let us assume that the subADC of the first stage (of resolution n_1) has a high INL $Q_1/2 \le |INL| \le (2^{s_i}-1)Q_1/2$. According to the previous section, the subranging can incorporate a design with redundancy of s_i bits between the two stages. In particular, to simplify the situation we will consider an ideal MDAC without gain block $(G_{0i}=1)$ and with a subADC, whose design follows the suggestions (41) and (42). If all the produced residues must be inside the conversion range $[-R_2, R_2]$ of the second stage (last quantizer in this case), $|r_1| \le 2^{s_1}(Q_1/2) \le R_2$, then the definition $R_2 = 2^{s_1}(Q_1/2) = R_1/2^{(n_1-s_1)}$ is compatible with it. Notice that this assignment becomes from (40) when the gain is equal to one. About the design of the Last Quantizer (LQ), it is not necessary to say much more, but to be a little more concrete we will assume that it has an accuracy of n_2 bits, that is to say, its INL is lower than 1/2LSB. The relationship among the references produces the relation of quantums, $Q_1 = 2^{(n_2-s_1)}Q_2$.

Any residue y_2 that arrives to the second stage can be expressed either as output of the MDAC: $y_2 = f_{\text{MDAC}}^{(1)}(x, c_1) = x - \tilde{p}_1(c_1)$, or as input of the last quantizer using, for example, the primitive form (10).

Equalizing both expressions, $y_2 = x - \tilde{p}_1(c_1) = l_2(c_2) + Q_2 INL_2(c_2) + \varepsilon(c_2)$,

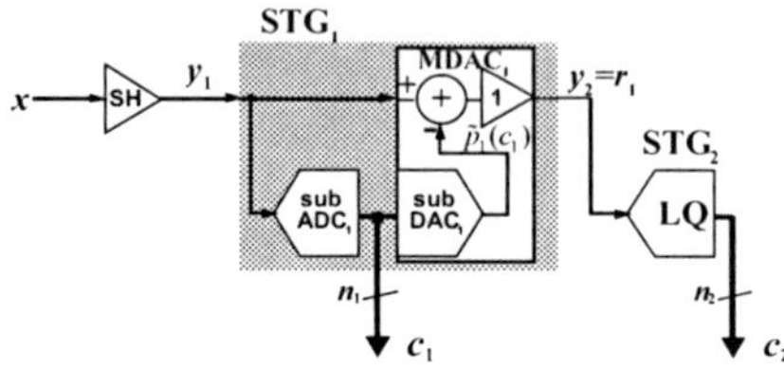

Figure 6-8. Subranging ADC without gain block between the stages

and using the expressions of the potencies, the ideal transitions and the relationship among the quantums, it results,

$$x = -R_2 + \delta_1 Q_1 - R_1 + (2^{(n_2 - s_1)} c_1 + c_2) Q_2 + Q_2 INL_2(c_2) + \varepsilon(c_2) \qquad (43)$$

If we take the combination of subcodes, $Z = 2^{(n_2 - s_1)} c_1 + c_2$ as the final code of the subranging, we will have that:

a) The evaluation of Z in binary logic requires performing the sum of the subcodes such that the s_1 least significant bits of c_1 are overlapped to the s_1 most significant bits of c_2. If s_1=0 (there is not redundancy) the subcodes are concatenated.
b) The final resolution of the subranging is $N = n_1 + n_2 - s_1$. If the redundancy level is lower than the resolution of the last quantizer, an increased resolution results in the total code.
c) The LSB of the subranging coincides with the LSB of the last quantizer, since if there is not residue amplification. It is the finest quantization that has been carried out.

$$q = (2R_1)/2^N = (2R_1)/(2^{n_1} 2^{n_2 - s_1}) = Q_1 / 2^{n_2 - s_1} = Q_2 \qquad (44)$$

As consequence, expression (43) remains as a pseudocanonical form with offset and non-linearity terms: $x + \beta_{os} = l(Z) + qINL_2(c_2) + \zeta$, being $l(Z)$ the transitions law of an ideal ADC with range $[-R_1, R_1]$ and resolution N.

The offset term $\beta_{os} = R_2 - \delta_1 Q_1 = 2^{(n_2 - s_1)}(2^{s_1 - 1} - \delta_1)q$ can be cancelled by design if the value $\delta_1 = 2^{(s_1 - 1)}$ is chosen for the offset of potencies that is compatible with (41). The non-linearity term coincides with the non-linearity term of the second stage. Since the latter is lower than at 1/2, the total ADC presents an accuracy of N bits. The residues of the subranging and of the last quantizer $\zeta = \varepsilon_2(c_2)$ are equal and are also close to the values of the LSB of the total ADC.

The conclusion is that the special structure of the subranging has produced an ADC better than the original, carrying out the full calibration of the first subADC at the expense of having a MDAC without imperfections and of incorporating redundancy. Although in a real situation there will be imperfections, the current state of the art allows, in general, a higher degree of control of the MDAC imperfections than it is possible to have in the design of the subADC. This makes the multistage ADC an efficient structure whenever the cost increment due to hardware resources, power and design effort are permissible.

It is easy to extend the previous results to the case the MDAC has gain different to the unit, but compatible with (40). All the results are identical except the value of the LSB q of the complete ADC and, consequently, the total residue of the conversion; it will be G_{o1} times lower than the last quantizer $q = Q_2 / G_{01}$.

6.2 Behavioral Model for Multistage ADC

6.2.1 General model with ideal MDACs

The generalization of the results of the previous section to a multistage ADC of L-stages is immediate. The non-ideality of any subADC will be corrected by the structure in cascade that follows it. At the end, only the INL corresponding to the last quantizer will remain, as well as an offset dependent on the potencies and redundancy level chosen for each stage. To obtain the pseudocanonical form of the L-stage ADC, it is enough to observe that its operation (26) follows the recurrence law $y_{n+1} = a_n y_n + b_n, (n \geq 1)$, where the coefficients are $a_n = G_{0n}$ and $b_n = -G_{0n}\tilde{p}_n(c_n)$.[23] Working out the value of the seed $y_1 = x$ (the ADC input) in the general solution[24] of this equation and substituting the expressions, already known, as the value of interstage gain (40) and the law of potencies (25), produces the following form:

$$x + \beta_{os} = I(Z) + qINL_L(c_L) + \zeta \tag{45}$$

where $I(Z)$ corresponds to an ideal ADC of range $[-R_1, R_1]$, resolution:

[23] Without lost of generality, we will always use positive interstage gains, $G_{0i} > 1$.

[24] The general solution of the finite-differences equation $x_{n+1} = a_n x_n + b_n$ for $n \geq 1$ is $x_{n+1} = \Pi_n \cdot \left\{ x_1 + \sum_{k=1}^{n} (b_k / \Pi_k) \right\}$ with $\Pi_j = \prod_{k=1}^{j} a_k$. The seed can be written as $x_1 = -\sum_{k=1}^{n} (b_k / \Pi_k) + (x_{n+1} / \Pi_n)$

$$N = \sum_{i=1}^{L-1}(n_i - s_i) + n_L = \sum_{i=1}^{L} n_i - \sum_{i=1}^{L-1} s_i \tag{46}$$

and a digital output that is a linear combination of the obtained subcodes:

$$Z = Z(c_1, c_2, \ldots, c_L) = \sum_{i=1}^{L} c_i 2^{m_i} \tag{47}$$

where the exponent m_i is the resolution associated to the significance (weight) of the subcode c_i inside the total code Z. Its value is:

$$m_i = N_i - n_i = \left(\sum_{j=i}^{L} n_j - \sum_{j=i}^{L-1} s_j\right) - n_i \tag{48}$$

A new parameter N_i is used in (48) for posterior developments. It is easy to identify N_i with the resolution that according to (46) would have a multistage converter formed by all the stages from the **STG$_i$** to the last quantizer. This latter sub-structure of the multistage converter will be denominated: *i-th order subconverter*. In this way, the first order coincides with the total structure with $N_1 = N$, the L-th order coincides with the last quantizer and $N_L = n_L$. A parameter that will be of great relevance is the LSB of i-th order subconverter:

$$q_i = \frac{2R_i}{2^{N_i}} \tag{49}$$

With this parameter the relation (40) for the interstage gain is $q_{i+1} = G_{0i} q_i$, that is to say, the LSB of each subconverter is G_{oi} times higher than the quantum of its precedent. The relationship with the stage-LSB of its more significant stage is $Q_i = 2^{m_i} q_i$.

The realization of Z in binary logic is obtained by means of a sum where each subcode c_i is overlapped in s_i bits with the following c_{i+1}. The overlapping disappears when the redundancy level is zero, then, the implicated subcodes will be concatenated.

The offset of (45) is:[25]

$$\beta_{os} = q \cdot \sum_{i=1}^{L-1} (2^{(s_i - 1)} - \delta_i) 2^{m_i} \tag{50}$$

[25] The offset is always positive or null, since for (41): $\delta_i \leq 2^{(s_i-1)}$.

This offset can be eliminated if it is incorporated to the total output in the way explained in Sect. 5.1, $(Z^{\dagger} = Z - \langle \beta_{os} / q \rangle)$. In that case it is not necessary to carry out any measure for its estimation because the value is known from the beginning of the multistage ADC design. We will say that it is a *by-design induced offset*, which is cancelled if for example all the stages use an offset of potencies equal to $\delta_i = 2^{(s_i - 1)}$. As the format of (50) is similar to the form of the total code (47), the offset correction is possible locally in the stage before passing it for its combination with the rest. A simple example that illustrates this is when the offsets of potencies are: $\delta_i = 1 - 2^{(s_i - 1)}$ (the minimum according to (41)); in that case: $Z^{\dagger} = Z - \langle \beta_{os} / q \rangle = \sum_{i=1}^{L} c_i 2^{m_i} - \sum_{i=1}^{L} 1 \cdot 2^{m_i} = \sum_{i=1}^{L} (c_i - 1) 2^{m_i} = \sum_{i=1}^{L} c_i' 2^{m_i}$ which means that the subcode that must give the stage for its combination is: $c_i' = c_i - 1$.

6.2.2 MDAC model with first order perturbations

Generalizing the result (45) for a more realistic situation requires a more real model of the MDAC block. The model adopted here includes additive first order perturbations of the inputs in the form:

$$y(x,c) = f_{\text{MDAC}}(x,c) = G_0(1+\varepsilon)(x - \tilde{p}(c)) + \gamma G_0 x + \lambda(c) + y_{os} \tag{51}$$

where ε represents the relative error of the output ideal value $y_{id} = G_0(x - \tilde{p}(c))$, y_{os} is an offset error, $\gamma G_0 x$ represents an additional gain error regarding the analog input, and, the error term $\lambda(c)$ considers a non-linear dependence of the output with the digital input. Habitually $\lambda(c)$ is a linear combination of the type $\lambda(c) \propto \sum_i \varepsilon_{b,i} Y(b_i)$, where $\varepsilon_{b,i}$ is the gain error of the path that transmits the signal $Y(b_i)$ associated to the bit b_i of the subcode c. If, for example, c is binary-coded with resolution n-bits, $\lambda(c)$ has n gain errors, while if it is thermometer-binary coded, the number of gain errors can increase up to 2^n. To interpret results it is better to associate each perturbative term to the degradation of the basic blocks forming the MDAC. In this way, ε is associated to the gain block of the MDAC, the factor γ to the direct path between x and y, the term $\lambda(c)$ is a combination of linear and non-linear effects corresponding to the subDAC, and the offset y_{os} is a combination of the offsets of each one of the subblocks.

As the real implementations of MDACs are usually carried out using discrete time design techniques, the model (51) can include perturbations associated to the finite time intervals in what the output is processed, for example:

$$y^{(n)}(x,c,T) = y_{\infty}^{(n)}(x,c;\varepsilon,\gamma,\underset{\sim}{\lambda},y_{os})\cdot(1-\psi(T)) + y_{0}^{(n)}(x,c,y^{(n-1)})\cdot\Phi(T) \qquad (52)$$

where the superscript n indicates the order of the sample in the discrete process, T is the duration of the process interval, the function $y_{\infty}(\cdot)$ is the proposal in (51), and $y_0(\cdot)$ determines the value that takes the output at the beginning of the n-th process (it can depend on the output value in the previous process). The functions $\psi(.)$ and $\Phi(.)$ evaluate the temporary perturbations and verify $\psi \sim \Phi \sim 1$ if $T \to 0$, and $\psi \sim \Phi \sim 0$ if $T \to \infty$.

Although models as (52) and other more complexes are deducible from the low-level study of the circuits implementing the MDAC, here we will only consider the proposal (51). Its simplicity is suitable for the symbolic development of the general expressions of multistage ADC models, but it is also effective for the description in a widespread fan of implementations where the second order effects can be obviated in phases of the design scanning. Therefore, if in a first moment we want to estimate the effects of second order, such as quadratic terms with the inputs ($\sim \alpha x^2$) or feedback terms ($\sim \beta y^{(n-1)}$), we will consider them contained in a term of additive noise $n_o(x,c,y^{(n-1)})$ in which the electronic noise is also included. Of course, the practical characterization of this noise must be carried out at low level and we will accept, in what follows, that it is possible to estimate its average power.

The expression (51) is linear for c fixed, but any electronic implementation will show saturation effects. To take into account this effect we can consider the following PWL transfer function for the output values in (51):

$$y_o = (1-\delta m_{sat})y(x,c) + \frac{\delta m_{sat}}{2}\left(\left|y(x,c)+y_{sat}\right| - \left|y(x,c)-y_{sat}\right|\right) \qquad (53)$$

This function preserves the values of $y(x,c)$ between the amplitude limits $\pm y_{sat}$, to which we consider the beginning of the saturation effect, and it reduces its value, when exceeding them, in a quantity equal to $\delta m_{sat}\cdot\left|y(x,c)-y_{sat}\right|$.

Finally and to complete this section with some results of physical design, we will show the values that the perturbative parameters of the model take in case the circuit chosen to implement the MDAC function is designed with SC techniques (**S**witched **C**apacitors). In the case of taking a block as the represented in the Fig. 6-9, the output voltage v_0 as function of the inputs $(v_{in}, c = \{b_j\})$ and the capacitances of the used capacitors is:

$$v_o = \alpha_A \left\{ v_{in} + \sum_j r_{jf} V_j - \sum_j r_{jf} R_j + V_{os} \right\} \tag{54}$$

where $r_{jf} = C_j / C_f = r_{jf}^0 (1 + \varepsilon_{rj})$ is the existent relationship between one of the capacitors C_j associated to the digital input and the feedback capacitor C_f. This relationship, which gives the gain of the path from the input associated to the bit b_j to the output of the block, is affected of a relative error ε_{rj} (*mismatch error*) with regard to the nominal value r_{jf}^0. In practice, the bound of this relative error is to a certain extent controllable by design. For that, the capacitors are built as parallel associations of unitary capacitors, which must have a capacitance value above certain critical value related with the required bound of ε_{rj}. In current CMOS technologies, without having to use excessively big values for the unit-capacitor (~1pF) and without resorting to a special process, it is possible to reduce the mismatch errors below 0.1%. The factor $\alpha_A \approx 1$ and the term $V_{os} \approx 0$ are related with the finite gain (A_0) and the offset v_{os} of the opamp, respectively.

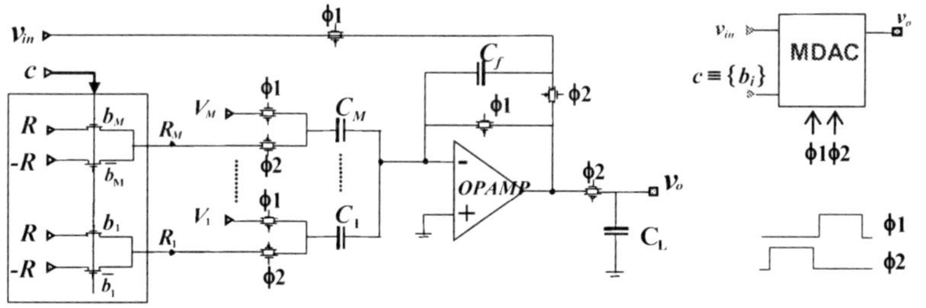

Figure 6-9. Non-overlapped two phases SC topology for MDAC.

If in (54) we take appropriate values for V_j and R_j, we will be able to find design solutions that correspond nominally with the behavior of a MDAC. For example, if the digital input to the block comes coded in a M-bit thermometer code[26], taking $R_j = b_j R - \bar{b}_j R = (b_j - \bar{b}_j)R = -m_j R$ (as it is indicated in Fig. 6-9) and $V_j = v_{in}$, expression (54) becomes:

[26] If the resolution of the subcode is n bits, the thermometer code has, as maximum, 2^n bits.

$$v_o = \alpha_A \left\{ \left(1 + \sum_{j \le M} r_{jf}^0 \right) v_{in} + R \sum_{j \le M} m_j r_{jf}^0 + \left(\sum_{j \le M} \varepsilon_{rj} r_{jf}^0 \right) v_{in} + R \sum_{j \le M} \varepsilon_{rj} m_j r_{jf}^0 + V_{os} \right\} \tag{55}$$

where R is the reference of the stage conversion range.
In this expression the parameters of (51) are identified: [27]

a) The interstage gain, $G_0 = 1 + \sum_j^M r_{jf}^0$

a) The position of potencies, $\tilde{p}(c) = \tilde{p}(\{b_1, \ldots, b_M\}) = (-R / G_0) \sum_j^M m_j r_{jf}^0$

b) $1 + \varepsilon = \alpha_A \approx 1 - G_0 / A_0$

c) $\gamma G_0 = \alpha_A \sum_j^M \varepsilon_{rj} r_{jf}^0$

d) $\lambda(c) = \lambda(\{b_1, \ldots, b_M\}) = \alpha_A R \sum_j^M \varepsilon_{rj} m_j r_{jf}^0$

e) $y_{os} = \alpha_A V_{os} \approx \alpha_A v_{os}$

For a design with unit redundancy $s = 1$, offset of potencies $\delta = 1$ and same ranges for input and output, there is a very simple compatible solution making $r_{jf}^0 = 1/2$ and $M = 2^n - 2$.

The expression (55) is also compatible with other designs, its form does not change essentially if the digital input comes n-bit binary-coded. Anyway, the previous expressions, in particular d) and e), reveal us an existent correlation, in this type of implementations, among the gain γ and the non-linearity $\lambda(c)$ parameters,

$$\gamma G_0 = \frac{\lambda(0)}{R} = -\frac{\lambda(M)}{R} \tag{56}$$

6.2.3 General model with real MDACs

Once we have defined the realistic model of MDAC that we will consider, we proceed to extend the results of the Sect. 6.2.1, and hence to obtain the pseudocanonical form for a real multistage ADC. But before, let us revise some of the hypotheses.

We will maintain the redundancy property and the condition of full calibration among stages (Sect. 6.1), but with the following remarks:

[27] In this single-ended topology, a coarse estimation of the average power of electronic thermal noise at its output is $n_0^2 \approx G_0 k_B T / C_f$, where k_B is the Boltzmann's constant and T the absolute temperature (contributions from opamp thermal noise and switch on-resistors have been underestimated).

a) We take by definition the interstage gain expression (40) .
b) The residue is redefined by (51), and hence, it is the input signal fraction that amplified by the nominal gain produces such output. That is: $r_i = y_{i+1} / G_{0i}$.
c) If the parameter $y_{sat,i}$ that defines the separation between the linear and the saturation range of the i-th stage according to (53), is smaller than R_{i+1}, then the stage output must be strictly inside to the linearity range, $|y_{i+1}| \leq y_{sat,i}$.

With these assumptions, the model proposed in Sect. 6.2.1 with ideal MDACs continues being valid.

The expression (39) used to define the redundancy level continues being valid, although in case of c) is verified, the maximum real residue must be limited by design to a fraction $y_{sat,i} / R_{i+1}$ of the value given by (39).

When requiring that the stage output be found in the linear range given by the model (51), then, the algorithm (26) of the multistage ADC still verifies the recurrent relationship $y_{n+1} = a_n y_n + b_n, (n \geq 1)$ that we use in Sect. 6.2.1, but now the coefficients are $a_n = G_{0n}(1+\varepsilon_n+\gamma_n)$ and $b_n = -G_{0n}(1+\varepsilon_n)\tilde{p}_n(c_n)+\lambda_n(c_n)+y_{os,n}+n_{o,n}$ (notice that we consider the noise term in the model). Working out the value of the seed $y_1 = x$ in the general solution, will provide us the expected functional form (see note 24). To get a result coherent with that obtained in the case of ideal MDACs, we must use the parameters there defined as (47) and (48). To describe the last quantizer we will use the canonical form (19) instead of the primitive one, that is:

$$g_L \cdot (y_L + n_{I,L}) + x_{os,L} = l_L(c_L) + Q_L enl_L(c_L) + \nu_L \tag{57}$$

where we will write its gain as $g_L = 1+\gamma_L$. In order to being uniform with the expression of a_n, $\varepsilon_L = 0$ is considered. The term $n_{I,L}$ is the total input referred noise of the last quantizer.

The final result after some algebra is:

$$\alpha \cdot (x + n_{Ix}) + \beta_{os} = l(Z) + qInL(c_1,\ldots,c_L) + \zeta(c_L) \tag{58}$$

Regarding the model (45), the essential differences are: a) the non-linearity term InL depends on all the subcodes, b) it appears a quasi-gain factor α, and, c) a noise term n_{Ix} is introduced.

Using,

$$\alpha_i = \prod_{j=i}^{L} (1+\varepsilon_j+\gamma_j) \qquad ,\mu_i = (1+\varepsilon_i)\alpha_{i+1} \tag{59}$$

the concrete values of the quasi parameters in the functional form (58) are:

– *Gain factor*:

$$\alpha = \alpha_1 \tag{60}$$

– *Non-linearity term*:

$$InL(c_1,\ldots c_L) = \sum_{i=1}^{L-1}(-1+\mu_i)c_i 2^{m_i} - \sum_{i=1}^{L-1}\alpha_{i+1}\frac{\lambda_i(c_i)}{q_{i+1}} + enl_L(c_L) \tag{61}$$

– *Offset term*:

$$\beta_{os} = q\sum_{i-1}^{L-1}(2^{(s_i-1)} - \delta_i)2^{m_i} + q\cdot\left(\sum_{i=1}^{L-1}(1-\mu_i)\frac{p_{0i}}{q_i} + \sum_{i=1}^{L-1}\alpha_{i+1}\frac{y_{os,i}}{q_{i+1}} + \frac{x_{os,L}}{Q_L}\right) \tag{62}$$

– *Noise term*:

$$n_{Ix} = q\cdot\sum_{i=1}^{L}\left(\frac{\alpha_i}{\alpha_1}\right)\frac{n_{I,i}}{q_i} \tag{63}$$

where we have used the input referred noise terms of each stage $(n_{I,i})$ obtained after dividing for the total gain, $n_{o,i} = G_{0i}(1+\varepsilon_i+\gamma_i)n_{I,i}$.

– *Residual term*: it is proportional to the corresponding of the last quantizer with the same scale of the LSB of the total ADC, $\zeta(c_L) = q\cdot(\nu_L/Q_L)$.

Notice that for each subconverter of the multistage ADC (for example the *k*-th order subconverter), it is possible to write a pseudocanonical form as (58) with parameters of the same form that the previously exposed but with a sequence of indexes that begins in $i=k$. This separation of effects will be frequently used in the digital self-calibration processes of the multistage ADC that we will show in the Sect. 6.3. The pseudocanonical form (58) does not have to be close to the canonical form (19) of the multistage ADC.

If we develop methods of digital calibration (Sect. 5.1), where the final code *Z* is corrected with an estimation of the value of the *InL*, just as it is in (61), then, it will remain only the gain (60) and the offset (62). If these errors are not tolerated, they can be more easily reduced by other techniques independent of the structure (Sect. 5.1.1 & 5.1.2).

6.2.4 Simplification of the general model

In the given expressions of the quasi-parameters there are second order interactions among the parameters of the MDACs, but the second order effects were underestimated or included as contribution of the noise term. Therefore, we must simplify the expressions keeping the first order effects. The expressions (59), (61), (62) and (63) become:

$$\begin{cases} \alpha_i \approx 1 + \sum_{k=i}^{L} \varepsilon_k + \sum_{k=i}^{L} \gamma_k = 1 + \Xi_i^{(L)} + \Psi_i^{(L)} \\ \mu_i = (1+\varepsilon_i)\cdot\alpha_{i+1} \approx 1 + \Xi_i^{(L)} + \Psi_{i+1}^{(L)} \end{cases} \tag{64}$$

$$InL(c_1,\ldots c_L) \approx \sum_{i=1}^{L-1} (\Xi_i^{(L)} + \Psi_{i+1}^{(L)}) c_i 2^{m_i} - \sum_{i=1}^{L-1} \frac{\lambda_i(c_i)}{q_{i+1}} + enl(c_L) \tag{65}$$

$$\beta_{os} \approx q \cdot \left(-\sum_{i=1}^{L-1} (\Xi_i^{(L)} + \Psi_{i+1}^{(L)}) \frac{p_{0i}}{q_i} + \sum_{i=1}^{L-1} \frac{y_{os,i}}{q_{i+1}} + \frac{x_{os,L}}{Q_L} \right) \tag{66}$$

$$n_{Ix} \approx q \cdot \sum_{i=1}^{L} \frac{n_{I,i}}{q_i} \tag{67}$$

It must be remarked that the first addend of the offset term in (62) coincides with (50) and, as we commented at the end of the Sect. 6.2.1, it can be eliminated without problems since it is known (induced) by design, then, we will incorporate it to the expression of the output code:

$$Z^* = Z - \sum_{i=1}^{L-1} (2^{(s_i-1)} - \delta_i) 2^{m_i} = Z - Zos \tag{68}$$

Hereinafter we will consider that the realization Z^* is the true output code of the multistage ADC.

If the MDACs were ideal, the non-linearity, gain and offset of the complete ADC correspond to last quantizer. For this reason, in the design of the multistage ADC, it is necessary to put special care of not making an enormous effort to control the errors of the MDACs and to forget the last quantizer design.

6.2.5 Gain of the multistage ADC

When expression (65) is not the non-linearity error *enl*(*Z*) of the complete converter, it is true, in general, that neither the gain factor (60) nor the offset term (66) represent the gain and offset of the multistage ADC. In practice, the second and third term of (65) not barely have linear correlation with *Z*, and therefore they have a little contribution to the offset and the gain of the final ADC. By the contrary, the first term in (65) has a strong linear trend with the final code, giving place to a substantial modification of the gain factor. This will be seen next, where we intend to obtain an approximate first order expression of the total gain of the multistage ADC. Before, we will rewrite the first sum term of (65) to make evident the expected correlation:

$$\sum_{i=1}^{L-1}(\Xi_i^{(L)} + \Psi_{i+1}^{(L)})c_i 2^{m_i} = \sum_{i=1}^{L-1}(\varepsilon_i + \gamma_{i+1})Z_{[1i]} \tag{69}$$

In this expression, the code $Z_{[1i]}$ is a truncated version of *Z* because it does not contain information from the subcodes following to the c_i,

$$Z_{[1i]} = Z(c_1, c_2, \ldots, c_i, 0, 0, \ldots, 0) = \sum_{k=1}^{i} c_k 2^{m_k} \tag{70}$$

Using this, the representation of $Z_{[1i]}$ respect of *Z* becomes a growing staircase of unit slope, in which each step corresponds with the set of values of *Z*, $Z \in \{(0,0,\ldots,0,c_{i+1},\ldots,c_L)\}$, where only the *L-i* least significant subcodes are changed. As consequence, each term of the right side of (69) is also a staircase with approximate slope of $(\varepsilon_i + \gamma_{i+1})$. Notice the strong correlation with *Z* in the case shown in the Fig. *6-10*a where $gl_2 = (0.07\%)Z_{[12]} = (0.07\%)(c_1 2^{12} + c_2 2^{10})$ and $c_i \in [0,6]$.
We will express each sum term as a linear term with *Z* (that contributes to the gain factor) and other no-linear $\varphi_i(Z)$ (see Fig. *6-10*b) that will have left as remanent *InL*:

$$(\varepsilon_i + \gamma_{i+1})Z_{[1i]} = (\varepsilon_i + \gamma_{i+1}) \cdot Z + \varphi_i(Z) \tag{71}$$

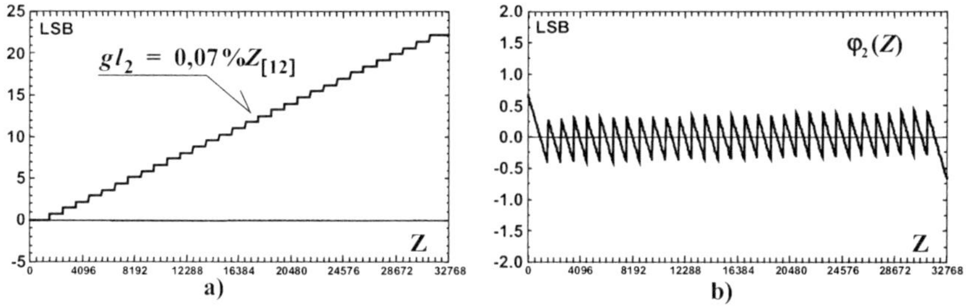

Figure 6-10. **a)** Graph of the second gain term with the final code
b) Non-linearity of **a)** after subtracting the correlation with Z

The term (69) becomes:

$$\sum_{i=1}^{L-1}(\Xi_i^{(L)}+\Psi_{i+1}^{(L)})c_i 2^{m_i} = Z\cdot\sum_{i=1}^{L-1}(\varepsilon_i+\gamma_{i+1})+\sum_{i=1}^{L-1}\varphi_i(Z) = (\Xi_1^{(L)}+\Psi_2^{(L)})\cdot Z+\sum_{i=1}^{L-1}\varphi_i(Z) \tag{72}$$

Passing the result (72) to the pseudocanonical form (58) and grouping all linear term with Z such as they result contained in the ideal law *l*(*Z*), it becomes that the whole form must be divided by the factor $1+(\Xi_1^{(L)}+\Psi_2^{(L)})$. Thus, the new gain factor (the gain of the multistage ADC if there are not more important linear correlations with *Z*) is:

$$g \approx \frac{\alpha_1}{1+(\Xi_1^{(L)}+\Psi_2^{(L)})} = \frac{1+(\Xi_1^{(L)}+\Psi_1^{(L)})}{1+(\Xi_1^{(L)}+\Psi_2^{(L)})} \approx 1+\gamma_1 \tag{73}$$

6.3 Digital Calibration in Multistage ADC

Along the Sect. 6 we have exposed a model for the static behavior description of multistage A/D converters. This model includes the effect of the redundancy that calibrates completely[28] the non-idealities of the subADC, as the expression (45) demonstrates. However, the degradation of the ADC functionality due to the non-idealities of the MDAC still remains. This is implicitly expressed in the pseudocanonical form (58). To reduce this degradation it is necessary to use extra calibration procedures. These procedures can be based on the modification of the internal structure of the

[28] Except those of the last quantizer.

MDAC ([26], [27]), in the modification of the residue evaluation ([28], [29], [30]), or in the direct correction of the output code ([31]-[39]); the two formers are known as analog calibration and the latter as digital calibration technique (Sect. 5.1).

To conclude this chapter we will expose two procedures based on a digital approach. Both procedures estimate the most significant part of (65), which we will call, $\langle InL_r(Z)\rangle$, corresponding to the r most significant stages of the multistage structure. Its value is then memorized in a register block, and during the normal operation of the ADC it is used to correct the final code, just as it was exposed in Sect. 5.1. In addition to this procedure, it would be necessary to apply others that calibrate the offset and/or the gain errors (Sect. 5.1.1 and 5.1.2).

The partial estimate of the quasi-no-linearity will be expressed as

$$\langle InL_r(Z)\rangle = \langle InL_r(c_1, c_2, \ldots, c_r)\rangle = \sum_{i=1}^{r<L} ze_i(c_i) \tag{74}$$

where the terms $ze_i(\cdot)$, depending only on the subcode value, are denominated *error codes* of the *i*-th stage, and the calibrated final code will be:

$$Z^{\dagger} = Z^{*} + \langle InL_r(Z)\rangle = Z(c_1, \ldots, c_r, \ldots c_L) - Zos + \sum_{i=1}^{r<L} ze_i(c_i) \tag{75}$$

The separation in error codes of the quasi-non-linearity estimation implicates that instead of having a block of 2^N registers addressed by Z (Fig. *6-5*a) we will have a set of r blocks of M_i registers addressed by the local value of each subcode.

Although there is a negative correlation[29] between the resolution of the stages and the number L of them in the multistage structure, the total number of registers to memorize, $\sum_{i=1}^{r} M_i \le \sum_{i=1}^{r} 2^{n_i}$, is always much smaller than the necessary in the full calibration.

The Fig. *6-11* shows the form in what the digital calibration of the output code is made using error codes. The error codes are stored in the register blocks $\mathbf{RB_i}$ and its content is accessed using as address the subcode value[30].

[29] To smaller chosen resolution for the stages, more number of stages are necessary to get a given resolution and, consequently, more number r of stages have to be included in the calibration.

[30] If the values of the subcode c_i don't begin in zero $(c_{i0} \ne 0)$, the address of the register block $\mathbf{RB_i}$ is in fact: $(c_i - c_{i0})$.

The offset register *Zos* can be incorporated into the block of subcode combinations.

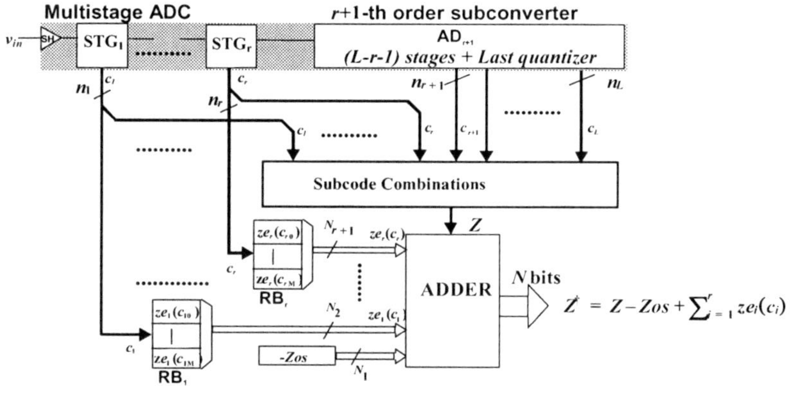

Figure 6-11. Topology of multistage ADC calibrated with error codes.

The rest of terms of (65) not included in the estimate will contribute to a remanent non-linearity. The calibration procedure is efficient to a certain resolution when the not estimated part is considered as insignificant and it is included in the noise term. As we have already argued in Sect. 5.1, to reduce the effect of the truncation errors in the results of the arithmetic operations it is necessary to add extra resolution (*p* bits) to the code of the ADC. This implicates that in the design of multistage ADC with digital calibration it is necessary to include more stages to obtain a given resolution *N*. In this case, the additional resolution plays a similar role to the extra resolution that corrects the redundancy of the design.

If the calibration procedure of the ADC is based explicitly on the pseudocanonical functional form (65) and, hence, it cancels the InL value of that expression, then the pseudocanonical form of the calibrated multistage ADC will not have the non-linearity term. Consequently, this form corresponds to a perfect ADC and its gain factor is, in fact, the gain of the multistage ADC. From (64), if we consider the r+1-th order subconverter as the last quantizer:

$$g^{\dagger} \approx \alpha_1 = 1 + \Xi_1^{(r)} + \Psi_1^{(r+1)} \tag{76}$$

If there is not a specific compensation, the gain of the calibrated ADC can be higher than for the multistage ADC without calibration (73).

The two procedures we will introduce are of self-calibration, that is to say, the converter is designed so that it incorporates in its operation a control and a reconfiguration that allow to estimate in an autonomous way the error codes. In both, direct measures are carried out of the **MDAC$_i$** output whose error code $ze_i(\cdot)$ is going to be estimated, by means of the next $i+1$-th order subconverter. That subconverter should be as perfect as possible. The process of self-calibration of the multistage ADC is in fact a recurrent self-calibration procedure of all the subconverters with order lower than the first one that can be considered linear, the $r+1$ in our case. Once the $i+1$-th order subconverter is self-calibrated (linearized), it can be already used as device for measure the i-th MDAC errors.

The resolution N_{r+1} of the $r+1$-th order subconverter can be decided a priori knowing the technological limits of the process where the ADC will be implemented (for example 8 or 9 bits in a conventional CMOS technology). It is necessary to take the biggest possible resolution to reduce to the maximum the cost of the calibration procedure and also the propagation errors due to the recurrence of the process.

Of course, in both methods it is necessary to have calibration stimuli for which a certain response is expected, however this is not problematic since we will use DC signals already existent in the design of the ADC without calibration.

6.3.1 KLB self-calibration procedure[31]

Before of introducing a way for realizing the error estimation, let us see how the error codes of this procedure can be expressed using our closed forms.

According to (65), the r terms of the quasi-non-linearity to be estimated have the form

$$InL_r(c_1,c_2,\ldots,c_r)=\sum_{i=1}^{r}(-1+\mu_i)c_i 2^{m_i}-\sum_{i=1}^{r}\frac{\lambda_i(c_i)}{q_{i+1}}=\sum_{i=1}^{r}\left((-1+\mu_i)c_i 2^{m_i}-\frac{\lambda_i(c_i)}{q_{i+1}}\right)=\sum_{i=1}^{r}e_i(c_i) \tag{77}$$

[31] We denominate KLB to this procedure in honor of the authors of [33] that reported this self-calibration technique for the case of pipeline ADCs with stages of 1 bit. Although our model proposes a generalization of this procedure to stages of more than 1 bit, we maintain the KLB name. Also, it is possible to show using our model that the method published in [36] is equivalent to the KLB method.

That is, the error code of the *i*-th stage should be: $ze_i(c) = \langle e_i(c) \rangle$ $c \in [c_{i0}, c_{iM}] \subseteq [0, 2^{n_i} - 1]$, where the subcode interval is given by (42).
Using the first-difference operator for the parameters $\lambda_i(\cdot)$, $\Delta_1\lambda_i(k) = \lambda_i(k+1) - \lambda_i(k)$, and for the subcode values, $\Delta_1 k = (k+1) - k = 1$, we can write:

$$\begin{cases} \lambda_i(c) = \lambda_i(c_{i0}) + \sum_{k \geq c_{i0}}^{c-1} \Delta_1\lambda_i(k) \\ c = c_{i0} + \sum_{k \geq c_{i0}}^{c-1} \Delta_1 k \end{cases} \tag{78}$$

thus obtaining recurrent relations as:

$$e_i(c) = e_i(c_{i0}) + \sum_{k \geq c_{i0}}^{c-1} \left((-1+\mu_i)2^{m_i} - \frac{\Delta_1\lambda_i(k)}{q_{i+1}} \right) = e_i(c-1) + W_i(c-1) \tag{79}$$

with $W_i(c) = (-1+\mu_i)2^{m_i} - \left(\Delta_1\lambda_i(c)/q_{i+1}\right), \quad c \in [c_{i0}, c_{iM})$.
Substituting (79) into equation (77) it results:

$$InL_r(c_1, c_2, \ldots, c_r) = \sum_{i=1}^{r} e_i(c_{i0}) + \sum_{i=1}^{r} \sum_{k \geq c_{i0}}^{c_i - 1} W_i(k) \tag{80}$$

The first term in (80) does not vary with the output code *Z*, and hence, it does not contribute to the non-linearity; it can be incorporated inside the quasi-offset term in the pseudocanonical form. The second term is the InL we are going to estimate, being the error codes the quantized values of its summation terms:

$$ze_i(c) = \left\langle \sum_{k \geq c_{i0}}^{c-1} W_i(k) \right\rangle = \sum_{k \geq c_{i0}}^{c-1} (\Omega_i(k) - 2^{m_i}), \quad c \in [c_{i0}, c_{iM}] \tag{81}$$

where Ω_i represents the *i-th* stage *relative transition codes* defined as:

$$\Omega_i(c) = \left\langle \mu_i 2^{m_i} - \frac{\Delta_1\lambda_i(c)}{q_{i+1}} \right\rangle, \quad c \in [c_{i0}, c_{iM}) \tag{82}$$

The reason of expressing the error codes as functions of the relative transition codes is that in this self-calibration procedure these latter values are directly obtained from the **MDAC$_\mathbf{i}$** response to the calibration stimuli.

Using (68) and (81), the calibrated output code can be expressed in the following form, which suggests the alternative implementation illustrated in Fig. *6-12:*

$$Z^{\dagger} = Z^{*} + \langle InL_r(Z)\rangle = Z^{*} + \sum_{i=1}^{r} ze_i(c_i) = \sum_{i=1}^{r} E_i(c_i) + Z_{r+1} - Zos_r \qquad (83)$$

where $E_i(c_i) = \sum_{c\ge c_{i0}}^{c_i-1} \Omega_i(c)$ is the accumulative sum of the relative transition codes, Z_{r+1} is the output code of the r+1-th order subconverter, and Zos_r is an offset term produced by the calibration:[32]

$$Zos_r = \sum_{i=1}^{L-1}\left(2^{s_i-1} - \delta_i\right)2^{m_i} - \sum_{i=1}^{r} c_{i0}2^{m_i} \qquad (84)$$

However, if the system incorporates an offset calibration procedure as that explained in Sect. 5.1.1, it is not necessary to consider Zos_r in (83) since it results included in the total offset estimation $\langle x_{os}/q\rangle$.

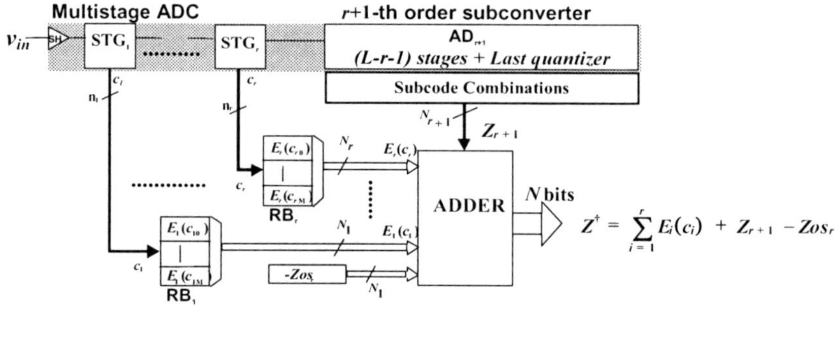

Figure 6-12. Alternative topology for the implementation of error code based digital calibration in multistage ADCs

Summarizing, the generalized KLB self-calibration technique can be applied using the following procedure:

– From stage i=r to stage i=1:

32 The consideration $E_i(c_{i0}) = 0$ implicates that we must modify the value of the induced offset from (50) to (84).

a) Estimation of the *i-th* stage transition codes, $\{\Omega_i(c)\}, c \in [c_{i0}, c_{iM})$ as it will be explained in a next section (Sect. 6.3.2).
b) Evaluation and storage of the *i-th* stage error codes using one of the two following alternatives:
   - 1st alternative:

$$\begin{cases} ze_i(c_{i0}) = 0 \\ ze_i(c+1) = ze_i(c) + \Omega_i(c) - 2^{m_i}, \quad c \in [c_{i0}, c_{iM}) \end{cases} \tag{85}$$

   - 2nd alternative:

$$\begin{cases} E_i(c_{i0}) = 0 \\ E_i(c+1) = E_i(c) + \Omega_i(c), \quad c \in [c_{i0}, c_{iM}) \end{cases} \tag{86}$$

– The resulting final calibrated code will be the given by (75) for the first alternative, and by (83) for the second.

6.3.2 Transition codes for the KLB procedure

This section explains the method for obtaining the transition codes needed for the application of the KLB self-calibration procedure. Figure *6-13* illustrates the method in the case of being the first stage the stage under calibration (SUC). The same configuration much be considerate for the *r* most significant stages.

Each transition code of the SUC is estimated applying to its MDAC (**MDAC$_1$** in the Fig. *6-13* example) a DC input signal, x_t, and making two measures at the MDAC output. These output values are digitized by the ADC composed by the rest of the stages (AD$_2$ in Fig. *6-13*). Since calibration is performed stage to stage starting from the least significant to the most significant, the AD$_2$ subconverter is already calibrated when used for calibration of the first stage.

The values of the DC input signal much be very closed to the transition codes of the subADC in the SUC; that is, the medium value of two contiguous potencies $x_t = t_{i,j} = (\tilde{p}_i(c_{ij} - 1) + \tilde{p}_i(c_{ij}))/2$, $c_{ij} \in [c_{i1}, c_{iM}]$. These stimuli can be easily obtained, for example, with a resistive network and applied through an analog multiplexer as it is indicated in Fig. *6-13*. In the case the subADC is of the Flash type [1], these stimulus values can be the same used as subADC references.

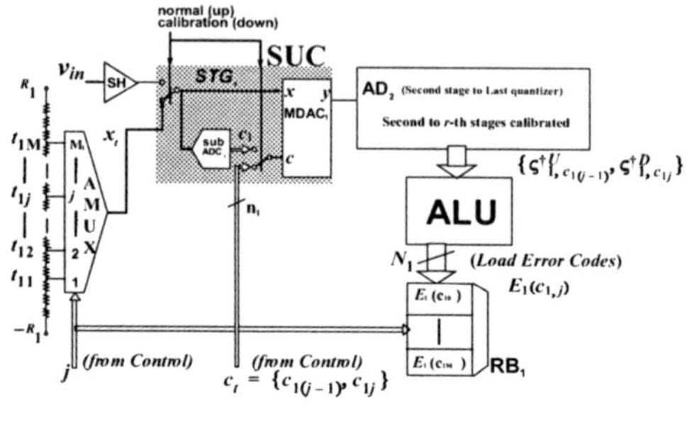

Figure 6-13. Reconfiguration of a stage and calibration stimuli to carry out the estimations of the relative transition codes

For each analog value x_t, two MDAC output values are taken. They are the response of the MDAC to the analog input x_t and to two different MDAC digital input subcodes c_t. These subcodes have to be those associated to x_t; for the *j*-th transition of the **STG**$_\mathbf{i}$ that is: $x_t = t_{ij}$, $c_t = \{c_{i(j-1)}, c_{ij}\} = \{c_{ij} - 1, c_{ij}\}$.

It is important to notice that the effectiveness of this calibration method is not dependent on the strict equality $x_t = t_{ij}$, but on the fact that this value must be held constant during the measurement of the two outputs corresponding to the two digital inputs c_t.

The role of the ALU block in Fig. *6-13* is to obtain the set of relative transition codes $\Omega_i(c)$ from the set of two measurements, and to determine using (85) or (86) the corresponding error codes which are then stored in the registers **RB**$_\mathbf{1}$.

In the following, the way to obtain the relative transition codes is explained. Let us consider the input stimulus $x_{t\kappa}$ corresponding to the transition between subcodes $c_i = \kappa$ and $c_i = \kappa + 1$. According to (51), we have the following two outputs,

$$\begin{cases} \chi^U_{i,\kappa} = f_{\text{MDAC}_i}(x_{t\kappa}, \kappa) = G_{0i}(1+\varepsilon_i)(x_{t\kappa} - \tilde{p}_i(\kappa)) + \gamma_i G_{0i} x_{t\kappa} + \lambda_i(\kappa) + y_{os,i} \\ \chi^D_{i,\kappa} = f_{\text{MDAC}_i}(x_{t\kappa}, \kappa+1) = G_{0i}(1+\varepsilon_i)(x_{t\kappa} - \tilde{p}_i(\kappa+1)) + \gamma_i G_{0i} x_{t\kappa} + \lambda_i(\kappa+1) + y_{os,i} \end{cases} \tag{87}$$

Subtracting both outputs, a value very similar to $W_i(\kappa)$ in (79) results:

$$\begin{aligned} \chi^U_{i,\kappa} - \chi^D_{i,\kappa} &= G_{0i}(1+\varepsilon_i)(\tilde{p}_i(\kappa+1) - \tilde{p}_i(\kappa)) - (\lambda_i(\kappa+1) - \lambda_i(\kappa)) = \ldots \\ \ldots &= G_{0i}(1+\varepsilon_i)Q_i - \Delta_1\lambda_i(c) = \left[(1+\varepsilon_i)2^{m_i} - \frac{\Delta_1\lambda_i(c)}{q_{i+1}}\right] q_{i+1} \end{aligned} \tag{88}$$

Let us consider now $\varsigma^{\dagger U}_{i+1,\kappa} = f^{(i+1)}_{AD}(\chi^{U}_{i,\kappa})$ and $\varsigma^{\dagger D}_{i+1,\kappa} = f^{(i+1)}_{AD}(\chi^{D}_{i,\kappa})$ as the two digital output codes (*absolute transition codes*) corresponding to the values in (87) obtained by the *i*+1-*th* order subconverter (AD_2 in Fig. *6-13*).[33] Using the pseudocanonical form for this converter, the relationship between these codes and the values in (87) is:

$$\begin{cases} \alpha_{i+1}\chi^{U}_{i,\kappa} + \beta_{os,i+1} = l(\varsigma^{\dagger U}_{i+1,\kappa}) + \xi^{U}_{i+1} \\ \alpha_{i+1}\chi^{D}_{i,\kappa} + \beta_{os,i+1} = l(\varsigma^{\dagger D}_{i+1,\kappa}) + \xi^{D}_{i+1} \end{cases} \tag{89}$$

and subtracting both expressions results:

$$\frac{\alpha_{i+1}(\chi^{U}_{i,\kappa} - \chi^{D}_{i,\kappa})}{q_{i+1}} = (\varsigma^{\dagger U}_{i+1,\kappa} - \varsigma^{\dagger D}_{i+1,\kappa}) + \frac{(\xi^{U}_{i+1} - \xi^{D}_{i+1})}{q_{i+1}} \tag{90}$$

If the residual contribution in (90) is negligible after a truncation of the *p* least significant bits, $\left|\xi^{U}_{i+1} - \xi^{D}_{i+1}\right| \ll q_{i+1}2^{p}$, the estimation of the first term in (90) will give the wanted relative transition code,

$$\left\langle \frac{\alpha_{i+1}(\chi^{U}_{i,\kappa} - \chi^{D}_{i,\kappa})}{q_{i+1}} \right\rangle = \varsigma^{\dagger U}_{i+1,\kappa} - \varsigma^{\dagger D}_{i+1,\kappa} \tag{91}$$

To show that this obtained value is the relative transition code it is enough to use (88):

$$\left\langle \frac{\alpha_{i+1}(\chi^{U}_{i,\kappa} - \chi^{D}_{i,\kappa})}{q_{i+1}} \right\rangle = \left\langle (1+\varepsilon_i)\alpha_{i+1}2^{m_i} - \alpha_{i+1}\frac{\Delta_1\lambda_i(\kappa)}{q_{i+1}} \right\rangle \cong \left\langle \mu_i 2^{m_i} - \frac{\Delta_1\lambda_i(\kappa)}{q_{i+1}} \right\rangle = \Omega_i(\kappa) \tag{92}$$

6.3.3 LS self-calibration procedure[34]

Although the self-calibration procedure published in [32] is less general than the KLB, it is easier to apply because in some specific ADC designs it is not necessary to carry out the recurrent self-calibration of the subconverters. In addition, a reduced number of stimuli is required (*r*+1 in the general case: one per stage, and one more, common to all the stages). By

[33] The dagger reminds us that these codes can be the result of a calibration.

[34] We call LS to this procedure in honor of the authors that reported it in [32].

the contrary, the non-linear terms of the MDACs gain errors are not calibrated, and a reconfiguration of the MDAC structure is necessary for the error code estimation.

In this subsection, we will proceed as in the previous one using our model to explain the general operation of this self-calibration technique. First of all, the general form of the error codes is shown, and after that, the procedure for the estimation of these codes is explained.

We start considering that either the gain error of the $r+1$-*th* subconverter, γ_{r+1}, and the gain errors ε_i of the first r MDACs, are all included in the terms of gain and noise of the pseudocanonical form (65).[35] This is equivalent to write the InL as,[36]

$$InL_r(c_1,c_2,\ldots,c_r)=\sum_{i=1}^{r-1}\Psi_{i+1}^{(r)}c_i 2^{m_i}-\sum_{i=1}^{r}\frac{\lambda_i(c_i)}{q_{i+1}} \tag{93}$$

Although the results expressed from (77) to (80) apply also to this case, we define here the *relative transition codes* of the *i-th* stage as,

$$\Omega_i(c)=-\left\langle\frac{\Delta_1\lambda_i(c)}{q_{i+1}}\right\rangle,\quad c\in[c_{i0},c_{iM}) \tag{94}$$

The contribution of the γ_i gain errors is indirectly calculated assuming that there is a known correlation between $\lambda_i(\cdot)$ and γ_i [34]. Using the correlation given in (56),

$$2\gamma_j G_{0j}=\frac{\lambda_j(c_{j0})-\lambda_j(c_{jM})}{R_j} \tag{95}$$

and the relationships between the interstage gains G_{0j} and the quantum Q_j and q_j results:

$$\gamma_j 2^{N_j}=-\frac{\lambda_j(c_{jM})-\lambda_j(c_{j0})}{q_{j+1}} \tag{96}$$

[35] Notice that in implementations where (56) is fulfilled, the ε_i values can be reduced increasing the opamp gain.

[36] This self-calibration procedure gives a final gain $g^{\dagger}\approx 1+\Psi_1^{(r)}$, according to the hypotheses in Sect. 6.2.5.

Comparing the second terms of (96) with the $\lambda_i(\cdot)$ terms in (93), which are now estimated through the relative transition codes (94), gives:

$$\left\langle \gamma_j 2^{N_j} \right\rangle = -\sum_{c \ge c_{i0}}^{c_{iM}-1} \left\langle \Delta_1 \lambda_i(c)/q_{i+1} \right\rangle = \sum_{c \ge c_{i0}}^{c_{iM}-1} \Omega_i(c) = \pi_i \tag{97}$$

The error code of the i-th stage results:

$$ze_i(c) = \left\langle \sum_{k \ge c_{i0}}^{c-1} W_i(k) \right\rangle = \sum_{k \ge c_{i0}}^{c-1} \left(\left\langle \Psi_{i+1}^{(r)} 2^{m_i} \right\rangle + \Omega_i(k) \right) = \sum_{k \ge c_{i0}}^{c-1} \left(\Gamma_i^{(r)} + \Omega_i(k) \right), \quad c \in [c_{i0}, c_{iM}] \tag{98}$$

where $\Gamma_i^{(r)}$ is an integer value obtained from (97) (and hence, from the relative transition codes) using the definition (64) and considering as last quantizer the $r+1$-th subconverter:

$$\Gamma_i^{(r)} = \left\langle \Psi_{i+1}^{(r)} 2^{m_i} \right\rangle = \sum_{j=i+1}^{r} \gamma_j 2^{m_i} \approx \sum_{j=i+1}^{r} \left\langle \gamma_j 2^{N_j} \right\rangle 2^{(m_i - N_j)} = \sum_{j=i+1}^{r} \pi_j 2^{(m_i - N_j)} \tag{99}$$

Notice that the value of this integer number is zero for the last calibrated stage ($i=r$). Hence, this correction is not necessary when only the first stage is calibrated, $r=1$. For $r=2$, it has the value: $\Gamma_1^{(2)} = \pi_2 2^{(m_1 - N_2)} = \pi_2 / 2^{s_1}$.

Summarizing, the LS self-calibration procedure can be defined as follows:

– From the stage $i=r$ to $i=1$:

a) Estimate the transition codes of the i-th stage, $\{\Omega_i(c)\}, c \in [c_{i0}, c_{iM})$, from direct measurements at the **MDAC$_i$** as we will see in Sect. 6.3.4.

a) Calculate π_i using (97).

b) Calculate $\Gamma_i^{(r)}$ using (99) if $i<r$; $(\Gamma_r^{(r)} = 0)$.

c) Calculate and store the error codes using (98) as:

$$\begin{cases} ze_i(c_{i0}) = 0 \\ ze_i(c+1) = ze_i(c) + \Omega_i(c) + \Gamma_i^{(r)}, \quad c \in [c_{i0}, c_{iM}) \end{cases} \tag{100}$$

– During the normal operation of the multistage ADC, the final calibrated code is obtained using (75).

6.3.4 Transition codes for the LS procedure

As for the KLB procedure, the MDACs of the stages to be calibrated (the r most significant stages) have to be designed to admit input stimuli different to the normal input. For the LS procedure it is also necessary to allow the reconfiguration of the MDACs in such a way that they exhibit a different input-output characteristic. This characteristic must permit to obtain values closed to the given in (94). A particular case already implemented in [32] is:

$$y = f_{\text{MDAC}}^{(\text{cal})}(x_i, \Delta c_i) = (1+\varepsilon)(x_i - G_0 Q \Delta c_i) + \Delta\lambda(c_i) + y_{os} \tag{101}$$

where, Δc represents the difference in the subcode from the beginning to the end of the interval in which the MDAC output is processed.

As in the KLB procedure, the estimate of each relative transition code requires two independent measurements of the **MDAC$_i$** output. For each $\Omega_i(\kappa)$, two analog stimuli are applied: $x_i = G_{0i} Q_i = R_{i+1} / 2^{s_i - 1}$ and $x_i = 0$. After the first one is applied, the value of the subcode c is incremented in one unit: $\Delta c_i = (\kappa + 1) - \kappa = 1$. For the zero stimulus this increment must be zero. The output values are quantized using the following subconverter (the $i+1$-th order) which must be linear or has been linearized by a self-calibration procedure.

For the two above stimuli, the two corresponding output values of the i-th stage are according to (101):

$$\begin{cases} \chi_{i,\kappa} = f_{\text{MDAC}}^{(\text{cal})}(G_{0i} Q_i, \Delta_1 \kappa) = \Delta_1 \lambda_i(\kappa) + y_{os,i} \\ \chi_{i,0} = f_{\text{MDAC}}^{(\text{cal})}(0_i, 0) = y_{os,i} \end{cases} \tag{102}$$

To prove now that the value of the relative transition code is the given by (94), we can follow the same process defined from (88) to (92).

ACKNOWLEDGEMENT

This work has been partially supported by the *Toolip Medea+* Pr.A-511.

REFERENCES

[1]. B. Razavi: "*Principles of Data Conversion System Design*". IEEE Press, 1995.

[2]. M. Gustavsson, J.J. Wikner, N.N. Tan: "*CMOS Data Converters for Communications*", Kluwer Academic Publishers, 2000.

[3]. P.G.A. Jespers: "*Integrated Converters. D to A and A to D Architectures, Analysis and Simulation*", Oxford University Press, 2001.

[4]. S.H. Lewis: "Optimizing the Stage Resolution in Pipelined, Multistage, Analog-to-Digital Converters for Video-Rate Applications", *IEEE Trans. on Circuits and Systems-II*, vol. 39, no. 8, pp. 516-523, August 1992.

[5]. D.W. Cline, P.R. Gray: "A Power Optimized 13-b 5Msamples/s Pipelined Analog-to-Digital Converter in 1.2μm CMOS", *IEEE Journal of Solid-State Circuits*, vol. 31, no. 3, pp. 294-303, March 1996.

[6]. J. Goes, J.C. Vital and J.E. Franca: "Systematic Design for Optimization of High-Speed Self-Calibrated Pipelined A/D Converters", *IEEE Transactions on Circuits and Systems-II*, vol. 45, no. 12, pp. 1513-1526, December 1998.

[7]. E. Peralías: "*Autocalibración Digital y Autotest en Convertidores A/D Multietapa*", Doctoral thesis (in Spanish). Universidad de Sevilla. July 1999.

[8]. J. Doernberg, H-S. Lee, D.A. Hodges: "Full-Speed Testing of A/D Converters", *IEEE J. Solid-State Circuits*, vol. SC-19, pp. 820-827, December 1984.

[9]. M. Mahoney: "*DSP-Based Testing of Analog and Mixed-Signal Circuits*", Computer Society Press of the IEEE, 1987.

[10]. S. Max: "Fast Accurate and Complete ADC Testing", *Proc. International Test Conference*, Paper 5.1, pp. 111-117, Washington, August 1989.

[11]. *IEEE Std 1241-2000*, "IEEE Standard for Terminology and Test Methods for Analog-to-Digital Converters", New York, USA, Dec. 2000.

[12]. U. Moon, B. Song: "Background Digital Calibration Techniques for Pipelined ADC's", *IEEE Transactions on Circuits and Systems-II*, vol. 44, No. 2, pp. 102-109, Feb. 1997.

[13]. S. Kwak, B. Song, K. Bacrania: "A 15-b, 5-Msample/s Low-Spurious CMOS ADC", *IEEE J. Solid-State Circuits*, vol. 32, no. 12, pp. 1866-1875, December 1997.

[14]. J.M. Ingino, B.A. Wooley: "A Continuously Calibrated 12-bit 10-MS/, 3.3-V A/D Converter", *IEEE J. Solid-State Circuits*, vol. 33, no. 12, pp. 1920-1931, December 1998.

[15]. A.J. Acosta, E. Peralías, A. Rueda, J.L. Huertas; "VHDL behavioral modelling of pipeline analog to digital converters", *Measurement*, vol. 31, pp. 47-60, Elsevier 2002.

[16]. A.J. Gines, E. Peralías, A. Rueda, N. Martínez, R. Seepold: "A Mixed-Signal Design Reuse Methodology Based on Parametric Behavioural Modeles with Non-Ideal Effects", *Proc. DATE*, pp. 310-314, Paris, March 2002.

[17]. E. Peralías, A. Rueda, J.L. Huertas: "New BIST for Structural Testing of Pipelined Analog to Digital Converters", *Journal of Electronic Testing: Theory and Applications (JETTA)*, vol. 17, nº 5, pp. 373-383, Kluwer Academic Publishers, Oct. 2001.

[18]. G.G. Gorbatenko: "High-Performance Parallel-Serial Analog to Digital Converter with Error Correction", *IEEE National Convention Record*, New York, March 1966.

[19]. S.H. Lewis, P. Gray: "A Pipelined 5-Ms/s 9-bit A/D converter", *IEEE J. Solid-State Circuits*, vol. SC-22, pp. 954-961, December 1987.

[20]. S.H. Lewis, H.S. Fetterman, G.F. Gross, R. Ramachandran, T.R. Viswanathan: "A 10-b 20Ms/s A/D Converter", *IEEE J. Solid-State Circuits*, vol. 27, no. 3, pp. 351-358, March 1992.

[21]. B. Ginetti, P.G.A. Jespers, A. Vandemeulebroecke: "A CMOS 13-b Cyclic RSD A/D Converter", *IEEE J. Solid-State Circuits*, vol. 27, no. 7, pp. 957-964, July 1992.

[22]. K. Hadidi, G.C. Temes: "Error Analysis in Pipeline A/D Converters and its Applications", *IEEE Trans. on Circuits and Systems-II*, vol. 39, no. 8, pp. 506-515, August 1992.

[23]. C. Mangelsdorf, S.H. Lee, M. Martin, H. Malik, T. Fukuda, H. Matsumoto: "Design for Testability in Digitally-Corrected ADCs", *Proc. Inter. Symp. on Solid-State Circuits*, Paper WP 4.7, pp. 70-71, 1993.

[24]. S.H. Lewis, R. Ramachandran, W.M. Snelgrove: "Indirect Testing of Digital-Correction Circuits in A/D Converters with Redundancy", *IEEE Transactions on Circuits and Systems-II*, vol. 42, no. 7, pp. 437-445, July 1995.

[25]. I.E. Opris, L.D. Lewicki, B.C. Wong: "A Single-Ended 12-bit 20 Msample/s Self-Calibrating Pipeline A/D Converter", *IEEE J. Solid-State Circuits*, vol. 33, no. 12, pp. 1898-1903, December 1998.

[26]. Y.M. Lin, B. Kim, P.R. Gray: "A 13-b 2.5Mhz Self-Calibrated Pipelined A/D Converter in 3-∝m CMOS", *IEEE J. Solid-State Circuits*, vol. 26, no. 4, pp. 628-636, April 1991.

[27]. J. Goes, J. Franca, N. Paulino, J. Grilo, G. Temes, "High-Linearity Calibration of Low-Resolution D/A Converters", *Proc. ISCAS*, vol. 5, pp. 345-348, London 1994.

[28]. B.S. Song, M.F: Tompsett, K.R. Lakshmikumar: "A 12-b 1-Msamples/s Capacitor Error-Averaging Pipelined A/D Converter", *IEEE J. Solid-State Circuits*, vol. 23, no. 6, pp. 1324-1333, Dec. 1988.

[29]. P.C.Yu, H.S.Lee: "A 2.5V, 12-b, 5-Msamples/s Pipelined CMOS ADC", *IEEE J. Solid- State Circuits*, vol. 31, no. 12, pp. 1854-1861, Dec. 1996.

[30]. H.S. Chen, B.S. Song, K. Bacrania: "A 14-b 20Msamples/s CMOS Pipelined ADC", *IEEE Journal of Solid-State Circuits*, vol. 36, no. 6, pp. 997-1001, June 2001.

[31]. H.S. Lee, D.A. Hodges, P.R. Gray: "A Self-Calibrating 15 Bit CMOS A/D Converter", *IEEE J. Solid-State Circuits*, vol. SC-19, no. 6, pp. 813-819, Dec. 1984.

[32]. S. Lee, B. Song: "Digital-Domain Calibration of Multistep Analog-to-Digital Converters", *IEEE J. Solid-State Circuits*, vol. 27, no. 12, pp. 1679-1688, December 1992.

[33]. A.N. Karanicolas, H.S. Lee, K.L. Bacrania: "A 15-b 1Msample/s Digitally Self-Calibrated Pipeline ADC", *IEEE J. Solid-State Circuits*, vol. 28, no. 12, pp. 1207-1215, December 1993.

[34]. S. Lee, B. Song: "Interstage Gain Proration Technique for Digital-Domain Multi-Step ADC Calibration", *IEEE Trans. on Circuits and Systems-II*, vol. 41, no. 1, pp. 12- 18, Jan. 1994.

[35]. H.S. Lee: "A 12-b 600 ks/s Digitally Self-Calibrated Pipelined Algorithmic ADC", *IEEE J. Solid-State Circuits*, vol. 29, no. 4, pp. 509-515, April 1994.

[36]. E.G. Soenen, R.L. Geiger: "An Architecture and An Algorithm for Fully Digital Correction of Monolithic Pipelined ADC's", *IEEE Transactions on Circuits and Systems-II*, vol.42, no. 3, pp. 143-153, March 1995.

[37]. M.K. Mayes, S.W. Chin: "A 200mW, 1Msample/s, 16-b Pipelined A/D Converter with 32-b Microcontroller", *IEEE J. of Solid-State Circuits*, vol.31, no.12, pp. 1862-1872, Dec. 1996.

[38]. G. Ahn, H. Choi, S. Lim, S. Lee, C. Lee: "A 12-b, 10MHz, 250mW CMOS A/D Converter", *IEEE Journal of Solid-State Circuits*, vol. 31, no. 12, pp. 2030-2035, December 1996.

[39]. E. Peralías, A. Rueda, J.A. Prieto, J.L. Huertas: "DFT & On-line Test of High-Performance Data Converters: A Practical Case", *Proc. International Test Conference*, Paper 21.3, pp. 534-540, Washington (USA), October 1998.

Chapter 7

DFT AND BIST TECHNIQUES FOR EMBEDDED ANALOG INTEGRATED FILTERS

Diego Vázquez-García de la Vega
IMSE, CNM, CSIC / University of Sevilla

Abstract: This chapter focuses on both Design-for-Test and BIST approaches for analog circuits testing, but with special emphasis in the case of embedded Analog Integrated filters.

Key words: Design-for-Test (DfT), Built-In-Self-Test (BIST), Analog Testing, Active Filters Testing.

1. INTRODUCTION

From the conception to the placement of an Integrated Circuit into its field operation environment is a laborious process that, unfortunately, is not perfect. Consequently, there is a possibility of creating non perfect circuits, and thus, subject to imperfections which can make the IC to failure. Such lack of perfection makes necessary to test the circuits to ensure the correctness of the final product to be shipped. That is, the aim of testing can be understood, in a wide sense, as the strategy applied to separate good and bad ICs.

The evolution of nano-technologies, the improvements made in the field of CAD tools and the growing demand for cheaper and more efficient electronic products has lead to produce larger and more complex mixed-signal ICs on a single chip. In this way, complete systems (the so-called SoCs, or System-on-Chip) are actually being integrated on a single silicon-piece. But the more complex an IC is, the more necessary, difficult and

J.L. Huertas (ed.), Test and Design-for-Testability in Mixed-Signal Integrated Circuits, 215–258.

costly is to verify the final product outcome. In this way, a bottleneck for the development of actual and future complex circuits is test. The major test challenges are mainly due to the analog portions of the mixed-signal ICs. Even more, the cost to produce mixed-signal devices is being dominated by the test of the analog part. Although testing a circuit, no matter if analog or digital, in a stand alone manner is a difficult task, this is even worst in the case of mixed-signal ICs, where there may be some blocks that are not externally accessible through I/O pins. In this sense, the IEEE 1149.4 boundary scan standard [1] has mean and important advance to provide external access to internal points with a reduced number of extra pins.

Although design and test, specially in the case of analog circuits, have been traditionally separated tasks, now days there is a need in linking the two process more and more coupled through the so-called Design-for-Test (DfT). A Design-for-Test (DfT) technique aims to reduce the test complexity and associated cost by taking into account the test needs of a system along the design process. It is achieved by altering the circuits to improve their testability. It includes extra facilities to provide external access to the embedded blocks and/or the use of specific test strategies to reduce the number of functional specifications to be verified, But a DfT technique in this sense does not eliminate the need of generating the test signals and observe them. Accordingly to the increased performances of the actual and future ICs it is demanded a more complex, accurate and, consequently, costly test equipments (known as ATE from Automatic Test Equipment). The rising cost of ATEs can be controlled by the use of Built-In-Self-test (BIST) methods which consists on moving some of the required test resources (signal generation and/or evaluation) on to the chip. In fact, BIST techniques are in essence DfT approaches in the sense of that they must be contemplated during the design phase of the IC to include the required extra hardware for signal generation and/or evaluation.

Analog filters are one of the most used components in complex mixed-signal ICs. Traditional test techniques are based on functional specifications [2, 3], and it includes the verification of parameters like frequency response, dynamic range, offset, distortion, voltage range, output noise, power consumption, etc. But measuring such parameters is a time consuming task and requires costly ATEs when high precision is required. The situation is even worst when the filter is embedded into a more complex system where the external access may be restricted by the limited number of I/O pins. In addition, there is a need to cut the cost of testing to maintain the global cost of the IC into reasonable values [4, 5].

This chapter will be focused on both DfT and BIST approaches for the particular case of embedded Analog Integrated Filters. First, the main test concepts and difficulties of analog and mixed-signal testing are briefly reviewed in Sections 2, 3 and 4. After that, relevant DfT schemes proposed during the last few years for off-line (Section 5), on-line (Section 6) and BIST (Section 7) testing of analog circuits in general but with emphasis in filter applications are revised. Finally, a unified DfT architecture incorporating several DfT approaches is discussed in section 8.

2. GENERAL TEST ISSUES

In essence, the test of an IC consists on the application of a given stimulus to the Circuit Under Test (CUT) and the processing of the response to extract some parameter that is used to check the correctness of the circuit behavior accordingly to a given set of specifications. Independently on the way the test stimuli are selected, there are some factors driving the selection: The stage development phase, the test procedures and the type of circuit.

2.1 Test phases in an IC

Testing is required along different phases in the life of an IC. During the circuit conception phase, design validation and prototype verification are essential before qualifying a chip for mass-production. Normally, the test procedures during these early phases include diagnosis in case of failure and are mainly ad-hoc actions driven by design knowledge, not so much influenced by the time and resources to be used, and performed by the designers.

The next phases are basically the circuit fabrication, shipping, on-board installation and on-field operation. In all these phases the role of testing is different:

- Post fabrication test. The invested time should be minimal and the required equipment as standard as possible. The cost in terms of time and external test resources (ATEs) is critical and demands systematic yet efficient procedures.
- Field diagnosis or maintenance test. Fast yet efficient tests are needed, ideally based on the test strategies developed in the post-fabrication test step. Test is usually performed at a subsystem level, and time constraints are not as critical as in the previous case.
- Start-up diagnosis or basic functional testing. Applied before starting operation (typically in aerospace and automotive applications, and expanding to other areas); this kind of test prevents erroneous operation

by checking the status of the on-board electronics. Again should be valuable to use procedures relying on those developed in the post-fabrication test, but signals have to be automatically provided by the system under test or by additional supporting circuitry.

- Concurrent test. Needs are also increasing and coming up from biomedical applications, non-stop robot applications, etc. In this case, there is an interest in using the actual signals driving the circuit to perform a basic operational test.

2.2 Structural and Functional Testing

Although there exists a wide terminology that pretends to distinguish and classify the test strategies in base to different attributes, they can be easily classified in two categories: structural and functional [6, 7]. Both classes are related to different abstraction levels of an IC as shown in Figure 7-1. Structural testing or also called defect-oriented testing is based on a set of available fault models, which are an abstraction of the electrical impact of the defects on the corresponding level of design description (layout, logic, etc.). On the basis of that the structure connection is corrupted by the presence of a fault, the test stimuli can be conveniently selected. The availability of the faults models gives the opportunity to define efficient algorithms to determine the optimum test set and to derive a metric to measure its efficiency. However, structural testing has some important limitations or disadvantages. Firstly, its effectiveness depends on the accuracy of the fault models. Secondly, it requires extensive simulations and computational resources that in the case of large and complex circuits may be a serious limitation. Finally, it does not guarantee the correct functional behavior of the IC. Structural testing is usually applied during post-production test, as they usually have the ability to detect a large number of manufacturing defects.

On the other hand, functional testing or also called specification-driven test is not based on fault models but in the functional specifications of the IC. That is, the test set are selected, normally by the designers and based on their experience, accordingly to the targeted specifications. This approach has also some limitations. For example, a practical functional test can not take into account all possible operating conditions of the IC (otherwise the test set would be infinite). It makes possible that, although the IC passes the test, it can be affected by faults that lead to IC failures under some circumstances. Moreover, ICs have normally a large number of specifications to be accomplished with different accuracy requirements, which makes functional testing a large and costly consuming task. For these reasons, Functional testing is normally applied at the end of the testing phases as there is a need to guarantee that the chip to be shipped meets the targeted specifications.

It should be clear at this point that functional and structural testing are complementary between them. For this reason, a test set is normally composed in practice by structural and functional sub-sets.

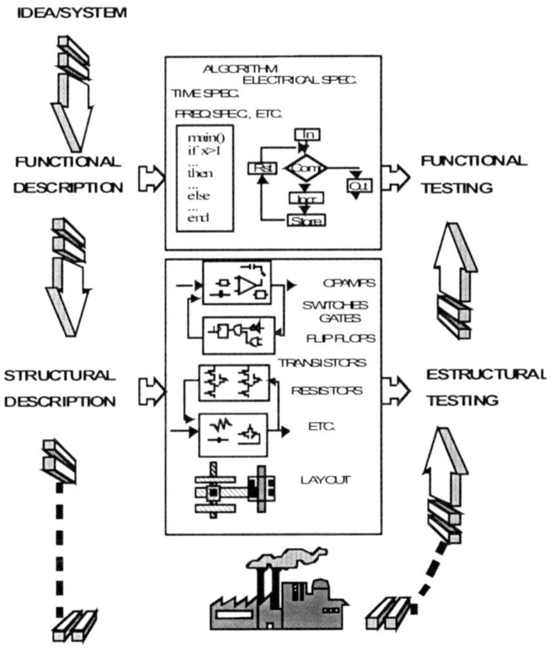

Figure 7-1. IC abstraction levels and test categories.

2.3 Test procedures: Off-Line, On-Line and BIST

Another key point refers to the generation and application of the test stimuli and how the responses to them are observed and processed. The answer from this point of view defines the test procedure. In this sense, the test process can be carried out using, basically, three procedures: Off-line, On-Line and BIST (Built-In-Self-Test).

Off-line test involves the application of specific test stimuli and the processing of the response to check the correctness of the circuit behavior. The test stimuli obey to a given test strategy based on checking functional specifications and/or a strategy to detect faults. Both the generation and the evaluation of the test response are performed out of chip using external test equipment. This is the usual procedure during the early test phases (prototypes and post-production). It requires to have access to those points where the test stimuli should be applied and the response to them observed. In this sense, DfT strategies for off-line testing are normally focused on providing accessibility from the I/O pins to internal points. It requires some extra circuitry to be added and/or the reconfiguration of the circuit during the test mode to provide such access.

On the contrary, in On-Line procedures the input stimuli do not must be known in advance; the object consists not in the response itself but of some property of the response. Its interest relies on three features:

a) System operation does not need to be interrupted during the test application process.
b) Test can be performed using realistic signals rather than specific ones just generated for testing.
c) Test is applied at full operation speed.

Such features make the on-line strategies specially suited for on-field testing. Moreover, it can be exploited during any of the test phases of the IC. The use of encoded signals and hardware redundancy are the main elements used as the basis to perform a basic operation checking of the circuit. So, on-line testing is in essence a DfT technique (it requires extra hardware added during the design phase) and the test performed is structural as no circuit's specifications are measured.

Finally, BIST procedures involve the on-chip generation of the test stimuli and the evaluation of the response. This is an increased interest in the development of BIST techniques as the ICs increase in complexity due to the limitations for testing imposed by the number of pins. BIST techniques aim to avoid the use of expensive ATE and the need of external accessing for testing purposes.

Any of the above procedures can be used in different test phases. However, it should be clear that the BIST and On-line procedures are most suitable for the start-up diagnosis and on-field operation testing.

3. ANALOG AND DIGITAL TEST CONTRASTS

The state-of-the-art of digital systems is quite different if compared to the analog and mixed-signal counterparts. While digital design is practically (if not fully in some cases) automated even for very large systems, analog is far from it. The same situation holds for the test as there are a variety of well-known and accepted rules and design methodologies to facilitate the test of digital circuits. The main reasons for this situation are mainly due to the nature of analog circuits and signals. To understand the contrast between states-of-the-art, several considerations can be made [8]:

Size and accuracy: Analog circuits normally contain relatively few devices compared to digital ones. In addition, the relative number of inputs and outputs are also small. So, unlike in the case of digital circuits, the difficulties of testing analog circuits do not reside in their size, but in the precision and accuracy they must achieve.

Modeling: Modeling is far more difficult than in digital circuits. The function of analog circuits can not be described by closed-form expressions such as Boolean equations. Instead of that, the behavior of an analog circuit depends on the exact behavior of a transistor which model requires a set of complex equations containing a lot number of parameters. It makes difficult the mapping of defects to suitable models. But, at the same time, fault modeling accuracy is crucial for fault simulation. So, it represents a big contrast with respect to the widely accepted digital stuck-at and path-delay fault models.

Signal tolerances: Digital signals have discrete values. Analog signals however have an infinite range, and good signal values are referred to certain tolerance margins that depend on process variations and measurements inaccuracies.

Component tolerances: Absolute tolerances in analog components can be very large (around 20%), but relative matching is normally very good (0.1% in some cases). Although multiple faults can occur, faults in multiple components can cancel each other's effect, so not every multiple fault is a real fault. In addition, the list of possible multiple faults is normally too large and prohibitive in terms of simulation time.

Sensitivity: In contrast to the robustness of digital circuits, analog are very sensitive to loading effects. The simple transportation of an analog signal to the output pin may have an important impact on the circuit topology and behavior.

Once all the above considerations are kept in mind, it is easy to understand why analog testing has been traditionally a functional-based test, in contrast with digital circuits where structural testing is a mature area. The robustness, discrete signal levels, high level description facilities and efficient fault models in the case of digital circuits has allowed the development of feasible DfT and structural test techniques together with the automation of designs with test facilities included.

4. DESIGN-FOR-TEST: BASIC CONSIDERATIONS, RULES AND GUIDELINES

Design-for-Test (DfT) was conceived in digital [9] and adapted to analog and mixed-signal circuits [10, 11]. There are different ways to define DfT in an analogue context [10, 11, 12]. In a broad sense, it can be defined as *"any technique, procedure or strategy leading to an improvement in off- and/or on-line tests (including BIST)"* [11, 12]. A particular view (as is commonly done in digital) is referred to strategies for increasing the accessibility of some nodes within an IC [10, 11, 12, 13, 14, 15].

To make a complex design more manageable, it is usual to fragment the overall system into a number of functional blocks (called macros [10] in the following), each of which performs a well-defined operation. In the same way, a common-sense approach for testing these complex ICs relies, accordingly to the design flow, on testing each macro or group of macros (partition in general) separately. Obviously, it requires external accessibility to both inputs (to inject the stimuli) and outputs (to observe the responses) of partitions that must be taken into account during the design phase.

With this view, providing access to some nodes of independent macros deeply embedded within an IC (i.e., not connected to the chip pin-out) is one of the main concern of Design-for-Test (DfT) for off-line testing. Even more, although on-line testing and BIST can be considered as DfT techniques, they need to be accompanied with DfT approaches which provide the indispensable accessing paths.

There are two major problems associated with making a node accessible:

a) The node must be isolated from its normal driving sources before injecting a test stimulus.
b) Any extra element used to make a node accessible must preserve the same performance levels of the original circuit. This is a difficult task due to the sensitivity of analog circuits in general to loading effects and structure modifications.

In general, to avoid performance degradation these ideas involve extra functionality of some devices which must be added during the design phase to provide the required accessibility to the input and output nodes of the subsystems (macros), since most of them are internal and not connected to external pins.

DfT in analog circuits was traditionally applied at an ad-hoc basis, what makes it difficult to use in a general design environment. The need of avoiding performance degradation normally requires a careful redesign of some parts if it was not taken into account along the design flow. For this reason, DfT strategies allowing design standardization are encouraged to reduce the cost of testing.

Several approaches have been reported in the technical literature about this topic. All of them fall into one of two basic methodologies: the multiplexing [10, 11, 15, 16, 17] and the by-passing [14, 18, 19, 20].

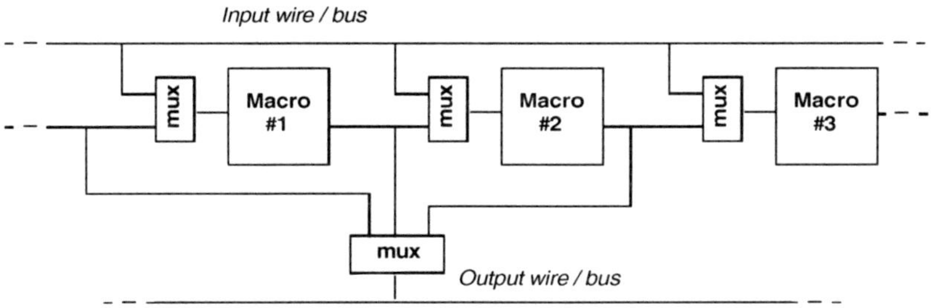

Figure 7-2. The Basic multiplexing methodology.

The multiplexing method (Figure 7-2) includes all these approaches where internal nodes are made accessible by connecting them either to a wire or a bus. Controllability (that is, the capability to inject stimuli) requires to disconnect the macro input from the main signal path and to re-connect that input to a test signal. Observability only requires an extra connection from the macro output to a wire. Unfortunately, the introduction of extra components within the main signal paths normally leads to performance degradation. Then, in order to minimize the influence of these extra components, several guidelines must be applied:

a) Minimize the number of accessible nodes (only observe nodes carrying "interesting" signals for testing).
b) Control (Observe) voltages instead of currents (except for current-mode circuits).
c) Control (Observe) DC, slow-varying, or low-frequency signals.

d) Select nodes with appropriate signal levels (not too small, not too large) and that are preceded by a buffering cell.
e) Try to displace the switching components away from the main signal path still preserving macro-level testability.
f) Minimize the number of buffers to/from the external environment. It has been an intense search for solutions based on a careful design of such extra components, but it requires a large design effort that cannot be easily reusable and depends on the particular application.

The by-passing methodology (Figure 7-3) rely in reconfiguring the macros which are not actually tested in such a way that a frequency-independent path (like a buffer) is established between the external and the macro input (equivalently, between the macro and the external output). In fact, the by-passing methodology can be considered an extension of the popular scan approach widely used in digital circuits. Again we need to address to possible performance degradation due to the extra components and/or the switching behavior of the different components.

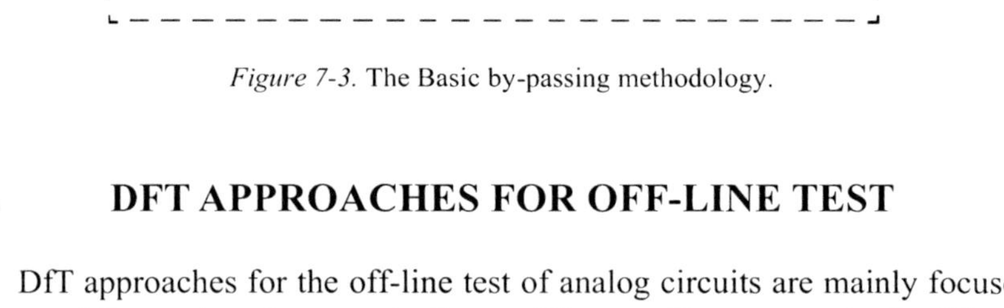

Figure 7-3. The Basic by-passing methodology.

5. DFT APPROACHES FOR OFF-LINE TEST

DfT approaches for the off-line test of analog circuits are mainly focused on providing external access to the partitions and/or macros in the Circuit Under Test (CUT). The un-intrusive and effective scan-path approach is widely used in digital circuits. In the analog counterpart, the multiplexing and by-passing conceptual architectures are feasible using conventional CMOS transmission gates (with or without buffers), but it normally requires significant design efforts to maintain the performance degradation into tolerable levels. A basic idea to preserve the original performance specifications of a circuit consists on introducing the switching components into active components, trying to ensure these switches only deal with small-signal level [22] or even to prevent them from being located within the internal signal path [19]. Practical implementations include analog shift registers, analog test buses and the so-called sw-opamp, together with other approaches based on circuit reconfiguration. This section will be dedicated to them.

5.1 Analog Shift Registers (ASRs)

Scan techniques are the most representative and widely accepted DfT in the case of digital circuits to increase the controllability and observability of internal points. For that, it is normal the efforts made to translate the same concept to the analog context.

One of the key points for off-line testing is the number of observable nodes when the number of I/O pins is restricted. It is well known that the number of test stimuli can be greater reduced by increasing the number of observable nodes in an IC. Reported approaches for this purpose are based on the so-called Analog Shift Registers (ASRs) [13, 23] which adapt the concept of digital shift registers to the analog context (Figure 7-4a). Two implementations have been reported: bucket-brigade-like devices and charge-coupled devices (CCDs). However these schemes are suited only for observability as the state of the circuit can not be controlled.

In bucket-brigade schemes (Figure 7-4b), the voltage at internal test points is captured by sample and hold circuits (it can be done in parallel). A proper switching scheme converts the sample and hold circuits to a chain of voltage followers that is used to pass, serially to the output, the voltage information. A buffer between the test point and the sampling capacitor is normally required to reduce the loading effects on the normal operation of the circuit.

Another well-known implementation of ASRs makes use of charge coupled devices (Figure 7-4c). Compact transistors and capacitors are used to shift the analog information in the form of charges. This solution is also valid for both voltage and current mode circuits. The requirements in terms of area and power are generally much smaller than in the bucket-brigade case, while provides small capacitive loads, high linearity and can operate at higher frequencies [23].

The major disadvantage for large circuits of these approaches is the need of a large number of clocking circuits and that large interconnection lines will induce significant measurement errors.

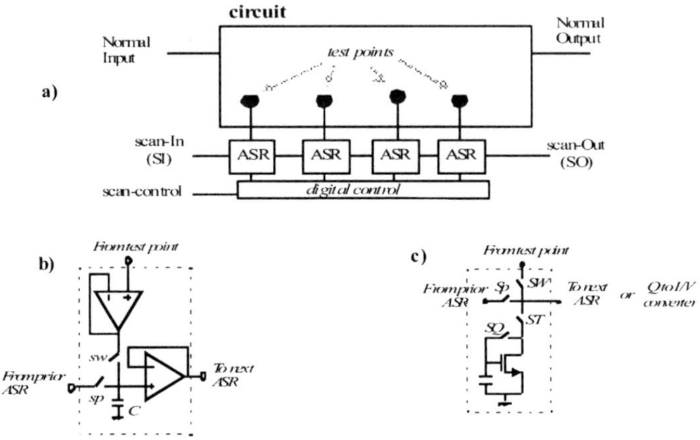

Figure 7-4. Analog Shift Registers. a) Use for observability. b) bucket-brigade devices implementation. c) charge-coupled devices implementation.

5.2 Analog Test Buses

Probably, the most used method to gain accessibility in analog circuits is through the use of analog test buses. Early analog buses were based on transmission gates connected to a common wire directly connected to an output pin [24, 25] or connected to and on-chip buffer amplifier which drives an output pin [26]. These approaches have however important drawbacks, as the limited bandwidth of the signals that can be monitored, the significant extra capacitance of the wires and the on-chip coupling with near sensitive analog signals, area requirements for the implementation of the on-chip buffers, etc. In [27], a high speed Analog Test Bus, illustrated in Figure 7-5, that overcomes these problems is reported. Its key features are: constant and small parasitic capacitances on nodes to be observed, large bandwidth (100MHz was achieved) and small silicon area. An internal node voltage is observed off-chip by first performing a voltage-to-current conversion using 3-state digital inverters and moving it off-chip. A high speed off-chip current-to-voltage converter converts the signal back to its original voltage. A calibration phase is needed to obtain high linearity. This scheme is a clear example where the original paths are preserved by embedding the virtual on-off switch activity into the design of active components (in this case the 3-state inverters).

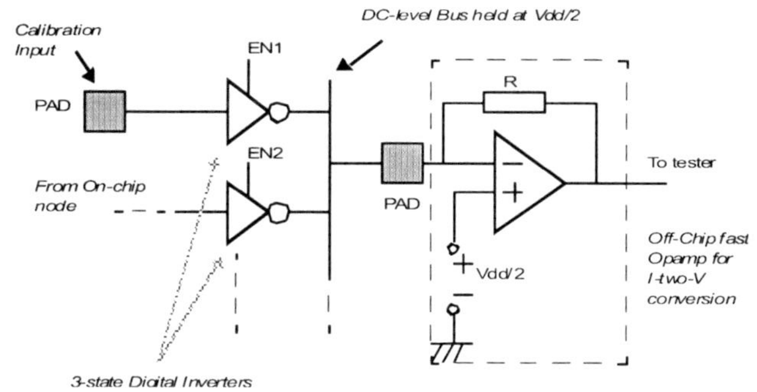

Figure 7-5. High Speed Analog Test Bus

5.3 Stage isolation by reconfiguration

The basic idea in the by-passing strategy shown in Figure 7-3 is to reconfigure the prior and succeeding blocks of that to be tested in such away that the test stimulus and the response can be transmitted to the primary input and output respectively without altering them. Undisturbed transmission can be done, in the case of analog signals, by widening the bandwidth of the blocks for which the signal will be passed through. This basic idea is exploited in the following schemes.

5.3.1 Switched MOS transistors

Soma suggested in [14] the use of analog switches in active analog filters to perform a dynamically broadening bandwidth in each filter stage. It is accomplished by associating conveniently MOS transistors, with the corresponding line control, to all passive components like resistors, capacitors and series and parallel combinations of them. Later, the same idea was extended to the case of switched capacitor circuits in [18] but using additional control of the switches.

The reported approaches in [14, 18] facilitate the detection of component deviations (parametric deviations) as well as catastrophic faults (opens and shorts). However, the main drawback from a practical point of view is the considerable design effort required to accommodate the extra circuitry (extra switches and a high number of additional control signals) in order to avoid a significant performance degradation.

5.3.2 Pole-zero cancellation

Another approach that exploits the bandwidth widening for testing purpose is reported in [28]. It is only valid for filters composed by a cascade of second order structures (biquads). In this case, the biquads are designed in such a way that, depending on a digital control signal, every section can behave either as a biquadratic (as specified) or as a constant transfer function. This is achieved by adding some passive components and switches and does not increase the design parameters (area, power, etc.) so much. Although the proposed approach will be described here in the s-domain, it is also valid in the *z*-domain.

In essence, programmable Universal Biquadratic Sections (UBSs) are used to realize the filter, each UBS being able to implement any transfer function $H_{UBS}(s)$ of the form,

$$H(s) = \frac{\alpha_2 s^2 + \alpha_1 s + \alpha_0}{\beta_2 s^2 + \beta_1 s + \beta_0} \tag{1}$$

Depending on the logical value of a control signal Φ_T, each biquad in the filter has two operation modes: Normal $H^N_{UBS}(s)$ and Transparent $H^T_{UBS}(s)$.

a) *Normal Mode*: In this mode, the biquad is under normal operation and coefficients α_j and β_j have the nominal values derived from the filter specifications:

$$\begin{aligned} \alpha_j &= \alpha_j^N \\ \beta_j &= \beta_j^N \end{aligned} \tag{2}$$

b) *Transparent Mode*: In this mode, the biquad is out of normal operation and coefficients α_j and β_j are changed by electrical programming to α_j^T *and* β_j^T respectively, so that the new values fulfill,

$$\frac{\alpha_j^T}{\beta_j^T} = k = cte \tag{3}$$

Hence, the transfer function realized by the biquad in this mode is,

$$H_{UBS}^{T}(s) = k = cte \tag{4}$$

So, the transparent mode can be used to pass signals without altering them.

Although the programming of the coefficients can be made in many ways, it is necessary to take into account some considerations:

1. The roots of the polynomial $\beta_2^T s^2 + \beta_1^T s + \beta_0$ must be inside the left-hand side of the s-plane. Otherwise, the section would be unstable in the transparent mode.
2. Internal nodes in the filter must remain scaled (maximum gain of k) for proper dynamic range.
3. Components and connections to implement the normal operation mode should not be altered to obtain the transparent mode. It is convenient to add extra circuitry, that can be disconnected and turned off during the normal mode (avoiding performance degradation), to implement the transparent mode. In addition, the definition of such extra circuitry must be done with care in order to reduce the design efforts.

Taken into account the above considerations, the best way to proceed is to make,

$$\begin{aligned} \beta_j^T &= \beta_j^N \\ \alpha_j^T &= k\beta_j^N \end{aligned} \tag{5}$$

The convenience to proceed in this way can be justified on the following considerations:

1. The poles location of the filter does not need to be modified. It means that the core structure which implements the section is maintained, and hence, design efforts are reduced. In addition, as the poles locations are not altered and they are the main responsible of the frequency behavior of the section during normal mode, performance degradation is minimized.
2. The proposed solution reduces the design efforts to accommodate the extra circuitry needed for the implementation of the transparent mode. It is because only the zeros of the section need to be relocated.

It leads to conclude that the extra hardware required to implement the transparent mode with respect to the normal mode is reduced to provide extra paths from the input to the internal sum nodes of the section and/or simply add extra components to the existing paths. In this sense, the relative area overhead for the implementation of the approach depends on the filter itself; in particular, on the number and dimensions of the extra components required. Obviously, components must obey the bond equations resulting from Eq(3) to obtain a transfer function of the form indicated in Eq(4). In general, it would be necessary to modify as many components as number of independent equations result from applying Eq(3) and scaling internal nodes. However, this area overhead may be small in many practical situations.

The main drawback of this concept is the difficulty of an exact pole-zero cancellation; of course, this is impossible in practice, but depending on sensitivity considerations, an approximate cancellation may lead to effectiveness more than reasonable. In addition, although this technique may seem relatively complex, practical implementations may be very simple. To illustrate it, let us consider the switched-capacitor (SC) structure shown in Figure 7-6. It is a UBS structure where only capacitors connected (directly or via switches) to the input (V_{in}) need to be substituted by a 1-bit programmable capacitor to implement the strategy (the programmable components of the filter must have two values: one for Normal and another for the Transparent mode).

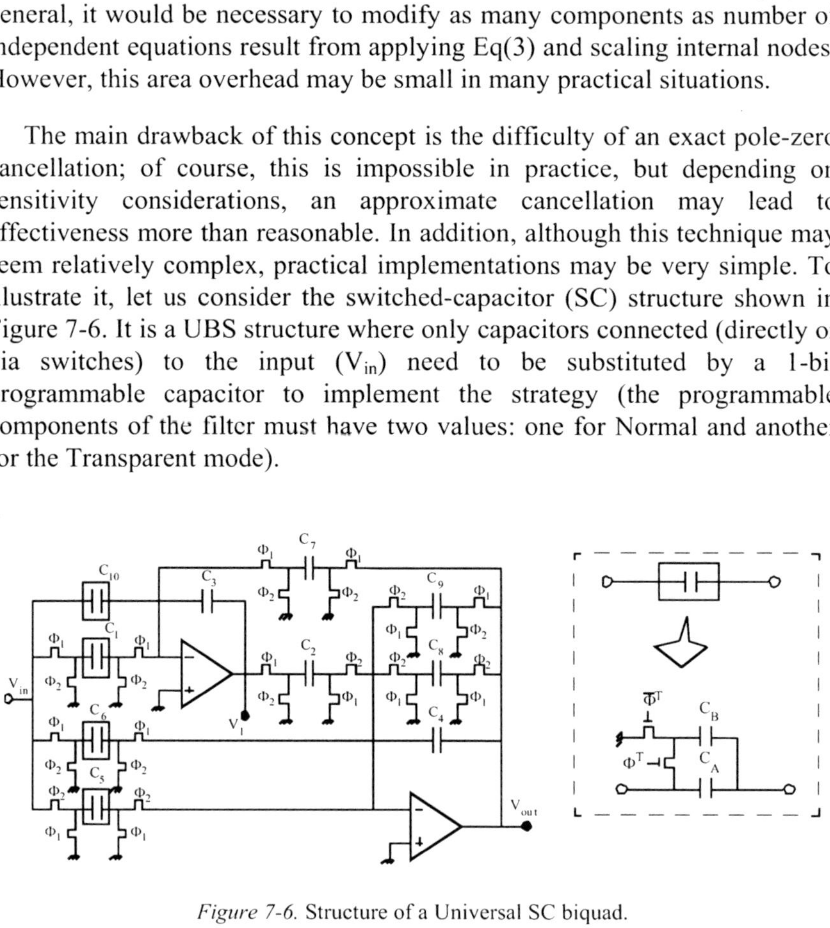

Figure 7-6. Structure of a Universal SC biquad.

An integrated prototype with the present approach was reported in [28]. The area overhead for a gain factor in the Transparent Mode k=1/1.189 was around 25%. A set of design considerations from a practical point of view for both SC and MOSFET-C implementations can be found in [28] and [29] respectively.

But a different use can be given to the pole-zero cancellation technique. In particular, it was exploited in [28] to perform a basic structural checking of the filter by setting all the biquads in the transparent mode and using a simple substracting circuit to generate an Error Signal $\varepsilon(t)$ in the form,

$$\varepsilon(t) = v_o(t) - Kv_{in}(t) \qquad (6)$$

where $v_o(t)$ is the output of the total filter, $v_{in}(t)$ the input and $K = \prod_{j=1}^{m} k_j$. Because circuit modifications are minimums, a maximum efficiency of the approach is guaranteed. Even more, it can be also modified to identify the section where the fault originates (if any). This can be done by multiplexing the input and output of each USB to a programmable gain adder to generate for each UBS an error signal $\varepsilon_j(t) = v_{out,j}(t) - k_j v_{in,j}(t)$, where $v_{out,j}(t)$ is the output of the jth-UBS and $v_{in,j}(t)$, its input.

5.4 sw-opamp

5.4.1 Basic concept and its use for accessibility

The sw-opamp concept is illustrated in Figure 7-7. It consists of a functional block with two operation modes depending on a digital control signal Φ:

1. Normal Mode: the circuit operates as an opamp (Opamp Mode), where terminals V+ and V- are the input and V_{out} the output.
2. Buffer Mode: the circuit acts as a buffer (Buffer Mode), where the signal present at terminal named VT is passed to the output.

The application of the sw-opamp concept to implement a multiplexing DfT strategy is illustrated in Figure 7-8. The last active element in each macro of the IC has been substituted by a sw-opamp. The inputs of the macros are made controllable through the sw-opamp in the previous macro.

The use of the sw-opam to implement the by-passing strategy is illustrated in Figure 7-9. The input of any macro can be controlled from the primary by passing the signal through the sw-opamps in the preceding macros (in buffer mode). In the same way, the output can be observed at the primary output through the sw-opamps in the succeeding macros. Consequently, each stage is fully controllable and observable from the primary input (V_{in}) and the primary output (V_{out}), respectively.

It should be clear at this point that there is a difference between the two basic schemes in Figure 7-8 and Figure 7-9 in terms of observability. In the by-passing structure of Figure 7-9, any macro can be observed at the primary output since every opamp in the system is a sw-opamp, and then observability is guaranteed by construction. On the other hand, the multiplexing structure needs some extra hardware (as shown in Figure 7-8) to support observability that must be added with care in order to avoid loading effects (for example, using buffers). However, s combination of both basic strategies in Figure 7-8 and Figure 7-9 can solve the observation problem in a very efficient manner. Let us consider Figure 7-10 where a modification of the multiplexing strategy is shown. The test input terminal of the sw-opamp can be connected either to an external test signal or to the output terminal of the previous macro. In this case, only one sw-opamp per macro is required (as in the basic multiplexing architecture), and a conventional multiplexer (out of the main signal path) connecting low and high impedance nodes. In this modified scheme, controllability is provided by a direct access through a sw-opamp, and observability is given through the primary output node.

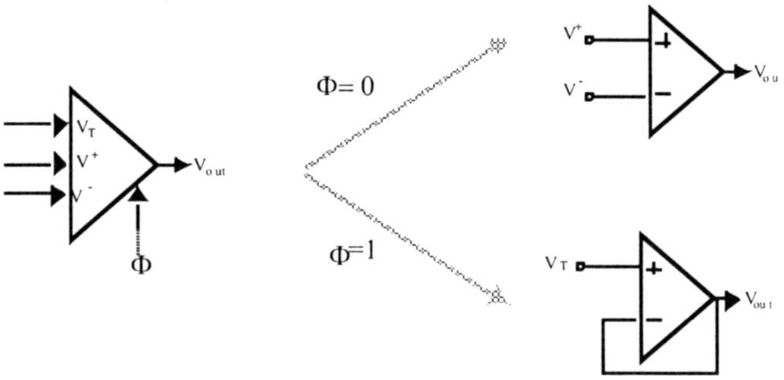

Figure 7-7. The sw-opamp concept.

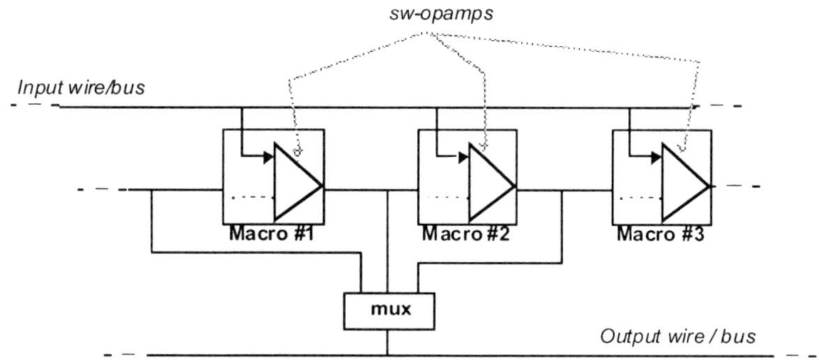

Figure 7-8. Implementation of the multiplexing strategy using sw-opamps.

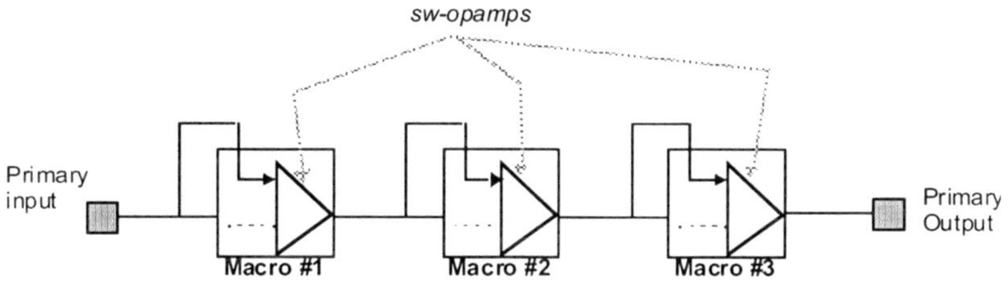

Figure 7-9. Implementation of the by-passing strategy using sw-opamps.

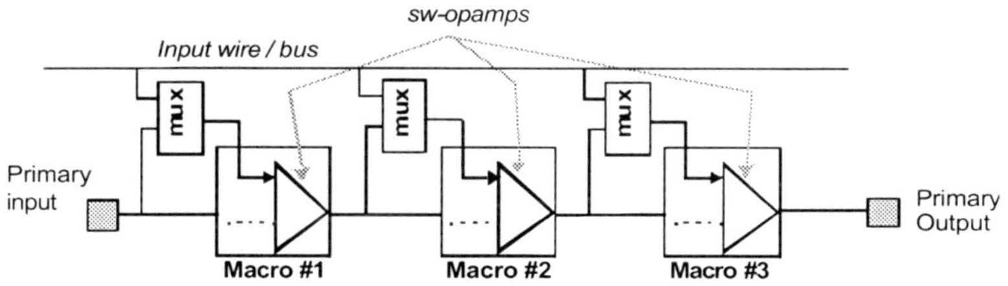

Figure 7-10. Improved sw-opamp based DfT scheme for off-line testing.

5.4.2 Sw-Opamp Implementations

A critical issue is the efficient design of the sw-opamp, since this is the key point to maintain tolerable signal degradation. Let us consider the typical two-stage opamp structure shown in Figure 7-11a. It is composed of three blocks: the input differential-pair (I), the active load (L) and the output stage (O).

Several implementations have been reported with different performance characteristics and requirements in terms of area and power overhead. However, all them can be derived from a reference opamp but maintaining practically the same performance specifications. Bratt et al. report in [22] the implementation shown in Figure 7-11b. It is based on a complete duplication of the input stage and the use of two minimum size non-complementary switches, NMOS or PMOS accordingly, to connect either of the input stages to the output stage accordingly to a digital control Φ. An alternative implementation [21, 30] consists on the use of some switches as shown in Figure 7-11c to make the necessary connections to configure the required operation. A third implementation (Figure 7-11d) [30] combines the two above cited approaches. In this case, only the input stage duplicated (only the input differential pair), while some switches are used to disconnect such blocks from the active load as necessary. In this configuration, switches can be NMOS or PMOS accordingly, but not of minimum size in general; they should be sized, simultaneously with the rest of components, to reduce their ON resistance and avoid undesired effects on the performance of the circuit.

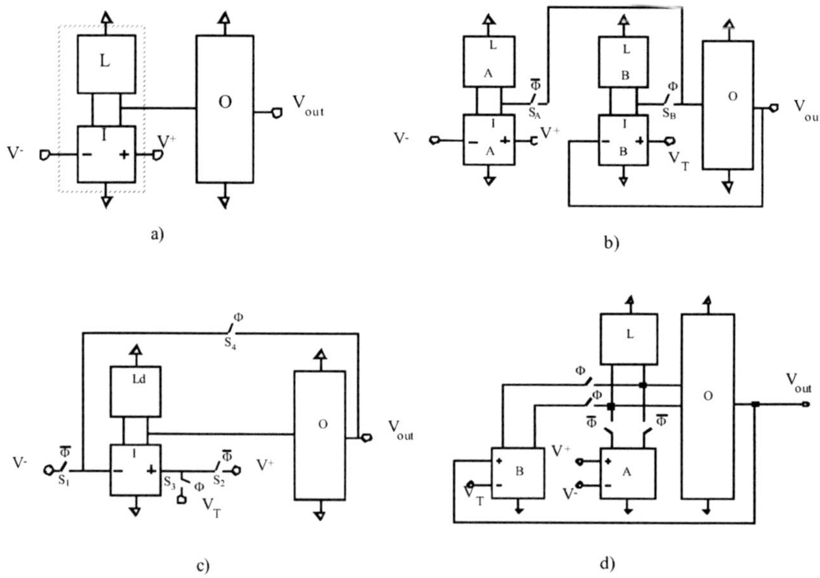

Figure 7-11. Block diagram of a typical two-stage opamp (a) and different implementations of the sw-opamp (c-d).

Practical CMOS implementations have been reported elsewhere. In particular, [30, 31] are good references for both single-ended and fully differential applications, where a comparison to non-testable structures is carried out, including experimental data confirming that in many practical situations the expected performance degradation can be tolerated without troubles.

6. DFT FOR ON-LINE TEST

The design of integrated circuits embedding on-line error detection and correction (fault tolerance) has been widely addressed in the digital field [32, 33]. In this line, the use of hardware redundancy has been extensively used to produce encoded signals [9]; under the assumption of that the presence of an error produce an out-of-code signal, a simple way to perform an on-chip on-line test consists on checking if the signals present in the circuit belongs to the permitted set of code words. The idea is illustrated in Figure 7-12. Moreover, the discrete natures of digital signals in both time and value have also been exploited to develop coding strategies that allow error correction, and thus, fault tolerance.

The required extra hardware for redundancy must be taken into account during the design phase of the circuit, what makes the code-based on-line test a DfT strategy in essence. In addition, the type of the test performed is structural as no circuit specifications are checked.

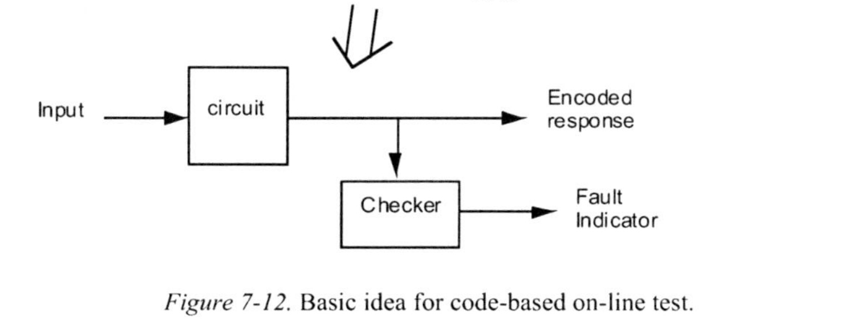

Figure 7-12. Basic idea for code-based on-line test.

Similarly to the digital world, on-line analog test strategies based on some degree of hardware redundancy to produce encoded signals have been proposed: fully [34] and partially duplicated analog circuits [16, 35], the checksum approach for state variable systems [36], balance checking of fully differential and balanced circuits [37, 38], etc. All them respond to the basic strategy illustrated in Figure 7-12 but based on different analog codes. In this sense, the checker design must be adapted to the code in use and to cope with the intrinsic characteristics of analog signals. In the case of analog signals, an "exact" comparison is not adequate as there is always a certain tolerance margin for which the circuit is considered to be operative.

A suite and well accepted comparison mechanism for analog signals that takes into account the required flexibility can be found in [34]: *Two analog signals V_1 and V_2 are considered "identical" if the absolute value of the difference between them is lesser than a given tolerance margin δ.*

$$V_1 = V_2 \Leftrightarrow |V_1 - V_2| < \delta \tag{7}$$

The δ value is determined by the maximum allowable amplitude and phase errors for a given application.

6.1 Replication and comparison

One of the simplest ways to provide a system with on-line testing is illustrated in Figure 7-13. The system architecture is denominated duplication and comparison. Two copies of the same circuit are driven by the same input signal. This scheme produces a encoded output signal represented by the pair (O1, O2). The checker object is to compare O1 and O2 and to decide whether they can be considered "identical" (O1= O2) or not (O1≠O2).

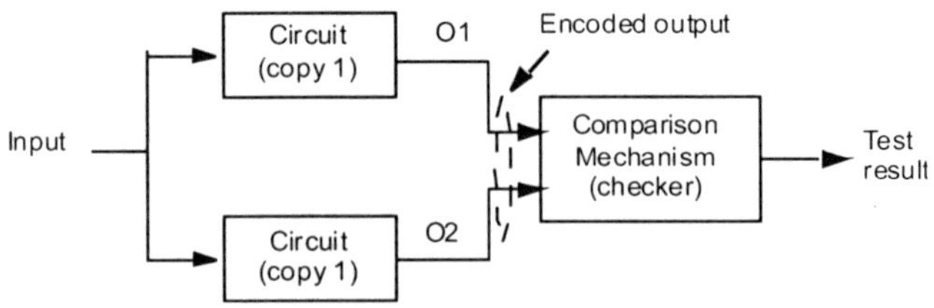

Figure 7-13. Duplication and comparison scheme.

The only extra effort required to extend this simple strategy to the analog field is the design of a comparison mechanism suitable for processing analog signals. Switched capacitor application examples can be found in [34], where a flexible comparison mechanism implementing Eq(7) is reported. Even more, the strategy is extended to provide fault tolerance to switched-capacitor circuits by taking advantage of the discrete nature (in time) of such kind of systems. But although simple, the fully duplication has the disadvantage of the extra area and power overhead (>100%) that in many cases may be unpractical. In addition, it is very sensitive to common mode faults, that is, faults affecting in the same way to both copies of the circuit.

6.2 Partial replication

To alleviate the inherent problems of the fully duplication scheme, the use of a programmable block was proposed in [16, 29, 35]. The proposed scheme is particularized in Figure 7-14 for the case of a filter composed by a cascade of biquads. The system architecture is composed by the following blocks:

- The original filter, where blocks labeled Bqj (j= 1, 2, ..., n) are the biquads.
- A Programmable Biquad (PBq) able to emulate, by electrical programming, the operation of each Bqj.
- A flexible comparison mechanism implemented by a window comparator where the limits are the maximum allowable disparity between "copies" of analog signals.
- A couple of multiplexing threes to connect the inputs of each Bqj to the input of the PBq (MUXRI) and their outputs to the comparison mechanism (MUXCI).
- A control logic which controls the operation of the PBq and the multiplexers.

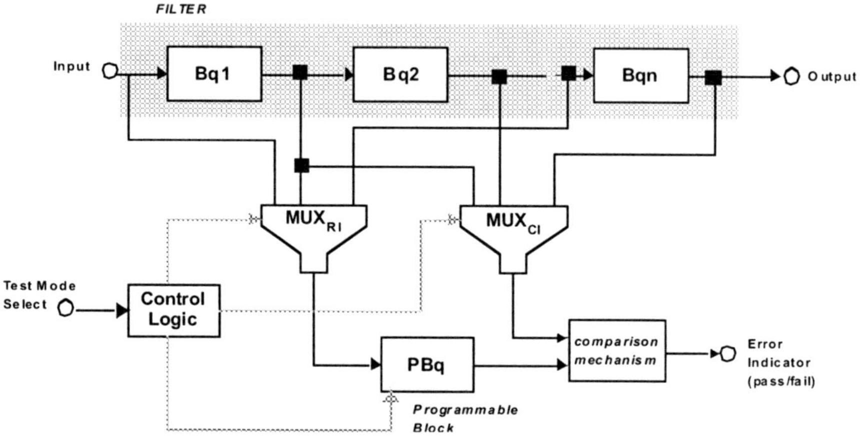

Figure 7-14. Basic system architecture for on-line testing using a programmable block.

The test process can be briefly described as follow: The control logic and the multiplexers select a biquaq (Bqj), connect its output to the comparison mechanism, its input to the input of the PBq and program it to simulate the operation of the selected biquad. The whole filter can be tested by testing each biquad in the form described above separately. Therefore, the test strategy is conceptually equivalent to the duplication and comparison scheme. But the present technique greatly reduces the risk of common mode faults with respect to the fully duplicated approach as no exact copies are being used. In addition, the PBq resources (active and passive elements) are shared in the test of all the biquads in the filter, what may mean important savings in terms of extra area and power.

The technique has been integrated and fabricated in the cases of a 4th-order switched-capacitor [35] and MOSFET-C [29] filters. In both cases, the multiplexers were integrated into the design of the blocks lowering the performance loss to marginal levels. Concerning the passive devices (capacitors and transistors acting as resistors) to be programmed in the PBq, an incremental strategy was used to reduce the area overhead.

This technique can been extended to other kind of filter design styles and circuits in general whenever a partitioning of the original system and the implementation of a programmable block (PRB) able to emulate the operation of the partitions be possible, as for example A/D converters [39]. However, in systems with multiple feedback loops (a occur in ladder filters) any parasitic in the PRB implementation, together with the fact that the PRB output does not enter in the complete system, could cause stability problems that invalidate real-time comparison, which is the basis of the fault detection of the approach. A solution that overcomes this problem is described in [40], where a simple feedback branch is added in the implementation of the PRB to stabilize its operation.

6.3 Checksums

In digital systems, the use of checksum encoding has been exploited for on-line error detection and correction in matrix multiplication. The technique has been extended to linear analog filters (independently of the design style used: ladders, cascade of biquads, etc.) and circuits that can be described as a linear state variable system (matrix representation) in [36]. The basic idea can be explained using the basic example illustrated in Figure 7-15. Let be a second order filter described in terms of the state variable matrices as,

$$\begin{bmatrix} x_1 \\ x_2 \end{bmatrix} = \begin{bmatrix} a_{11} & a_{12} \\ a_{21} & a_{22} \end{bmatrix} \begin{bmatrix} x_1/s \\ x_2/s \end{bmatrix} + \begin{bmatrix} b_1 \\ b_2 \end{bmatrix} \left(u/s \right) \tag{8}$$

where x_1 and x_2 are the state variables and u the input. The idea consists on creating a checksum encoded new state variable as follow,

$$x_t = \begin{bmatrix} r_1 & r_2 \end{bmatrix} \begin{bmatrix} x_1/s \\ x_2/s \end{bmatrix} + q\left(u/s \right) \tag{9}$$

where,

$$\begin{aligned} \begin{bmatrix} r_1 & r_2 \end{bmatrix} &= \begin{bmatrix} CV_1 & CV_2 \end{bmatrix} \begin{bmatrix} a_{11} & a_{12} \\ a_{21} & a_{22} \end{bmatrix} \\ q &= \begin{bmatrix} CV_1 & CV_2 \end{bmatrix} \begin{bmatrix} b_1 \\ b_2 \end{bmatrix} \end{aligned} \tag{10}$$

Vector $CV = [CV_1 \quad CV_2]$ is called the code vector. Notice that Eq(9) is equivalent to the weighted integration of the actual state variables and the input. The checksum properties of matrix multiplication say that the weighted sum of the state variables,

$$x'_t = [CV_1 \quad CV_2]\begin{bmatrix} x_1 \\ x_2 \end{bmatrix} \tag{11}$$

has to be equal to the state variable obtained as described in Eq(9) ($x_t = x'_t$). The presence of a fault corrupting the checksum code can then be detected through a signal error generated as,

$$\varepsilon = x_t - x'_t \tag{12}$$

Since signals are analog, the checking mechanism has to incorporate tolerances.

The overhead to be paid is equivalent to the circuitry required to implement the extra state variable, and the checker. In terms of fault detection, only those faults corrupting the checksum code can be detected. The technique, although described here in the continuous domain, it is also valid for discrete-time applications.

Figure 7-15. Example of checksum-based test for a second order filter .

6.4 Fully-differential filters

Fully differential (FD) architectures have been very popular in the design of circuits (filters, A/D converters, etc.) for high performance applications. Their properties to cancel many second order effects together with the good matching properties of integrated devices are the main reason for such popularity. In the case of filters, applications include switched-capacitor, Gm-C, and MOSFET-C implementations.

Figure 7-16 shows a typical opamp-based FD filter structure. In order to keep up the opamps operating correctly, the signals level at their inputs and outputs must be maintained close to a proper DC value (V_{cm}), called the common-mode voltage. For that, a specific circuitry (omitted in Figure 7-16 for simplicity) must be added to the design. The differential signal $V_{k,diff}$ is obtained by differencing the pair of signals (V_{k+}, V_{k-}) generated at the corresponding nodes in the paths. The main property of such pairs is that signals are virtually balanced with respect to the common mode voltage V_{cm}. It can be expressed as follows:

$$\begin{aligned} V_{k+} &= \left(\frac{V_{k,diff}}{2}\right) + V_{cm} \\ V_{k+} &= -\left(\frac{V_{k,diff}}{2}\right) + V_{cm} \end{aligned} \tag{13}$$

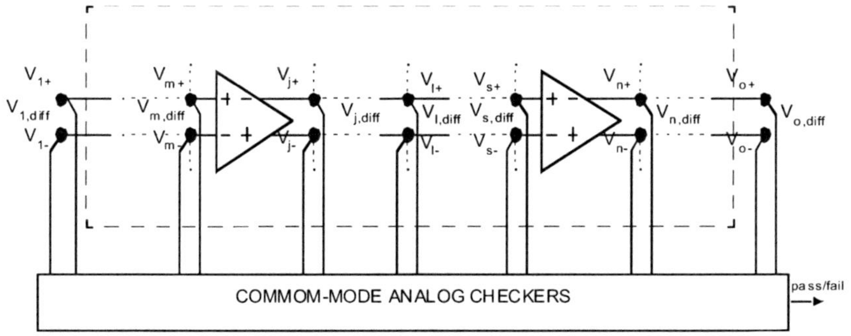

Figure 7-16. Typical opamp-based fully-differential structure.

From the test point of view, FD circuits are then very attractive because signals are inherently encoded as described in Eq(13), in such a way that,

$$\frac{V_{k+} + V_{k-}}{2} = V_{cm} \tag{14}$$

The code checker object is to detect intolerable shifts in the value of the common mode voltage at any of the signal pairs. Single faults (opens, shorts, component value deviations, etc.) in the opamps or in any of the signal branches causing signal unbalance, would be detected.

A complete discussion of this technique including application examples and fault detection capabilities can be found in [41, 42, 38, 37].

7. BUILT-IN-SELF-TEST

Built-In-Self-Test (BIST) techniques are DfT strategies very attractive to overcome the problems of testing embedded blocks to which external accessibility through the I/O pins is limited. They reduce the need of external testing by moving tester functions (stimuli generation and/or test evaluation) on to the chip. Typically, BIST techniques provide a digital pass/fail output.

Because BIST requires extra hardware, and thus, the major cost associated is the area, performance impact and yield loss. But on the other hand, testing automation, better accuracy (as signals are not routed off-chip and loading effects are lowered), reduced dependency on expensive ATE and its reuse at different test phases (from production to on-board and on-field testing) must be considered as a trade-off for its use. Even more, overhead can be cut down by reconfiguring the existing circuitry during the test process [43], and/or exploiting the availability of DSP, A/D and D/A converters in complex mixed-signal ICs to perform the test of the analog part [17].

A BIST scheme for functional testing normally requires accurate signal generators and the processing of a large amount of data. In the case of filters, a precise test in the frequency domain demands the generation and application of calibrated sinewaves at different frequencies together with a complex processing of the response (FFT, DFT, sinewave correlation, power metering, etc.) that is normally made externally or internally using complex DSP cores (when available). Professor Gordon Roberts has performed a large and good work in this line [44].

BIST schemes based on structural testing alleviate the complexity of signal generation and evaluation. Simple signals that do not need to be accurately calibrated (DC, ramps, square waves, etc.), or even, signals generated by the circuit itself are used to excite faults and observe their effects at the test points even through a very simple on-chip circuitry [43, 45, 46]. The so-called Oscillation-Based-Test is a clear example in this sense [47, 48, 49, 52].

It is not the intention of this section to perform a review of all BIST approaches proposed in the literature suite for the particular case of filters. Instead of that, it will be restricted to two particular approaches that focused the interest of a large part of the test community during the last years: the Spectrum-Based and the Oscillation-based BIST strategies.

7.1 Spectrum-Based BIST

Functional testing of analog filters demands the measure of specification-related frequency parameters (bandpass performance and stop-band edges, distortion, signal-to-noise-ratio, dynamic range, etc.). Spectrum-Based testing deals with this task.

The test problem can be divided into two tasks: the generation of the stimuli and the evaluation of the response. Spectrum-Based testing requires the application of specific yet precise sinewaves; the evaluation of the responses using appropriated processing algorithms allows the extraction of parameters like the frequency response, gain, harmonic and intermodulation distortion. As can be deduced, they are typical parameters related to filter specifications.

7.1.1 On-chip signal generators

Most of the proposed strategies for on-chip signal generation of analog test signals make use of digital resources [51, 44]. It has the advantage of that digital test methods can be applied and/or adapted for the generation and evaluation of the test signals. In this section, three basic methods will be discussed. Their differences are mainly in their requirements in terms of hardware, the quality of the generated signals and in their flexibility.

For example, analog signals can be generated using a digital-frequency synthesizer followed by a D/A converter and a reconstruction filter. The ROM-Based implementation shown in Figure 7-17 is a typical example. The accuracy of signals generated in this way is limited by several factors: in terms of frequency by the number of samples stored in the ROM and in terms of quality either by the word length of each sample and the characteristics of the D/A converter. But anyway, the main drawback of this approach is the large silicon overhead if the required circuitry is not available on the system; for example, the ROM function can be substituted by an existing RAM during the test phase of the analog part.

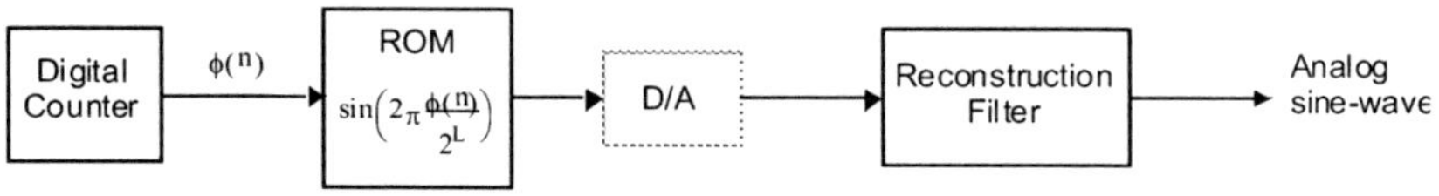

Figure 7-17. ROM-based frequency synthesizer.

Alternative approaches that overcome the problems of area and quality of the ROM-based synthesizers in Figure 7-17 are based on the use of ΣΔ encoding schemes [62], where the noise shaping characteristics of these kind of encoding methods are exploited. The basic idea consists basically on generating a 1-bit stream ΣΔ encoded version of an N-bit digital signal. Matching the noise shaping characteristics of the encoding ΣΔ scheme with the reconstruction filter, a precise and high quality analog signal can be obtained. This is illustrated in Figure 7-18.

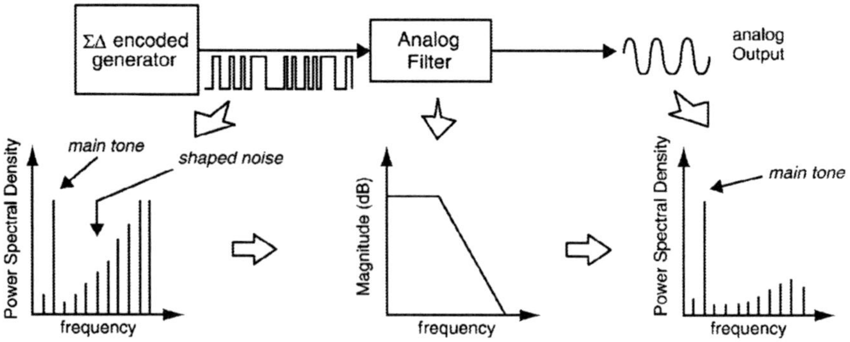

Figure 7-18. Use of ΣΔ-encoded bit streams to generate precise analog signals

There are two basic approaches to generate the 1-bit ΣΔ encoded pattern: a ΣΔ oscillator and a memory-based signal generator. Both approaches are feasible to be implemented on-chip. However, while the first achieve better accuracy, the second is more area-efficient.

A ΣΔ oscillator (Figure 7-19) is basically composed by a digital resonator and a ΣΔ modulator in a loop gain configuration as shown in Figure 7-19. The resonator creates an N-bit sinusoidal signal (it has a pair of poles placed on the unit circle), which amplitude and phase is determined by the initial conditions set in the multibit registers. The ΣΔ modulator creates the 1-bit pattern of the signal created by the resonator. The noise shaping characteristics of the 1-bit stream is determined by the encoding method. In this sense, low-pass ΣΔ modulators are more suitable for low frequencies while bandpass modulators are better for high frequencies. The frequency of the generated signal is set by the loop gain (K) and the sampling frequency. This approach offers high precision and high frequency resolution; up to 90dB spurious-free dynamic range signals has been reported [53, 54] with this method. However, design of this type of oscillators is not straightforward and its application is limited to single tone signals.

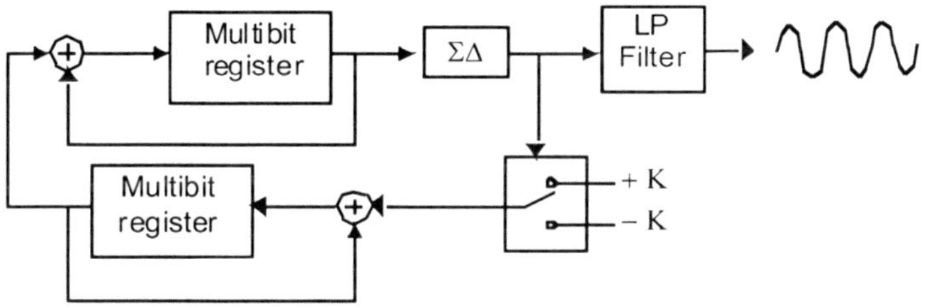

Figure 7-19. ΣΔ oscillator.

A second approach based on memory elements is feasible [44]. The basic idea is to truncate the 1-bit output of the ΣΔ oscillator and reproduce it periodically using a feedback chain of memory elements as shown in Figure 7-20 Single tones with spurious-free dynamic range as high as 80dB has been obtained for sequences as short as 100bits. But this technique is not restricted to single tones. On the contrary, arbitrary and multi-tone waveforms can be easily generated. Moreover, optimized methods for generating the bit stream improve the quality of the generated signals [44]. Concerning the area overhead, it can be reduced to near zero if existing digital structures in the system (boundary scan chains, RAMs, etc.) are exploited to implement the technique.

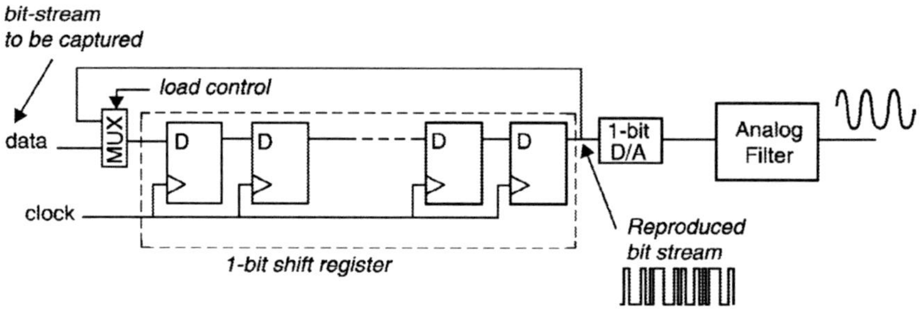

Figure 7-20. Basic memory-based oscillator circuit.

A complete discussion on this topic together with silicon demonstrator results can be found in [44].

7.1.2 Parameter extraction

The extraction of frequency-based parameter specifications requires a way of separating sinusoidal and noise components in the output signals. This is a complex task that is normally done in the digital domain; thus, an A/D conversion is required in this sense. Otherwise, analog signals should be transported off-chip to perform the extraction using appropriated software tools.

The Fast Fourier Transform (FFT) is the standard tool for this purpose and gives very accurate results. It allows the decomposition of finite length (N) signal into a set of harmonic frequency components. The on-chip implementation would necessitate N memory elements (plus the A/D converter) to store the samples of the signal and would require on the order of $N\log(N)$ arithmetic operations. As the accuracy of the measures depends on the number of samples, this extraction procedure is normally prohibitive for on-chip implementation if no complex DSP core that can be exploited for this purpose is present into the chip.

When only one frequency is of interest, a sinewave fitting algorithm can be used. In terms of computational efforts, this method is less expensive than the FFT, what makes it more suitable for on-chip implementation. However, it requires memory elements to store sine functions what means large area requirements for a good accuracy.

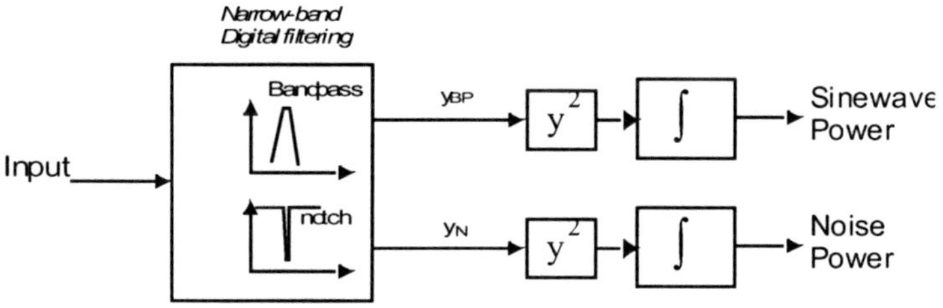

Figure 7-21. Use of bandpass and notch filtering for parameter extraction.

A combination of bandpass and reject-band narrow digital filters is another possibility [55]. Figure 7-21 shows a block diagram of the approach. A digital filter provides two outputs: a bandpass function ant its complementary notch function. For a given frequency, the power of the signal and the noise can be obtained by squaring and adding the appropriated output (the bandpass and the notch respectively). The accuracy obtained with this approach depends on the selectivity of the filters and the number of samples; the larger the selectivity the larger the accuracy. However, it must be taken into account that the time required to obtain the measurement is larger while larger is the selectivity of the filter.

7.2 Oscillation-Based-Test[1]

Experience gained from the development of digital circuits invites to move towards structural or fault-driven Built-In Self-Test (BIST) alternatives. An emerging solution in this line is the so-called Oscillation-Based-Test (OBT). It consists on converting the circuit into an oscillator during the test phase in such a way of that oscillation parameters be directly related to the circuit behavior and performance during normal operation. The presence of faults can be detected through the measure of the cited oscillation parameters.

The main advantages of the OBT technique are:

- It is conceptually simple. No strong circuit modifications are needed.
- No dedicated resources are needed for the generation of the test stimuli as they are generated by the circuit itself.

[1] An extended study on Oscillation-based techniques can be found in the last Chapter of this book.

- It is applicable to many kind of circuits with a variety of different specifications (filters, A/D and D/A converters, PLLs, etc). Notice that the test of analog circuits would be standardized in this sense.
- Test evaluation is reduced to the extraction of the oscillation parameters. It facilitates its extension to BIST (OBIST in this case).

Although the OBT is mainly a fault-driven approach, oscillation parameters are closely related to the behavior and performance of the circuit under normal operation [47, 56], and even functional specifications can be extracted from them [57].

But a feasible application of the OBT technique is not straightforward and several factors must be taken into consideration:

1. How to build the oscillator. A robust oscillator has to be configured in order to guarantee self-sustained oscillation and predict the oscillation parameters accurately. In addition, the circuit structure and components should not be altered. Fault detection capability and performance loss will depend directly of that.
2. A start-up mechanism. A mechanism to force the oscillation to begin is normally required whenever an oscillator is involved. Even more, test is not valid until the configured oscillator reaches the steady-state; a short transient time is thus needed to cut the test time.
3. Support for parameter extraction. This is a requisite to be implemented on-chip in order to extend the technique to BIST.

7.2.1 Building the oscillator

Converting a filter into an oscillator requires a mechanism to force a displacement of (at least) a pair of poles to the imaginary axis in the s-domain (on the unit circle in the z-domain). It is well known that it can be achieved in the case of filters by adding a feedback branch as illustrated in Figure 7-22. Two modifications are required: a) the regular filter input has to be disabled, and b) the feedback loop has to be closed.

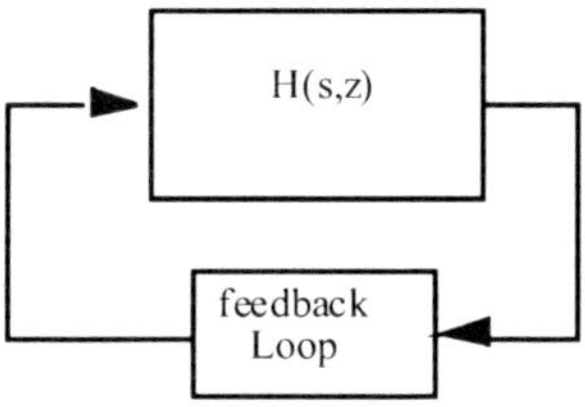

Figure 7-22. Block diagram to convert a filter into an oscillator.

In principle, linear and non-linear feedback loops can be considered. However, it is well known [58] that, in practice, linear feedbacks are not adequate and a non-linear mechanism is needed as a feedback element to guarantee sustained oscillations. In the case of active filters, the non-linear behavior of the active components can be exploited for this purpose as done in some implementations reported in [47, 48, 59, 60]. However, this strategy is not suitable because oscillation parameters will depend strongly of such non-linear behavior while the interest of the filter during normal operation is in its linear behavior. It leads to a reduced dependence of the oscillations with respect to the filter performance under normal operation, and consequently, the fault detection capability is limited by this fact.

The saturated function depicted in Figure 7-23 has been demonstrated to be very suitable for the case of second order sections (biquad-based implementations) [56]. Its main advantages are that it is easy to implement, almost all biquad structures can be forced to oscillate, and oscillation parameters can be estimated or accurately predicted using simple linear analysis. Appropriated cases for its use are frequency discrimination applications, where there is normally a filter preceding a zero-crossing detector acting as a front-end with a digital circuitry. In such cases, the overhead to implement the OBT technique can be greatly reduced [61].

Figure 7-23. Saturated function: a) negative, b) positive.

In the case of second order oscillators, formed by a second order filter section and the saturation function as the feedback element, closed-form expressions can be derived using the describing-function approach [56, 58]. Important facts for filters can be summarized as follows:

- Most second-order transfer function can be forced to oscillate just adding a feedback branch incorporating a saturation function.
- The frequency of oscillation (f_{osc}) is determined by the coefficients of the transfer function.
- The amplitude of the oscillations can be controlled by the saturation level, V, of the saturation function. In this way, fault detection capability may be maximized.
- Although approximate, the describing function approach gives usually reasonable fitting when the real signal is close to a pure sinewave. Otherwise, harmonic distortion and other factors may lead to certain inaccuracies. But anyway, this error is not an important problem since the derived closed-form formulae are good enough to be a starting point for behavioral-level simulation, thus reducing (easing) the computational cost of using a device-level simulator.
- Finally, it is convenient to remark that the considerations above are valid for both discrete- and continuous-time circuits.

Concerning the oscillation parameters to be measured, some authors have paid attention only to the frequency [47, 48, 59, 60]. However, it has been demonstrated elsewhere that the measures of the amplitude do not only increases drastically the fault coverage, but it also allows to relaxing the accuracy of the measures [56, 61]. In general, any predictable parameter as the DC-level of the oscillation waveform, the distortion, etc. can be used.

In the case of high order filters, analytical analysis of the resulting system is complex to deal with, making difficult to predict the oscillation parameters, which is crucial. In these cases, it has been suggested to divide the filter into smaller blocks where the OBT strategy is applied separately [49, 52].

7.2.2 Start-up

The start-up of the oscillator is not a marginal problem. Otherwise, the feasibility of the OBT technique can be seriously compromised. First, it must be ensured that the configured oscillator will start to run (safe start-up). Secondly, the measurement of the oscillation parameters has to be performed once the steady-state has been reached; a mechanism to reduce the transient time (an consequently the test time) is of vital importance. Finally, it has also important consequences from the fault simulation point of view because it would contribute to short the simulation time. In principle, any of the start-up techniques proposed in the literature can be used [64]. However, the sw-opamp has been shown to be also very effective for this task.

7.2.3 Extension to BIST. On-chip signal evaluation.

One of the crucial issues to be considered is concerned on how to give support to the evaluation of the test signals, that is, to the measurement of the oscillation parameters.

There are many ways to proceed. In particular, any of the approaches described on section 7.1 are (in principle) valid. However, their use may require excessive extra on-chip resources (A/D, DSP core, etc.) if they are not already available in the system. Instead of that, a comprehensive evaluation scheme for OBIST applications must take into account the following considerations:

- The interest relies on just checking whether the main characteristics of the generated oscillation wave (amplitude, frequency, etc.) matches or not the expected ones under fault-free conditions. In filter applications, the generated wave is normally very close to a pure sinewave.
- A measurement circuit must be implemented on-chip with low hardware overhead. It should be robust and, at the same time, its accuracy should provide high fault coverage. The results are preferable to be digitally coded.
- A pass/fail decision mechanism must be conceived accordingly to the measurements performed. It must be flexible in order to cope with tolerances as always as analog signals are involved.

Indirect measurements by test signal compaction schemes and the analysis of the resulting signature has been addressed by some authors [43, 45, 46, 65]. To avoid the inaccuracies of analog measures, some works have proposed the use of a $\Sigma\Delta$ modulator, instead of a full A/D converter, to obtain a digital encoded version of the test signal [49, 66] which can be processed either internally using a digital algorithm or externally by a pure digital tester. In the second case, the evaluation of the test signal is solved partially as only the off-chip transportation of the test signals is facilitated. But the on-chip processing requires different algorithms to the conventional ones because, otherwise, the required hardware and the test time would be very large. In this line, a robust and low-cost mechanism based on a $\Sigma\Delta$

modulator was firstly reported in [67] and improved in [50]. It is specially suitable for OBIST applications as it is able to extract the main oscillation parameters (amplitude, frequency and DC-level) in an efficient manner and with reasonable accuracy.

8. UNIFIED TEST ARCHITECTURE

Any of the DfT approaches described in this chapter do not completely solve the problems during the different testing phases of an IC. For example, DfT for off-line testing are useful for production testing but, in general, they are not adequate for on-field operation testing and/or maintenance. On the other hand, on-line and BIST approaches requires extra hardware that normally has to be tested in the early phases using off-line techniques.

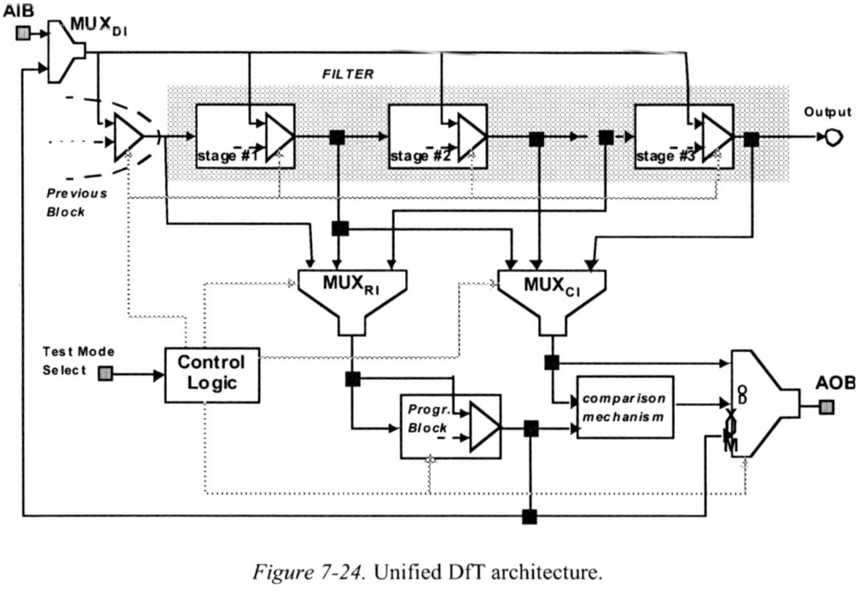

Figure 7-24. Unified DfT architecture.

The above considerations lead to a need of unified DfT schemes where off- and on-line DfT and BIST as well be integrated on a comprehensive manner. An example is the structure shown in Figure 7-29 and reported in [68]. It integrates the on-line DfT structure using a programmable block, the sw-opamp and analog test buses for off-line testing, and the OBIST approach described in the previous section. It assumes that the system is fully embedded so that both the primary input (AI) and the primary output (AO) are not externally accessible. However, it is considered that the input signal (AI) is provided by a buffering cell in the previous block. As can be deduced from the figure, the basic components are essentially those needed to implement the on-line testing approach in Figure 7-14. The main differences are:

a) Two external pins connected to buses, ATIB (Analog Test Input Bus) and ATOB (Analog Test Output Bus), which in turn are connected to the system through two multiplexers MUX_{DI} and MUX_{DO}.
b) Both additional multiplexers, MUX_{DI} and MUX_{DO}. The first is used to select the test input signal of the sw-opamps. It can be the signal injected through ATIB or the output of the PRB. The second is used to externally observe (through ATOB) the response of the stages, of the PRB or of the comparison mechanism.
c) Other differences are functional, as in the case of the PRB as will be described below.

The proposed architecture has basically three operating modes:

1. On-line mode. For on-line testing, the system is reconfigured as in Figure 7-14. To do so, the control circuitry (commanded by TMS) puts the sw-opamps in normal mode (opamp) and realizes the operations described in Section 6.
2. Off-line mode. The functional operations of the system, the PRB, and the Comparator are tested by interrupting the normal system operation and performing the following actions. For a stage and for the PRB testing, the sw-opamp in the previous stage is used to apply the same external test stimuli put at ATIB to the stage under test and to the PRB inputs. The respective outputs are then either observed separately at the ATOB or compared and the comparison result observed at TOB. Notice that the test of each stage and the PRB is performed in parallel (no extra time is required). The test of the Comparator block is performed by applying two known input signals and observing the response at ATOB. One of these two signals is generated by the PRB and the other is an external signal at ATIB entered through the sw-opamp of one of the stages and the MUX_{CI} multiplexer.
3. BIST Mode. This mode implements the Oscillation-Based-Test scheme. The procedure is the following: a) The PRB is programmed to emulate the operation of one of the stages (SUT). b) The feedback path from the PRB output to the sw-opamp test input through MUX_{DI} is enabled. This feedback is closed around the BRP which is converted in this way into an oscillator. The generated signal is injected to the stage under test through MUX_{DI} and the sw-opamp in the previous stage. c) The generated signal

and the response of the SUT can be externally observed MUX_{CI} and MUX_{DO}. Alternatively, as the BRP and the SUT implement the same function (under fault-free conditions) and they are excited by the same stimuli, both the BRP and the SUT outputs can be compared and the result observed at AOTB as done in the on-line mode.

Fortunately, there are some applications, especially in the case of filters, where this unified architecture can be implemented. An integrated and fabricated chip with a 6th-order high-Q fully differential filter for RDS applications [69] was reported in [68]. The most important contribution of this work is not only the architecture itself, but also the experimental results demonstrating that the proposed DfT strategies are feasible to be implemented without performance loss, what was one of the main arguments of the IC industry to not use them.

9. CONCLUSIONS

It has been reviewed several DfT approaches for off-line, on-line and BIST applications offering respectively to embedded analog IC filters: a) accessibility to internal nodes, b) on-line testing, and c) BIST capabilities. Some of them are extensions of existing techniques in the digital field. Other however copes with the difficulties imposed by the own nature of analog circuits and signals. Functional and structural DfT approaches are feasible in this way but, in general, functional testing requires large on-chip resources that may be prohibitive form a practical point of view if they are not already available in the chip. Structural DfT approaches offers on the other hand more efficient implementations, although they do not ensure the correctness of functional performances. In conclusion, structural and functional approaches are complementary and both must be taken into account to break the bottleneck of testing in actual complex mixed-signal ICs.

10. REFERENCES

1. K.Lofstrom, "A demonstration IC for the P1149.4 mixed-signal Test Standard", *Proc. of the International Test Conference*, pp. 92-98, 1996.
2. M. Ismail and T. Fiez, *"Analog VLSI: Signal and Information Processing"*, McGraw Hill, 1994.
3. R. Gregorian and G. Temes, *"Analog MOS Integrated Circuits for Signal Processing"*, Wiley Interscience, 1986.
4. T. M. Souders and G. N. Stenbakken, "Cutting the high cost of testing". *IEEE Spectrum*, pp. 48-51, March 1991.

5. R.Kapur , R. Chandramouli and T.W. Williams, " Strategies for Low-Cost Test", *IEEE Design & Test of Computers*, pp. 47-54, Nov-Dec 2001.
6. M. Sachdev: *"Defect Oriented Testing for CMOS Analog and Digital Circuits"*, Kluwer Academic Publishers, 1998
7. J. Pineda de Gyvez and D.K. Pradhan: "Integrated Circuit Manufacturability: The Art of Process and Design Integration", *IEEE Press*, 1999.
8. B. Vinnakota: "Analog and Mixed-Signal Test". Prentice Hall, 1998.
9. M. Abramovici, M. A. Breuer and A. D. Friedman, *"Digital System Testing and Testable Design"*. New York, Computer Science Press, 1990.
10. P. P. Fasang, D. Mullins and T. Wong: "Design for Testability for Mixed Analog/Digital ASICs", *Proc. IEEE 1988 Custom Integr. Circ. Conf.*, pp. 16.5.1-16.5.4, 1988.
11. K. D .Wagner, T. W. Williams: "Design for Testability of Mixed-Signal Integrated Circuits". *Proc. IEEE 1988 Int. Test Conf.*, pp 823-828, 1988.
12. J. L. Huertas, *"Test and Design for Testability of Analog and Mixed-signal Integrated Circuits: Theoretical Basis and Pragmatical Approaches"*. Selected Topics in Circuits and Systems, H. Dedieu Editor. Elsevier 1993.
13. C.L.Wey, "Built-in Self-test (BIST) Structure For Analog Circuit Fault Diagnosis" *IEEE Trans. on Instr. & Measur.*, Vol.39, 1990.
14. M. Soma, "A Design-For-Test Methodology for Active Analog Filters", *Proc. IEEE Int. Test Conf.*, pp 183-192, 1990.
15. G. Schafer, H. Sapotta and W. Denner: "Block-oriented Test Strategy For Analog Circuits" *Proc. of European Solid-State Circuits Conference*, pp.217-220, 1991.
16. J.L. Huertas, A.Rueda and D. Vázquez: "Improving the Testability of Switched-Capacitor Filters", *Int. Journal of Electronic Testing*, Vol. 4, No. 4, pp. 299-313, November 1993.
17. M. Ohletz, "Hybrid Built-in Self-test (HBIST) For Mixed Analogue/Digital Integrated Circuits", *Proc. European Test Conerence*, pp.307-316, 1991.
18. M. Soma and V. Kolarik, "A Design-For-Test Technique for Switched-Capacitor Filters", *Proc. IEEE Int. Test Conerence.*, pp 42-47, 1994.
19. D. Vázquez, A.Rueda, J.L.Huertas and A.M.D. Richardson,*"Practical DfT Strategy for Fault Diagnosis in Active Analogue Filters"*. IEE Elec. Letters, Vol. 31, N. 15, pp. 1221-1222, July 1995.
20. D. Vázquez, A. Rueda, J. L. Huertas, "A DFT Methodology for Fault Diagnosis in Active Analog Filters". *Proc. Int. Mixed Signal Testing Workshop'95*, Grenoble (France), pp. 124-129, June 1995.
21. M Renovell, F. Azais and Y. Bertrand, "A DFT Technique to Fully Access Embedded Modules in Analog Circuits Under Test". *Proc. Int. Mixed Signal Testing Workshop'95*, Grenoble (France), pp. 172-177, June 1995.
22. A.H. Bratt, R.J. Harvey, A.P. Dorey and A.M.D. Richardson, "Design-For-Test Structure to Facilitate Test Vector Application with Low Performance Loss in Non-Test Mode". *IEE Elec. Letters*, Vol. 29, No. 16, pp. 1438-1440, Aug. 1993.
23. C.L. Wey and S. Krishnan, "Built-In-Self-Test (BIST) Structures for Analog Circuit Fault Diagnosis with Current test Data", *IEEE Trans. on Instrumentation and Measurement*, Vol 41, NO, 4, August 1992.

24. Yunsheng Lu, et al, " Structure and Metrology for a Single-Wire Analog Testability Bus", *Proc. of IEEE International Test Conference*, pp. 919-928, 1994
25. K. Parker, J.E.McDermid and S.Oresjo, "Structure and metrology for an Analog testability Bus"; *Proc. of IEEE International Test Conference*, pp. 309-320, 1993.
26. P.M. Van PeteGhem and J.F. Duque, "Compact High-Frequency Output Buffer for Testing of Analog CMOS VLSI circuits", *IEEE Journal of Solid-State Circuits*, Vol. 24, pp. 540-542, 1989.
27. S. Sunter, "A Low Cost 100MHz Analog test Bus", *Proc. of VLSI Test Symposium*, pp. 60-65, 1995.
28. D. Vázquez, J.L.Huertas and A.Rueda, " A New Strategy for Testing Analog Filters", *Proc. IEEE VLSI Test Symposium*, pp. 36-41, April, 1994.
29. D. Vázquez, A. Rueda and J. L. Huertas, "On-Line Error Detection for Continuous-Time MOSFET-C Filters". *Proc. of European Solid-State Circuits*, pp. 206-209, Sevilla, September 1993.
30. D. Vázquez, J.L. Huertas and A. Rueda, "Reducing the impact of DfT on the Performance of Analog Integrated Circuits: Improved Sw-Opamp Design". *Proc. VLSI Test Syposium*, pp. 42-47, 1996.
31. D. Vazquez, A. Rueda and J. L. Huertas, "Fully-Differential SW-Opamp for Testing Analog Embeded Modules". *Proc. of Int. Mixed Signal Testing Workshop*, Quebec, 1996.
32. B.W.Johnson, *"Design and Analysis of Fault Tolerant Digital Systems"*, Addison Wesley, 1989.
33. P. K. Lala: *"Fault tolerant & fault testable hardware design"*. Prentice Hall, 1986.
34. D. Vazquez, A. Rueda, J. L. Huertas, "A Practical Implementation of Fault-Tolerant Switched Capacitor Circuits", *Proc. of Int. Symposium on Circuits and Systems (ISCAS)*, pp. 1565-1568, Singapore, June 1991
35. J. L. Huertas, A. Rueda, and D. Vazquez, "Testable switched-capacitor filters". *IEEE Jour. of Solid-State Circuits*, V. 28, n. 7, pp. 719-724. July 1993.
36. Chatterjee, "Concurrent Error Detection and Fault-tolerance in Linear Analog Circuits Using Continuous Checksums", *IEEE Trans. on VLSI Systems*, Vol. 1, NO 2, pp. 138-150, June 1993.
37. N.J. Stessman, B. Vinakkota and R.Harjani, "System-Level Design-for-Test in Analog Circuits", *IEEE Journal of Solid-State-Circuits*, pp. 1320-1326, October 1996.
38. S. Mir, V. Kolarik, M. Lubaszewsky, and B. Curtois, "Built-In-Self-Test and fault diagnosis in Fully Differential Circuits", *Proc. of Int. Conf. on Computer-Aided Design*, 1994.
39. E. Peralias, A. Rueda and J. L. Huertas, "An On-line testing approach for Pipelined A/D Converters". *Proc. of Int. Mixed-signal Test Workshop*, pp. 44-49, 1995.
40. D. Vázquez, A. Rueda and J. L. Huertas, "A solution for the on-line test of analog ladder filters". *Proc. VLSI Test Symposium*, pp. 48-53, 1995
41. V. Kolarik, M. Lubaszewsky, and B. Curtois, "Towards Self-Checking Mixed-Signal Integrated Circuits", *Proc. of European Solid-State Circuits Conference*, pp. 202-205, 1993.
42. V. Kolarik, M. Lubaszewsky, and B. Curtois, "Designing Self-Exercising Analogue Checkers", *Proc. of VLSI Test Symposium*, pp. 252-257, 1994.

43. M. Lubaszewsky, S. Mir and L. Pulz, "A Multifunctional Test Structure for Analog BIST", *Proc. of International Mixed-Signal test Workshop*, 1995.
44. B.Dufort and G.W.Roberts, "On-Chip Analog Signal Generation for Mixed-Signal Built-In-Self-Test", *IEEE Journal of Solid-State Circuits*, Vol. 33, No. 3, pp. 318-330, March 1990.
45. N. Nagi, A. Chatterjee and J.A. Abraham, "A Signature Analyzer for Analog and Mixed-Signal Circuits", *Proc. of Int. Conf. on Computer-Aided Design*, pp. 284-287, 1994.
46. M.Renovell, F. Azais and Y. Bertrand, "On-Chip ANalog Output Response Compaction", *Proc. of the European Design & Test Conference*, pp. 568-572, 1997.
47. K.Arabi,B.Kaminska."Oscillation- Test Methodology for Low-Cost Testing of Active Analog Filters" *Trans. on Instrumentation and Measurement*, vol. 48, no. 4, pp. 798-806, Aug. 1999.
48. K. Arabi and B. Kaminska."Oscillation-Test Strategy for Analog and Mixed-Signal Integrated Circuits" *14th VLSI Test Symposium*, 1996, pp. 476-482.
49. G. Huertas, D. Vázquez, E. J. Peralías, A. Rueda and J. L. Huertas , "Testing Mixed-Signal Cores: A Practical Oscillation-Based Test in an Analog Macrocell ", *IEEE Design & Test of Computers*, pp. 73-82, Vol 19, NO 6, pp. 64-72, 2002.
50. D. Vázquez, G. Huertas, G. Leger, A, Rueda and J.L. Huertas, "On-Chip Evaluation of Oscillation-Based-Test Output Signals for Switched-Capacitor Circuits", *Analog Integrated Circuits and Signal Processing*, Kluwer Academic Publishers, Vol 33, 2002.
51. M.V. Mahoney, "DSP-Based Testing of Analog and Mixed-Signal Circuits", *IEEE Computer Society Press*, 1987.
52. G. Huertas, D. Vázquez, E. J. Peralías, A. Rueda and J. L. Huertas , "Practical Oscillation-Based Test of Integrated Filters", *IEEE Design & Test of Computers*, Vol 19, NO 6, pp. 64-72, 2002.
53. A.K.Lu, G.W.Roberts,and D.Johns, " A high quality analog oscillator using oversampling D/A conversion techniques", *IEEE Trans. on Circuits and Systems II: Analog and Digital Processing*, pp. 437-444, 1994.
54. X.Haurie and G.W.Roberts, "Arbitrary-precision signal generation for bandlimited mixed-signal testing", *Proc. of the IEEE International Test Conference*, pp. 78-86, 1995.
55. M.F.Toner and .W.Roberts, "A BIST scheme for an SNR, gain tracking and frequency response test of a sigma-delta ADC", *Proc. of the VLSI Test Symposium*, pp. 60-65, 1994.
56. G. Huertas, D. Vazquez, A. Rueda and J.L. Huertas. "Effective Oscillation-Based Test for application to a DTMF Filter Bank". *Proc. IEEE International Test Conference*, September 1999, pp. 549-555.
57. Hassan IHS, K. Arabi,B. Kaminska. "Testing Digital to Analog Converters Based on Oscillation-Test Strategy Using Sigma-delta Modulation". *Int. Conference on Computer Design (ICCD 98),* Oct. 1998.
58. M. Vidyasagar, "Nonlinear Systems Analysis", Prentice-Hall, New Jersey 1978.
59. K. Arabi and B. Kaminska, "Testing Analog and Mixed-Signal Integrated Circuits Using Oscillation-Test Method". *IEEE Transactions on Computer-Aided Design of Integrated Circuits and Systems*, vol.16, no. 7, July 1997.

60. M. Santo-Zarnik, F.Novak, S.Macek, "Design of Oscillation-based Test Structures for Active RC Filters". *European Design & Test Conference*, pp 618, 1997.
61. G. Huertas, D. Vazquez, A. Rueda and J.L. Huertas, "Oscillation-based Test Experiments in Filters: A DTMF Example", *Proc. International Mixed-signal Test Workshop*, pp.249-253, Whistler, June 1999.
62. S. Norsworthy, R. Schreier, G. temes: *"Delta-sigma data converters"*, IEEE Press, 1997.
63. D. Vázquez, G. Huertas, A. Rueda and J.L. Huertas: *"DTMF Silicon Demonstrator Report"*, Internal Report of the European Sprit Project ASTERIS.
64. D. Vázquez, G. Huertas, A. Rueda and J.L. Huertas: "A Simple and Secure Start-Up Circuitry for Oscillation-Based-T Application", *Analog Integrated Circuits and Signal Processing*, Kluwer Academic Publishers, Vol 32, pp. 187-190, 2002.
65. J. Roh and J. Abraham. "A Comprehensive TDM Comparator Scheme for Effective Analysis of Oscillation-based Test". *Proc. IEEE VLSI Test Symposium*, April 2000, pp143-148.
66. G. Huertas, D. Vazquez, A. Rueda and J.L. Huertas. "Practical Methods for Reading Test Outcomes in Oscillation-Based Test". *Int. Mixed-signal Test Workshop*, June 2000.
67. D. Vázquez, G. Huertas, G. Leger, A, Rueda and J.L. Huertas, "Practical Solutions for the Application of the Oscillation-Based-Test: Start-Up and On-chip Evaluation", *Proc. of VLSI Test Symposium*, 2002.
68. D. Vázquez, A. Rueda, J. L. Huertas and E. Peralías: "A High-Q Bandpass Fully Differential SC Filter with Enhanced Testability", *IEEE Int. Journal of Solid-State Circuits*, 1998.
69. F.Alberti et al.: "3V Bicmos Analog Front-End for RDS Demodulation System". *Proc. of European Solid-State Circuits Conference*, pp. 112-115, 1994.

Chapter 8

OSCILLATION-BASED TEST STRATEGIES

G. Huertas, G. Leger, D. Vazquez, A. Rueda and J.L. Huertas
IMSE, CNM, CSIC / University of Sevilla

Abstract: This chapter aims to present a structural test methodology using the so-called OBT technique. The conceptual bases of the OBT approach are presented as well as many practical details on its application to practical integrated circuits.

Key words: Structural testing, Oscillation-based test, Test of analog and mixed-signal ICs

1. INTRODUCTION: MOVING FROM FUNCTIONAL TO STRUCTURAL TEST

Analog circuits are difficult to test because there is a wide variety of analog building blocks, their specifications are very broad, and there is a strong dependency of circuit parameters on component variations. For mixed-signal ICs, where analog circuits must coexist with digital components, testing difficulties increase substantially because the access to both analog and digital blocks is severely restricted. The consequences are a reduced fault coverage, a higher test application time and a longer test development time.

In mixed-signal IC's, the most difficult components to test are the analog cores, since analog test is based on checking functional specifications, what can be conflicting when test time has to be kept small, the number of available pins is reduced and full access to input/output core terminals can not be granted. Furthermore, functional test techniques greatly differ depending on the involved analog components and/or their application field,

J.L. Huertas (ed.), Test and Design-for-Testability in Mixed-Signal Integrated Circuits, 259–298.

turning almost impossible to define a general (functional) test methodology applicable to any analog block.

Experience forged from the test of digital circuits encourages researchers to try structural or fault-driven test methods for analog components and explore Built-In Self-Test (BIST) alternatives as well. This has to be done in a manner that increases accessibility to provide core isolation and test resources access, but it might have a high cost in terms of area overhead, power wasting, performance degradation and/or noise and parasitic penalties. But neither moving from functional to structural testing nor incorporating BIST are trivial issues in what analog circuitry is concerned, and are still far from a wide acceptance by the designer community. This acceptance will depend on several factors like compatibility with functional test approaches, test efficiency, test confidentiality and additional design effort.

In order to get an efficient, yet economically feasible test procedure for a complex mixed-signal ASIC, several factors have to be considered:

1.- ***stimuli generation***. A single signal should be preferred, specially a signal that can be easily supplied to a selected input or can be generated on-chip.

2.- ***access to several/many internal nodes*** that can be read either sequentially or in parallel, allowing us to select convenient test points.

3.- ***a single test output*** containing all the information required for performing an interpretation of the test signals. It should be preferred to have such an information coded in digital, since it would reduce the tester requirements.

4.- ***a simple measurement set*** giving enough information about the operational status of the Circuit Under Test (CUT)

5.- ***a system-level strategy*** to decompose the ASIC into meaningful parts that can be tested in a common way.

Among the emerging structural test solutions, the so-called Oscillation-Based Test (OBT) technique is very appealing. It is conceptually simple, does not demand strong circuit modifications during testing and can handle BIST without the penalty of dedicated, additional on-chip signal generation hardware. In broad terms, when OBT is employed no external test stimuli are required, some few simple measurements are used, and can be combined with a multiplexing scheme to probe internal nodes, thus complying with some of the factors above.

In this Chapter, the basic concepts underlying OBT are presented, as well as the principles for applying this test methodology to complex integrated circuits. Our aim is to provide the reader with an overview of the lights and shadows this test technique offers nowadays. Section 2 reviews the existing knowledge on this topic. Section 3 deals with the principles supporting the

OBT concept and a general manner to apply this technique to test a circuit. The OBT methodology can be used for testing from simple blocks (like opamps, filters, converters...) up to complete systems. Although the bases are not so different, the more complex a circuit, the more difficult to apply OBT to test it is. Then, practical strategies for coping with such a difficulty have been developed and are systematized in Section 4. The general OBT methodology is detailed by the means of an example along Section 5. There are some circuits for which the concept has to be applied with specific modifications. This is illustrated for oversampled modulators in Section 6. Finally, Section 7 is draws some conclusions.

2. A BRIEF HISTORICAL GLIMPSE ON OBT

Using built-in-self oscillations to determine faulty behaviors in systems is a relatively new test strategy dating from 1995 [2], when K. Arabi and B. Kaminska established the basic definitions of a test methodology they called "Oscillation-Based Test" or simply OBT. Since then, the OBT concept has been interpreted in different manners and its development can be found in the test literature [1]-[20]. In these references, the OBT approach has been successfully applied to several significant examples ranging from biquad filters to some basic data converter circuits [7]-[20].

Many researchers affirm that the OBT technique can be considered as a general test solution applicable to most basic blocks composing complex analog and mixed-signal circuits (amplifiers, operational amplifiers, comparators, Schmitt triggers, filters, voltage references, oscillators, phase-lock loops, analog-to-digital converters, etc...) or to a combination of these blocks. However, a more detailed study of the existing OBT strategies - examining all their difficulties and shortcomings- points that, until recently, there was not available an absolutely general OBT solution which can be employed for the totality of the analog and mixed-signal systems. This last assertion was based mainly on three facts. First of all, if the key literature is revised [1]-[20], along the nineties only some kinds of circuits are considered (several types of filters, operational amplifiers, and some analog-to-digital converters but not more). Moreover, for every case, the used algorithm and the required OBT mechanism are different and strongly determined by the specific features of the System Under Test (SUT). A second fact is that there was not enough practical evidence regarding the problems of the OBT technique when it is being applied to an embedded complex system. Only [10]-[11] try to be a first attempt to overcome the practical system-level problems. In these works, a whole complex SUT is considered and the difficulties related to the actual implementation of the OBT strategy are studied. And finally, the third fact, and possibly the most important, is that there is not enough experimental support to postulate that the OBT technique guarantees a complete fault coverage for all the examples and that it can be applied in any situation with the exception of those works reported elsewhere [12]. At most, discrete component realizations can be found in the literature to confirm the robustness of a specific OBT strategy [5]. However, following the work reported during the last three years, it seems that the referred three facts are not so insurmountable as they appear to be. Experimental evidence on a relatively complex integrated system, including a practical decomposition scheme, has been demonstrated, opening the door for other design experiments and proving the feasibility of the OBT method.

With the aim of setting this Chapter within a framework, we have reviewed all the main references dealing with the OBT strategy published in the last years. This set of principal references was chosen from a very subjective viewpoint. Obviously, such a set can be expanded and corrected in other contexts. However, the goal here simply is to outline the state-of-the-art of OBT concept which forms the backdrop to this Chapter.

Let us thus distinguish two stages in the evolution of the OBT concept:

-a first stage, *the prior-art of the OBT concept*:

From 1995 to 1997, when the OBT concept emerged [1]-[6]. Basically, in this stage the SUT is converted in a linear sinusoidal oscillator and, therefore, only a single oscillation frequency (whose nominal value is determined by linear mathematical analysis) is measured. Moreover, the only documented results were obtained by simulation or discrete circuit realizations. In this stage, researchers overlooked some important practical considerations, oscillatory behavior is made strongly dependent on the extra elements added for test purposes, and the only test parameter used was the oscillation frequency.

-a second stage, *the state-of-art of OBT concept*:

In subsequent years, some authors [18]-[19] continued with the same philosophy and the same line of reasoning than above. However, a new trend emerges as well. Another authors wonder whether only one test parameter (the oscillation frequency) was or not enough to both identify the fault locations and to obtain a high fault coverage [7]-[14], [20], [61].Therefore, a new concept of OBT arises where the SUT is transformed in a nonlinear oscillator which guarantees robust oscillations providing the control over other test parameters (such as the amplitude of the oscillations). Such new OBT concept achieves to solve with success most of the problems set out in the first stage of the OBT concept.

-a third stage, *the generalized OBT concept:*

The authors have worked out a generalization of OBT addressing its generalization to different mixed-signal circuits. This relies on measuring frequency but may also use other oscillation parameters for more accurate testing. Oscillator robustness is addressed during the design of the SUT, and nonlinear analysis techniques are employed to get an accurate model of the oscillator behavior [7]-[14]. This extended concept, introduced by the authors in these last years, is what will be further developed in this Chapter.

3. OSCILLATION-BASED TEST PRINCIPLES

3.1 OBT technique: the basic concepts

Basically, when Oscillation-Based Test (OBT) techniques are used, the System Under Test (SUT) is converted to a circuit that oscillates. Such an oscillator inherently produces a test signal whose oscillation frequency and/or other oscillation parameters, is/are related to the fault-free structure of the specific SUT in such a way that its/their exact value (in practice a window around the predicted nominal value) can be considered as a test signature. So, a fault in a component can be detected by measuring the oscillation parameters and by checking whether they deviate from the expected nominal values. In other words, the basic idea underlying OBT is to transform the SUT into an (almost) sinusoidal oscillator and to measure the oscillation frequency (or any other given parameter) to alert on potential faults.

Theoretically, converting any analog system into an oscillator requires a mechanism to force a displacement of (at least) a pair of poles either onto the unit circle in the z-plane (when dealing with a discrete-time linear system characterized in the z-domain) or on the s-plane imaginary axis (when a continuous-time linear system is involved). One of the most effective, yet general ways to do it, [7],[21]-[23], [26]-[28] consists in adding a feedback path to the circuit structure and then regulating the feedback elements until obtaining oscillations. This is illustrated by Fig. *8-1* where, at its left, a SUT is represented, and at the right the same system is shown converted in an oscillator.

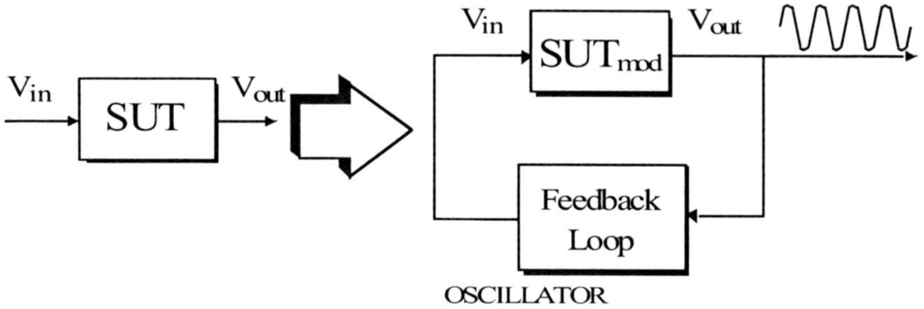

Figure 8-1. Basic idea of the OBT technique: converting the SUT to an oscillator

This method is useful provided that the SUT meets certain requirements and thus the closed-loop system satisfies some conditions to force oscillations. In practice, there are two important requirements to be taken into account. One is that we can effectively force oscillations in the fault-free SUT, and the other is that we can accurately model such oscillations. The former implies the method is applicable, the latter is needed in order to link the oscillator parameters to the SUT behavior (either fault-free or faulty), that is, to obtain a meaningful test output. Regarding the first point, for most of the second-order band-pass filters it can be possible to find a feedback loop that, during the test mode, makes the quality factor of the filter very high (which is equivalent to move its poles to the imaginary axis) at least from a theoretical viewpoint. But many times when any specific SUT structure is examined in detail it can be observed that it is not possible to fulfill the oscillation conditions by only connecting a feedback loop with a given function to the SUT. So, it can also be required to modify the circuit structure by either adding or removing some passive components.

Clearly, in this strategy two modes have to be distinguished: an ***operational mode*** when the system is connected to its regular input and all the additional components required for testing are removed, and a ***test mode*** where the feedback loop is closed around the SUT, its regular input is disconnected, and potentially some circuitry is removed and/or added in the SUT for test purpose. Obviously, adding extra elements to the SUT must be made carefully in order to prevent problems in the operational mode (taking care not to affect critical nodes and/or critical components), and the last option of removing parts of the SUT circuitry must be avoided because faults in those elements would not be covered. So, in a broad sense, the OBT proposal is to achieve oscillators formed by the SUT (or a part of it) plus (or minus) a set of extra components, and a feedback loop also formed by additional components. Obviously, the purpose is to establish the basis of doing that extra circuit more or less fixed and more or less independent on a particular SUT and thus extensible to many kinds of SUTs.

Fig. *8-2* shows a general block diagram corresponding to a system modified to apply the OBT concept. As can be seen from this figure, the only modification affecting the signal path is due to the fact that it is needed to switch between operational and test modes. As was proposed in [7]-[14], during the test mode, besides the feedback loop, some extra elements may be added or removed within the SUT, to produce self-sustained and well-established oscillations.

In theory, this procedure can be used no matter how simple or complex the SUT should be. However, there are some practical limitations that will be addressed later on in this Chapter. In order to get practical results, designers must enface the implementation of the system and the added circuitry as a global design problem. Besides achieving the system's functional specifications, they must strive to build a robust yet precise oscillator that exists around the system when the feedback loop is closed.

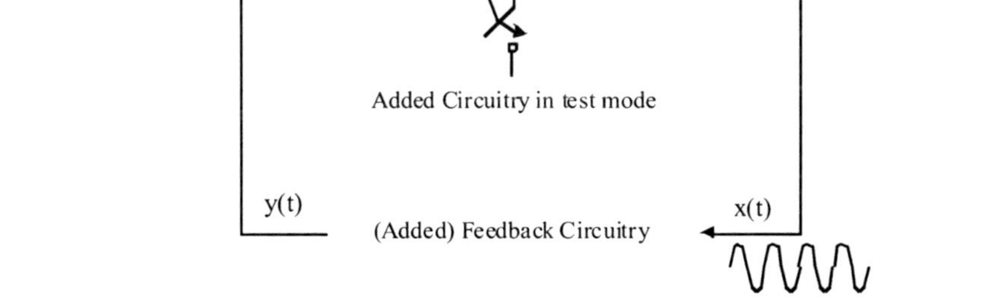

Figure 8-2. All the modifications required to convert any SUT in an oscillator

3.2 Systematic application of the method to complex systems

The basic OBT concept can be applied, in principle, to any linear or nonlinear circuit, no matter how complex it would be, including mixed-signal ICs incorporating digital components. However, although oscillations can be forced in most cases, they have to be modeled to accurately predict their parameters and their values linked to the behavior (both fault-free and faulty) of the original circuit. In practice, the more complex a circuit, the more difficult to establish a correlation between the resulting oscillations and the SUT behavior when working in its regular operation mode. The most meaningful results can be obtained when *H(z)* is a second-order function, as modeling a second-order almost sinusoidal oscillator is usually feasible and its frequency and amplitude are predictable. However, higher orders are difficult to deal with since there are no simple ways to explicitly solve equation the oscillation conditions.

In practice, the usual way to apply OBT [1]-[14] to a complex system is based on splitting the (complex) SUT in simpler functional building blocks which are separately tested. During the test mode, every one of these Blocks Under Test (BUT) is turned into an oscillator. Then, as can be seen in Fig. *8-3*-a, a given system has to be decomposed into as many blocks as necessary to guarantee an efficient implementation of the OBT test procedure, these blocks are named BUT #j to indicate that they will be separately tested. Fig. *8-3*-b shows the OBT approach: each BUT #j must be configured as an oscillator during the test mode.

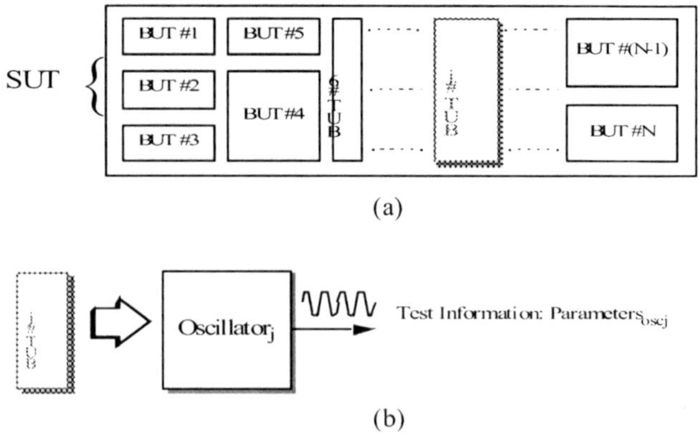

Figure 8-3. Scheme of OBT strategy for a complex system

From Fig. *8-3* it should be clear the actions to be taken in order to apply the OBT concept. These actions are at two different levels, namely, at system and at circuit (or block) levels. At system level, the first issue is to efficiently split the SUT into component BUTs. Second, a mechanism has to be devised to isolate the BUTs in such a way that every one can be "disconnected" from the system signal path and "re-connected" to a feedback loop without mutual interactions. In addition, a method has to be provided as well for reading every BUT test outcome. Finally, the test results for the whole system have to be interpreted from the measurements coming from every oscillator. It should be obvious that the BUT can be made to oscillate either simultaneously or sequentially, this conditions the overall test time for the SUT. Table*8-1* gives a summary of the actions to be undertaken when applying OBT to a system.

At block level, the actions to perform have been stated explicitly in the previous Section. These actions configure ***Step_3*** in Table*8-1*, what are detailed in Table *8-2*.

This test methodology is very appealing to ease the test problem mainly by the reasons already mentioned. First of all, test signals are internally generated, avoiding either specific hardware from a tester or dedicated on-chip resources, thus eluding the problem of generating optimal test vectors.

Indeed, as the complexity of the SUT increases, the problem of producing suitable test vectors in order to guarantee high fault coverage becomes crucial. Therefore, this vector-less test method, which does not require specialized input stimuli generators, represents an interesting improvement.

Apart from that, at least as far as the start-up time is under control, the test time may also be reduced due to the fact that only a limited number of oscillation frequencies has to be evaluated for each BUT. On the other hand, in the OBT strategy, the basic measurements can be translated into the digital domain, making easier the task of test signature evaluation and removing, likewise, most error sources related to the determination of precise voltage or current references.

4. PRACTICAL CONSIDERATIONS TO APPLY OBT

4.1 Block-level application

Self-starting and self-sustained oscillations are required with stable parameters (frequency, amplitude,...) in order to make OBT practical. It obliges to carefully think about a general feedback mechanism valid in any case. That is, how to make any system oscillating independently of its

Table 8-1. Steps to carry out when applying the OBT concept at system level

Sequence	Action
Step_1	Rearranging and Partitioning the SUT into blocks easier to test (into BUTs).
Step_2	Giving support to multiple oscillators. Every BUT must be "disconnected" from the system signal path and "re-connected" to a feedback loop.
Step_3	Transforming each BUT (or a combination of these BUTs) to an oscillator producing an oscillatory output signal whose frequency and amplitude values, f_{osc} and A_{osc}, are related to the BUT internal structure.
Step_4	Measuring the oscillation frequency and amplitude (and/or any other oscillation parameter) for every oscillator.
Step_5	Detecting a fault when the measured parameters for any BUT deviate from their nominal values.

Table 8-2. Steps to carry out when applying the OBT concept at block level

Sequence	Action
Step_3_a	Converting a BUT in a self-starting, robust oscillator
Step_3_b	Relating the BUT specifications to the oscillation parameters by analytical models and/or simulations
Step_3_c	Ensuring that all BUT components are tested

transfer function and (if possible) using a common feedback element. As was discussed somewhere else [7]-[14], that mechanism cannot be linear in practical circuits (because it is not so efficient for obtaining self-sustained oscillations) and must also take into account those non-linearities inherent to the operation (fault-free and faulty) of active components. This problem has been extensively considered in [7]-[14], where a general and practical solution for building up robust oscillators using a nonlinear block within the feedback loop was demonstrated. In that case, self-maintained oscillations can be guaranteed although the self-starting conditions would have to be dealt with and settled separately. This issue (start-up conditions) will be reintroduced later.

As was seen in [10], a symmetrical comparator provides a general, yet simple function to guarantee robust oscillations (at least in the case of filters). In fact, this nonlinear block is equivalent to a 1-bit Analog-to-Digital Converter (ADC) followed by a 1-bit Digital-to-Analog Converter (DAC). This is illustrated for a linear discrete-time circuit in Fig. *8-4*, where at the

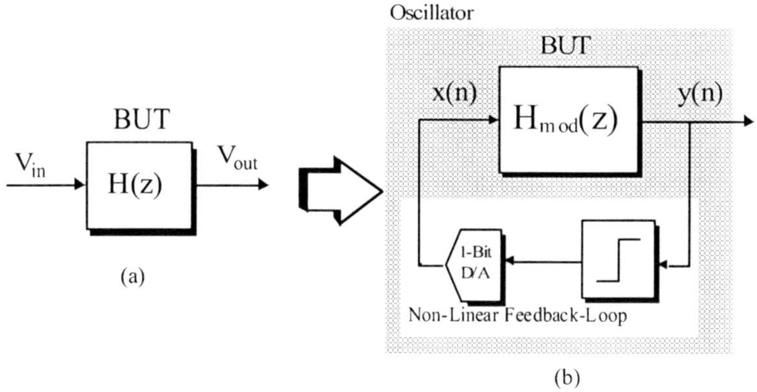

Figure 8-4. Converting the SUT/BUT in an oscillator

left-hand side, a system of transfer function *H(z)* is represented, which is assumed to be the BUT in this example, and at the right-hand side, the same system is shown converted in an oscillator after: a) disconnecting the actual block input from its regular input terminal, b) closing a feedback loop around the system input and output, and c) adding some passive components to the linear block, forming a modified transfer function, $H_{mod}(z)$.

An essential feature of this technique is to guarantee the spectral purity of the oscillator output, *y(n)* in Fig. *8-4*-b, at least to an extent that the fundamental clearly dominates the output signal. Since we are dealing with a nonlinear system, the associated dynamics can be very complex unless we can ensure that high-order harmonics of the basic oscillator frequency are filtered out. Doing this, a secondary advantage is that simplified, quasi-linear analysis methods (like the describing-function approach) can be used to model the oscillator response.

In this case, assuming the BUT be linear, we can determine the oscillation condition for the closed-loop system in Fig. *8-1*. Assuming *N(A)* is the describing-function [21] representing the transfer characteristic of the non-linear block as a function of the amplitude A of its input first harmonic, we can write the oscillation condition as,

$$1 - N(A)H(z) = 0 \tag{1}$$

From the solutions of this equation, we can calculate the frequency and amplitude of the oscillations. In general, *N(A)* may have different forms (depending on the nonlinear element used in the feedback loop), as well as

$H_{mod}(z)$, which is always a rational function but with a variable number of zeroes and poles. The concept can be extended to nonlinear circuits, where the added (embedded in the feedback loop) and the inherent (existing in the SUT) nonlinearities can be combined.

4.2 Start-up problems

As was explained in the previous section the basics of OBT approach is the adaptation of the system circuitry to transform it in an oscillator and then, to measure the oscillation parameters. Therefore, it should be obvious that the OBT feasibility can be seriously compromised if a fast and safe start-up is not ensured.

In fact, two main problems can be clearly highlighted when an OBT scheme is examined:

-First, it could occur that under certain conditions (initial state, offset values...) oscillations may either start very slowly or even not start at all [24]. In this sense, we have to provide a mechanism that guarantees ***safe*** start-up.

-Second, the oscillation parameter measurements have to be performed once the steady-state has been reached. But, since a short test time is a requirement of vital importance in most cases, a safe start-up mechanism is not enough. OBT requires also a start-up strategy which should reduce the transient time as much as possible.

In summary, for most of the cases we have to provide a ***fast*** and ***safe*** start-up strategy which ensures oscillations. Solutions have been considered in the literature, an example has been proposed in [52], [55] which is based on the versatility of the swopamp circuit.

4.3 System-level application

As has been discussed, one main point, in what is concerned when dealing with the application of OBT to a complex system, is to devise a procedure for separating the BUTs from the common signal path and facilitating their conversion into isolated oscillators. This can be done in many ways, though the one to be preferred will depend on the system itself, the extra resources that can be borrowed from other parts of the system, and the test time constraints.

Let us consider a system decomposed in several BUT[1]. Every BUT must be "disconnected" from the system signal path and "connected" to a nonlinear feedback loop. As was said above, depending on the system we have many possibilities. For the sake of illustration, let us consider the case of a cascade of BUTs. When this is the case, we have (at least) two options. One is to convert all BUTs simultaneously into oscillators as it is shown in Fig. *8-5* (parallel test), and another is to convert them sequentially (one after another) as illustrated in Fig. *8-6*. The former is based on closing a dedicated feedback path around every BUT after isolating this BUT from the rest of the circuit, and the latter is based on connecting the same feedback network to every BUT using a multiplexer/demutiplexer pair, provided the feedback network being adequate for all BUTs.

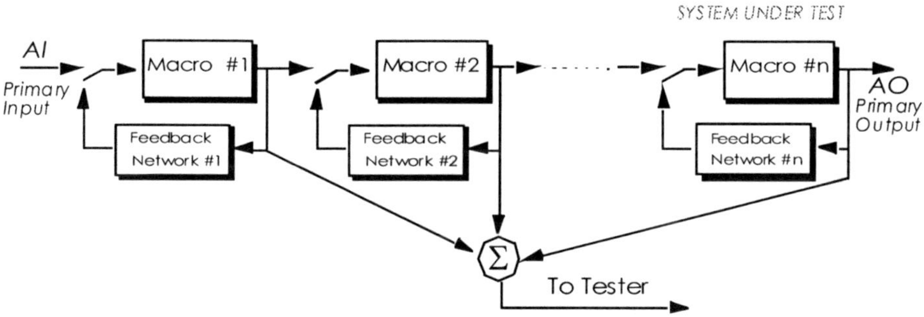

Figure 8-5. Supporting multiple oscillators in a SUT: the parallel case

A main difference between both schemes is in terms of test time. The architecture in Fig. *8-5* is inherently parallel as all the BUT can be forced to oscillate simultaneously. On the other hand, the architecture in Fig. *8-6* is sequential since every BUT is converted in an oscillator one by one.

From these figures, it should be clear that there is a remaining difficulty since now several switches are needed to connect and disconnect all the local feedback loops. Designing such switches is a critical issue since their inclusion may degrade the overall circuit performance. Later in this section we will comment how to minimize this problem by using, for example, the so-called swopamp concept. To decide among the many possibilities for partioning a system and applying OBT, one must consider this potential performance degradation as well as the area/power penalty that every approach has.

[1] Sometimes these blocks are referred as test ***macros***

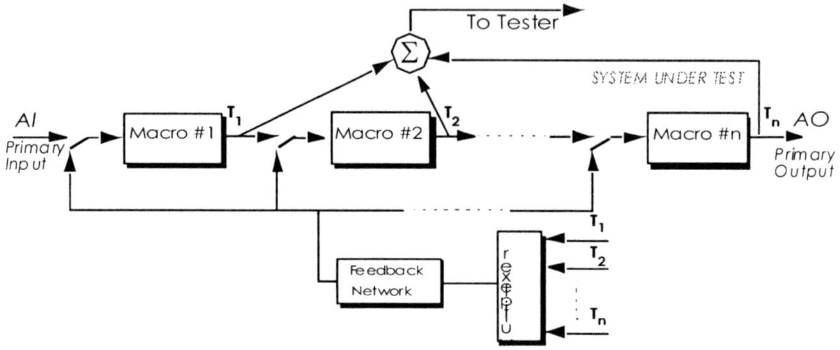

Figure 8-6. Supporting multiple oscillators in a SUT: the sequential case

The schemes of Fig. *8-5* and Fig. *8-6* do not exhaust all the possibilities. A third alternative that is also worth a mention is the one in Fig. *8-7*. There is a single feedback loop that can be shared by all BUTs. Around every BUT a by-pass wire is provided under the control of a 2-input 2-output switch. Whenever the position of the switches is adequately selected, only one stage (let us call it the j-th stage) is configured as an oscillator. The other BUTs does not play any role since there is a path (the external wire) either to pass the feedback signal to the j-th stage input or to pass the oscillator output to the filter primary output, depending respectively on whether they are before or after the block under test in the feedforward path. The control can be made quite simple: when the j-th stage is under test, all the switches, excluding the switch at the input of this particular stage, must enable the by-pass wires.

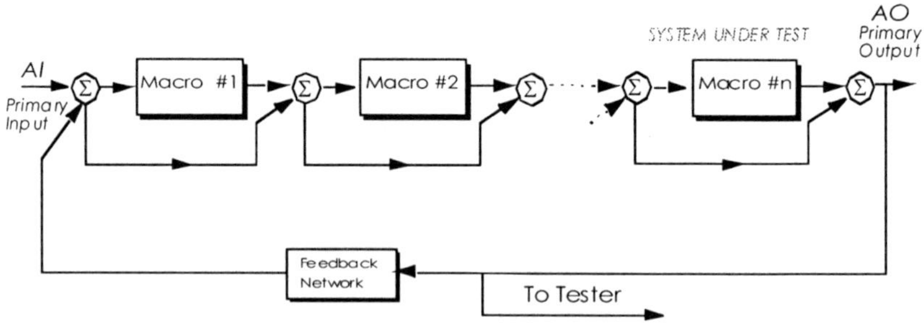

Figure 8-7. Supporting multiple oscillators in a SUT: the sequential by-pass case

5. OBT METHODOLOGY THROUGH AN EXAMPLE

In order to make easier the understanding of the OBT method, an example will be detailed in what follows. The aim is to provide the reader with the flavor of the pros and cons of this test technique.

5.1 A DTMF Receiver

Our example is a Dual-Tone Multi-Frequency (DTMF) receiver [53], [56]. This cell is broadly used in telephony and is gaining importance since it is required in paging systems, repeaters, mobile radio, credit card systems, remote control, personal computers, telephone answering machines, etc. The input of this receiver is a composite audio signal made of the superposition of two tones. The output of the receiver is a digital code which depends on input signals. The DTMF signal is processed first by a band splitting filter which separates the high and low band frequencies as it is shown in Fig. *8-8*. Each frequency is then square-shaped and decoded. The decoder task is to establish whether the present frequencies are recognized as a valid DTMF tone.

Fig. *8-8*-c illustrates the system configuration with the low- and high-group filter paths. The filter bank is formed by a cascade of several second-order functions where many of the typical transfer function types are present (LP, BP, HP, LP-notch, HP-notch). Two comparators make the interface to the digital part, which detects the presence of correct dialing tones and validates a tone accomplishing with the requirements of tone duration. In what follows we will address only the analog subsystem, and we will assume it is implemented using switched-capacitor techniques.

5.2 Step 1: Partitioning

From the structure shown in Fig. *8-8*, it should be clear that splitting the system into its component biquads is a convenient manner to apply OBT in this example. Then, we can decompose the DTMF in up to eight different biquads, corresponding to the components of both filter paths, as is represented in Fig. *8-9*.

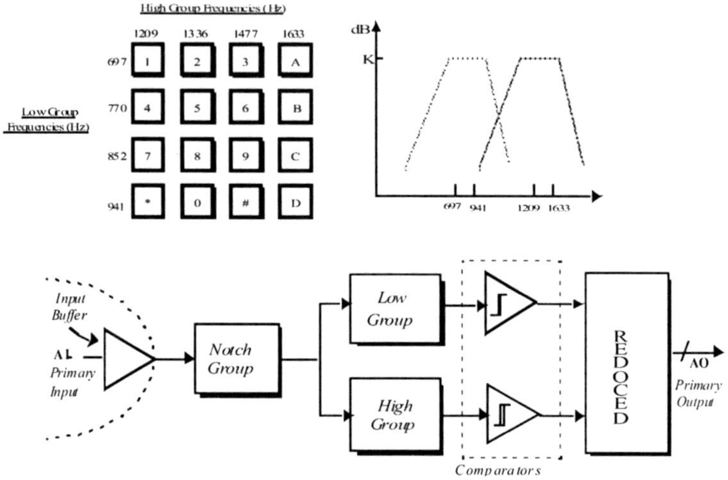

Figure 8-8. DTMF: a) keyboard associated frequencies, b) Band splitting, c) DTMF architecture

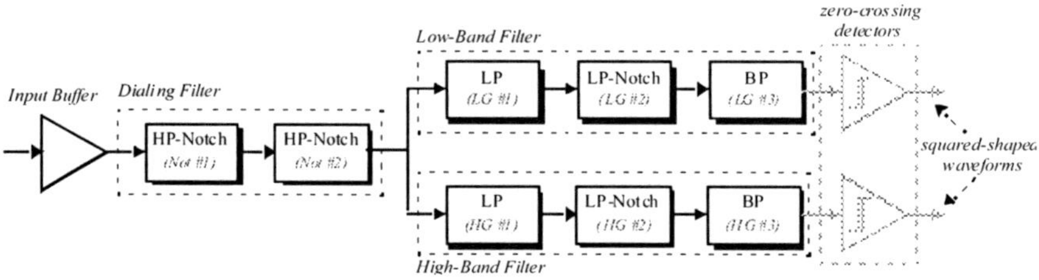

Figure 8-9. DTMF Filter Bank

5.3 Step 2: Giving support to multiple oscillators

Our next step is to provide the overall DTMF with a mechanism to separate every biquad during the test phase, in order to build one oscillator per biquad after closing a local feedback loop around every one and adding the required passive components. Such a mechanism has to separate a biquad from the system signal path and to isolate it from the others.

An efficient manner to isolate every biquad, forcing all the modified biquads to oscillate one after the other was depicted in Fig. *8-7*. As we discussed above, this is just one of the potential strategies we can employ but in this particular system is quite cost efficient.

In practice, steps 1 and 2 can be combined and may take advantage of the particular circuit technique to be used for implementing the SUT. In the case of the DTMF, when the biquads are implemented by relying to a two-opamp structure, the by-passing action in Fig. *8-7* can be implemented by resorting to swopamps [49]-[50] instead of conventional opamps and switches. As seen in Fig. *8-10*, a swopamp is a device able to perform either as a conventional operational amplifier or just as a buffer, controlled by a 2-level command signal.

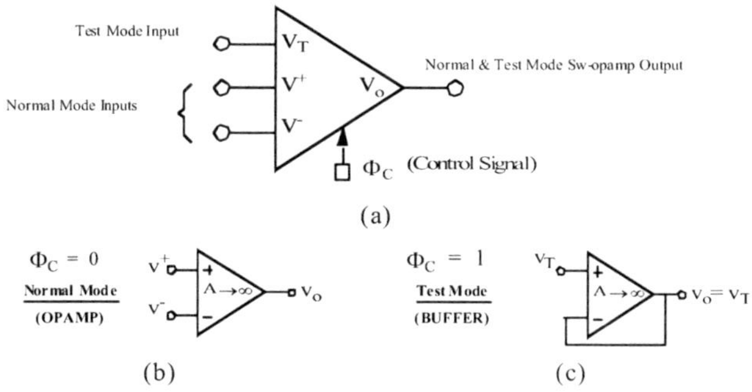

Figure 8-10. Swopamp concept

Then, a practical way to carry out the scheme of Fig. *8-7* is to substitute the opamp providing the output of every biquad by a swopamp. In this manner, a single feedback loop can be used as depicted in Fig. *8-11*. All the swopamps with the exception of the macro under test are programmed as buffers, thus by-passing the feedback signal. On the other hand, the BUT receives such a signal and may oscillate.

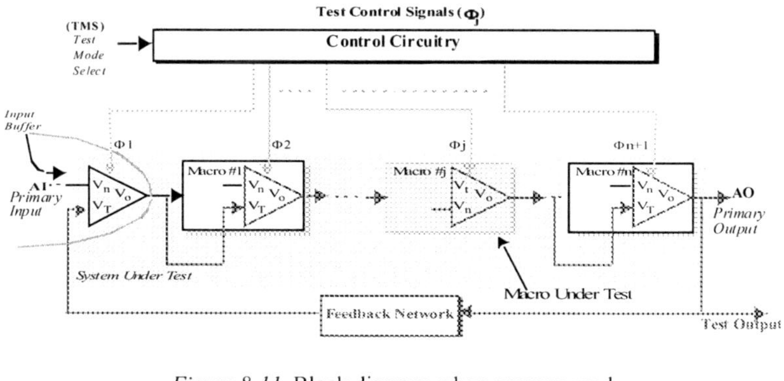

Figure 8-11. Block diagram when opamps used

Keeping in mind the use of swopamps in the DTMF, we can incorporate the structure of Fig. *8-11* to the core demonstrator. Since there are already two paths in this system, we can implement a double oscillation loop as shown in Fig. *8-12*. As can be seen from this figure we propose to use the comparators available in the circuit connected to a 1-bit D/A. Then, testing is achieved by closing the adequate loop. For instance, testing the biquads in the upper (alternatively, in the lower) path is carried out by closing the upper (respectively, in the lower) loop. On the other hand, testing any of the two biquads shared by both paths can be achieved by closing any of the two loops. It is worth noting that this implementation only requires the addition of a DAC in the common part of the feedback loops, plus a (simple) control circuit. In this way, the silicon area overhead is quite small. In terms of the design of those elements present in both the "conventional" and the OBT design, they are required to have a similar performance, what means the design effort is not so much increased.

In order to help understanding the proposed OBT-DTMF structure, in Fig. *8-12* we illustrate how the system may be programmed to test the low-pass notch located in the lower path. A loop is closed using the corresponding comparator and the BUT is left unaltered. All the remaining stages (in the loop) emulate a buffer through their swopamps, in accordance with Fig. *8-11*. Then, the overall closed-loop system (emphasized by thicker lines in this figure) corresponds to the oscillator associated to this biquad. Blocks placed in the upper path are represented by a discontinuous trace since they are not part of the oscillating loop in this case. These upper blocks may be used to read the test signal out from the BUT.

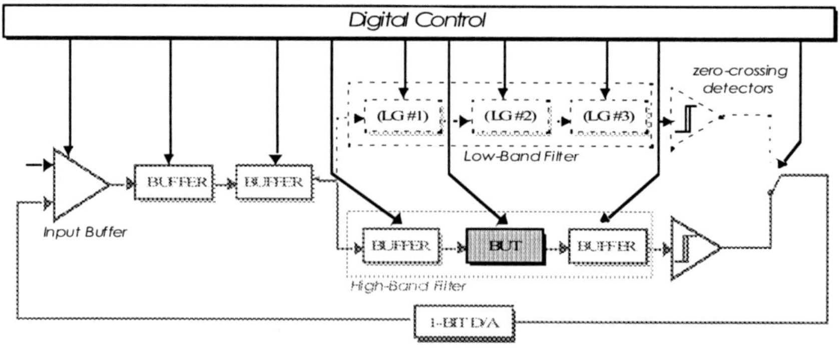

Figure 8-12. Modified DTMF illustrating the test of a biquad

There are practical limitations to this approach. First of all, every buffer in the chain (during test modes) introduces a delay at the test frequency. Nevertheless, for frequencies much lower that the unity-gain frequency, this delay is quite small and the accumulated loop delay can be neglected. If this is not the case, it can be estimated and taken into account.

There is second important problem to consider. We must guarantee that every component, either in the filter or in the feedback network, is tested. It should be clear from Fig. *8-12* and the test operation described above that every biquad is tested once along the test procedure. Additionally, all the swopamps are tested in both modes (operational and testing). In fact, the test path is tested n^{-1} times (being n the number of biquads), which can help in diagnosis, reducing the impact of the extra components on system testability. Only the input terminal and its connection to the signal path must be additionally checked out.

5.4 Step 3: Building oscillators

Consider the feedback structure shown in Fig. *8-4*-b, corresponding to a biquadratic discrete-time filter connected in an oscillatory mode. *H(z)* will be in this case the general biquadratic transfer function in the z-domain, which can be expressed by:

$$H(z) = \frac{x(z)}{y(z)} = \frac{k_2 z^2 + k_1 z + k_0}{z^2 + b_1 z + b_0} \tag{2}$$

where the coefficients are related to the particular circuit implementation of the filter.

As discussed, we use a nonlinear block implementing a saturation function, since then self-starting, self-maintained oscillations can be guaranteed. An advantage of using such a nonlinearity is that a closed-form oscillation condition can be derived, as well as approximations for the frequency and amplitude of the resulting oscillation. Furthermore, since amplitude is known, two measurements are now available for testing purposes. A linearized model for the closed-loop system in Fig. *8-4*-b (formed by a generic biquad and a comparator) can be derived using the describing-function approach [7]-[14]. Poles placement for the system in this figure would be determined by the roots of (1) where *N(A)* can be expressed (for a saturation function) as:

$$N(A) = \frac{2V_{ref}}{\pi A}$$

For a generic second-order section, closed-form expressions can be derived, and then used as design equations for the resulting oscillator. When the coefficients in (2) are taken into account, oscillations can be forced with an amplitude and frequency given by

$$f_{osc} = \frac{1}{2\pi T_s} \cdot \mathrm{acos}\left[\frac{1}{2} \cdot \frac{b_1(k_2-k_0)+k_1(b_0-1)}{k_0-b_0k_2}\right] \qquad A_{osc} = \frac{2V_{ref}}{\pi} \bullet \frac{k_2-k_0}{1-b_0} \tag{3}$$

being T_s the sampling period.

Only the following oscillation conditions have to be fulfilled

$$k_2 \neq k_0$$
$$sign(V_{ref}) = sign(k_2 - k_0) = sign(k_2 b_0 - k_0) \tag{4}$$

From these general expressions, the fault-free oscillation nominal values for any biquad can be determined.

Band-pass transfer functions are easily converted into oscillators just by adding some feedback. When this is not the case, it is still feasible to force oscillations without altering the filter transfer function if the biquad has (at least) two opamps. There exist structures that can simultaneously provide any transfer function for one output, and a transfer function for the other output capable to generate oscillations as indicated in Fig. *8-13* (see for example [10]). In particular, this is a property fulfilled by the biquad structure of Fig. *8-14*, which is a modified version of the Fleisher-Laker universal biquad [25][2]. It is an integrator-based second order universal biquadratic section, with two available outputs (one per integrator). It is possible to implement a given function taking V_{o1} or V_{o2} as the output. Then, either the normal output or a secondary output (originally an "internal" node) can be used to force oscillations [7]-[12]. This biquad can be customized to operate as required for any filter stage by adequately sizing the existing capacitors.

[2] This filter structure has been used to design every biquad from the DTMF case.

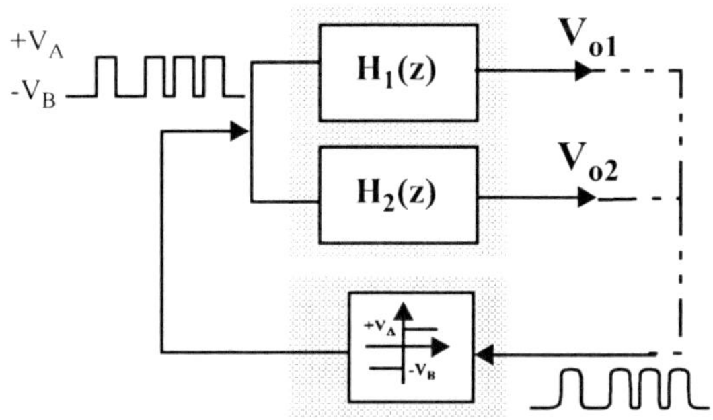

Figure 8-13. Amplitude-Controlled Oscillator

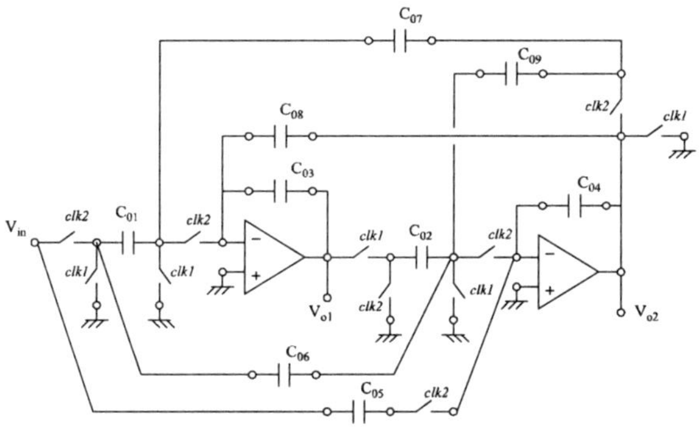

Figure 8-14. Switched-capacitor Biquadratic Filter

The coefficients in both the numerator and the denominator of (2) are related to the normalized capacitors in Fig. *8-14* as given in Table*8-3*.

In general, not all capacitors are present in the expression of the frequency in (3) which suggests that faults in such capacitors do not explicitly show up when only frequency is measured. On the other hand, it

Table 8-3. Biquad coefficients

$H_{01}(z) = \frac{V_{o1}}{Vin} = \frac{k_{21}z^2 + k_{11}z + k_{01}}{2 \ldots}$	$H_{02}(z) = \frac{V_{o2}}{Vin} = \frac{k_{22}z^2 + k_{12}z + k_{02}}{2 \ldots}$
$k_{21} = -C_1 + \frac{(C_5 + C_{56}) \bullet (C_7 + C_8)}{1 + C_9}$	$k_{22} = \frac{-(C_5 + C_{56})}{1 + C_9}$
$k_{11} = \frac{C_1 - C_8 \bullet (C_5 + 2C_{56}) - C_{56}C_7}{1 + C_9}$	$k_{12} = \frac{C_5 + 2C_{56} - C_1 C_2}{1 + C_9}$
$k_{01} = \frac{C_{56}C_8}{1 + C_9}$	$k_{02} = \frac{-C_{56}}{1 + C_9}$
$b_1 = \frac{-2 - C_9 + C_2 \bullet (C_7 + C_8)}{1 + C_9}$	$b_0 = \frac{1 - C_2 C_8}{1 + C_9}$

can be proved that, when both amplitude and frequency are simultaneously considered, all capacitors have an influence on this equation system. Since this is not a general property for every biquad, it is a key issue to develop Design-for-Testability (DfT) guidelines for selecting a particular biquad structure for filters.

For the sake of clarity, (3)-(4) were derived assuming no delay in the feedback path. However, usually this is not the case, and such extra delays must be taken into account in (3) to derive more accurate values of amplitude and frequency.

5.5 Step 4: Measuring and interpreting oscillation parameters

Several alternatives can be used to perform the measurements required for evaluating the oscillation parameters. Obviously, conventional peak detector circuits can be employed to measure the oscillation amplitude [65]. Instead, we can resort to more complex solutions, like the TDM comparator scheme reported in [62], which can be used as a signature analyzer of oscillation signals. However, we prefer the evaluation and interpretation scheme in Fig. *8-15*, first presented in [54], [57]-[60], which makes use of a first-order sigma-delta modulator and a few simple counters. This modulator encodes the oscillation waveform into a digital bitstream, and the non-linear feedback loop around the SUT provides a digital signal giving its sign, q(n).

The Evaluation Unit in Fig. *8-15* consists of three counters which provide a coded digital version of the frequency, the amplitude and the DC level of the oscillatory signal. The frequency counter measures the oscillation frequency by counting the number of system clock periods between two positive (alternatively negative) edges of the square signal, q(n). Then, the state of the counter at the end of each period of q(n) can be formally expressed as,

$$\left\lceil \frac{T_{osc}}{T_s} - 1 \right\rceil \le count\text{-}freq = N \le \left\lfloor \frac{T_{osc}}{T_s} + 1 \right\rfloor \tag{5}$$

being T_{osc} the oscillation period and T_s the sampling period of the BUT.

The accuracy is obviously determined by the oversampling ratio (T_{osc}/T_s) For example, for an oversampling ratio as low as 30, the accuracy of the frequency measurement is of 3% approximately. This precision is large enough for industry [52]-[55], where a typical required accuracy can be around 10%.

The DC level counter and the amplitude counter principle of operation is based on synchronous integration. The square wave q(n) giving the sign of the oscillation waveform is used to trigger the counters which perform the sum (integration) of the sigma-delta bitstream. A "1" increments the counter and a "0" decrements the counter. For the DC level counter this operation is realized over one period, defined by two positive edges of q(n). The authors in [54], [57]-[60] show that the counter output at the end of an evaluation period is,

$$countDC = BN \pm 2 \tag{6}$$

Where B is the DC level value, with respect to the modulator full-scale, and N is the oversampling ratio (T_{osc}/T_s) as defined before.

The amplitude counter performs the same operation, but the sigma-delta bitstream is inverted on the negative lobe of the signal. That is, the square wave triggers the counter on a positive edge of q(n) and inverts the bitstream on a negative edge of q(n). This is conceptually equivalent to integrate the rectified waveform over one period. The counter output at the end of an evaluation period is,

$$coutamp = \frac{2}{\pi}AN \pm 4 \tag{7}$$

Where A is the waveform amplitude with respect to the modulator full-scale and N is the oversampling ratio.

The counter outputs are thus directly proportional to the oscillation parameters with a precision that is mainly driven by the oversampling ratio. It has been shown in [57]-[60] that it is possible to improve the precision by a factor of two at a very low cost by averaging out the bitstream values around the zero crossing instants defined by q(n). Then, the called Interpretation Unit would only consists in a digital comparison with a nominal window stored on-chip. More details about this strategy are given in [54], [57]-[60].

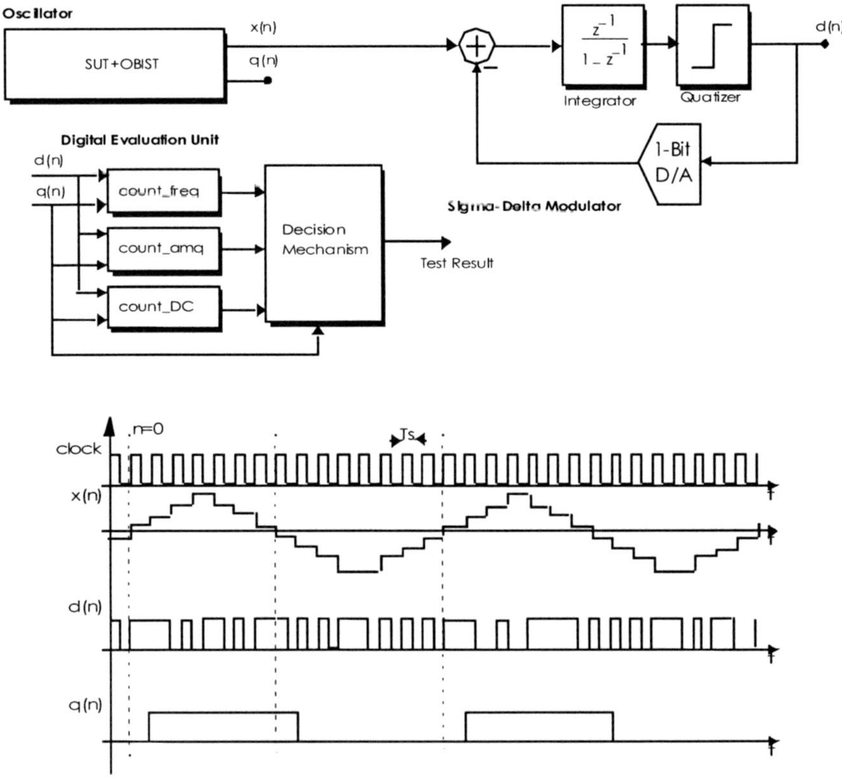

Figure 8-15. On-chip evaluation scheme

5.6 Modelling Accuracy

An important issue when applying OBT is validating the accuracy of the oscillation parameters as predicted by the describing-function approach. To do this we have simulated every section of the overall filter bank forming the DTMF, by using SIMULINK [36]; then, it makes sense to compare the value of the oscillation parameters given by this simulator to those predicted by the describing-function approach as done in [10]. There is a good agreement between the theoretical and simulation results for both amplitude and frequency, which are within a 10% bound as shown in Table*8-4*.

Table 8-4. Comparison between oscillation parameters given by SIMULINK and by the DF approach.

		DF approach		SIMULINK		Error (%)	
Group	Biquad	f_{osc} (Hz)	A_{osc}(Vp)	f_{osc} (Hz)	A_{osc}(Vp)	f_{osc}	A_{osc}
Dialing Filter	Not. #1	503.66	6.545	503.66	6.523	0.00	0.34
	Not. #2	666.34	1.378	629.62	1.532	5.51	9.99
Low-Band Filter	LG # 1	816.19	1.272	796.51	1.276	2.41	0.34
	LG # 2	975.68	1.927	967.51	1.926	0.84	0.06
	LG # 3	616.89	5.064	603.90	5.072	2.11	0.12
High-Band Filter	HG # 1	1363.12	1.306	1345.49	1.310	1.29	0.31
	HG # 2	1188.59	1.526	1181.89	1.533	0.56	0.45
	HG # 3	1728.93	7.130	1710.38	7.23	1.07	1.36

It can be argued that these quantitative results apply to the particular demonstrator we are targeting in. However, although approximate, the describing-function method gives usually a reasonable fitting. In any case, this error is not an important problem since the derived closed-form formulae are good enough to be a starting point for simulation and OBT design, thus alleviating the computational cost of using a device-level simulator. Therefore, generally speaking, the proposed OBT closed-loop systems can be analyzed by using the describing-function approach, where the nonlinear block is replaced by an "equivalent" linear block.

5.7 Fault analysis

The goal of the fault analysis [37],[39]-[40] is to qualify the use of OBT and study how faults translate into changes in either the frequency or the amplitude of the oscillations (or even in both). In order to determine if the OBT approach is a feasible test technique for achieving a high fault coverage one needs a previous theoretical fault analysis where three main steps must be worked out:

1.- Evaluating the changes which must be made in the original BUT structure in order to implement the oscillator architecture.

2.- Fixing the values of the test extra elements in order to satisfy not only the oscillation conditions but also to accurately predict the nominal value of the oscillation parameters by the linearized model (describing-function approach).

3.-Checking by analytical equations and/or simulations how faults (or deviations) in all the elements both those from the BUT structure and those included for test, affect the oscillation parameters nominal value.

5.8 Test quality

On the other hand, an estimation of the overall test quality of the OBT technique is required to support its use. In terms of fault coverage, there is a lack of a widely accepted criterion. Alternatively, we will try to give an assessment by injecting meaningful faults into every biquad of Fig. *8-9* and proving that all of these faults can be detected. This assessment will be complemented by showing that a fault-free biquad is not graded as faulty by using this approach. These two issues will be discussed at simulation level. However, in addition, we will also report experimental results obtained in the laboratory of a silicon prototype corresponding to the use of OBT-OBIST when applied to the DTMF of Fig. *8-9*.

As a first step, an acceptability region has to be defined as the place where all the good circuits must lie. During the design process of an analog IC, designers use the worst-case processing parameters given by the silicon foundry, to find a robust solution which optimizes production yield. If a filter is well-designed and correctly centered in the technology, the influence of worst-case electrical parameters should be minimized. On the other hand, capacitor mismatchings may have an important influence as well. In practice, centering a design is equivalent to defining a region in the design space which must be within the test acceptability region for the application specifications. For OBT we need to map such an inner zone into our test space, where the coordinates are deviations in the oscillation parameters with respect to the nominal ones. To determine (in the DMTF example) where the fault-free circuits are within this test space, a Monte Carlo analysis was performed for every biquad considering process parameter variations and a 0.2% mismatching in capacitor ratios. The black box in the lower-left corner of the plot in Fig. *8-16* shows a typical situation corresponding to one of the biquads. Although Monte Carlo's predictions are always pessimistic, all fault-free circuits exhibit a rather small deviation in the oscillation parameters. The results for all of the biquads are shown in Table*8-5*. It can be seen from this Table that the frequency is very stable in all the cases, while the amplitude varies slightly more. These values will be used as the minimum tolerance margin allowed to decide whether a fault may or may not be detected, and to define an inner region within the acceptability region of the SUT.

Table 8-5. Maximum parameter deviations for a 0.2% of mismatch in capacitor ratios and process parameter variations.

(%)	NOT #1	NOT #2	LG #1	LG #2	LG #3	HG #1	HG #2	HG #3
ΔA_o	4.7	4.3	1.7	3.9	1.9	2.2	3.4	4.4
Δf_{osc}	2.5	0.7	0.8	0.8	0.9	1.1	0.9	1.2

For all of the biquads in the above Table, we undertook an exhaustive validation of the OBT technique. We have considered both small parametric changes which can potentially take the circuit out of specs and catastrophic faults causing great changes or even change the circuit topologies. Fault simulations have been carried out with SWITTEST [40]-[47] for every biquad. Shorts and opens for switches and capacitors have been considered, as well as the influence of a permanent stuck-at ON of switches. Additionally, deviations from 5% up to 50% in capacitor values have also been injected. Therefore, both hard and soft faults have been considered in order to cover a wide range of possibilities. From the simulator results, both the frequency and amplitude of the first harmonic of the oscillations have been measured. The results corresponding to one of the biquads are shown in Fig. *8-16*, where only those faults causing a deviation below 50% have been displayed.

The results depicted in Fig. *8-16* corresponds to one of the biquads in the DTMF. In this Figure, the design data gathered through extensive Montecarlo simulations and the test empirical results are combined to show the robustness of the OBT approach. We use the prototype of the DTMF receiver designed in a 0.6μm double-poly technology shown in Fig. *8-17*. In Fig. *8-16*, a tolerance window has been drawn in gray illustrating how the acceptability region can be found. The selected window allows us to discriminate all faulty circuits in this example.

In essence, Fig. *8-16*, depicts a typical case for one of the filter biquadratic sections representing the design acceptability region and several faulty situations in the oscillator parameter space, and also data corresponding to five tested samples for one biquad. From this Figure it should be clear that *a priori* simulation predictions give a good estimate for the experimental oscillation values.

Similar results have been obtained for the remaining biquads. Although every biquad presents different plots, the one shown in Fig. *8-16* is representative of all of them. The main conclusion we can draw is that when both frequency and amplitude are considered, most of the injected faults are easily detected. Even in the most pessimistic situation an accuracy of around 5% in frequency and 10% in amplitude was required for separating good from bad circuits in any of the biquads forming the DTMF. When this is the case, the percentage of detected faults (from those injected by simulation) ranges from 98.3 to 100 % (depending on the biquad).

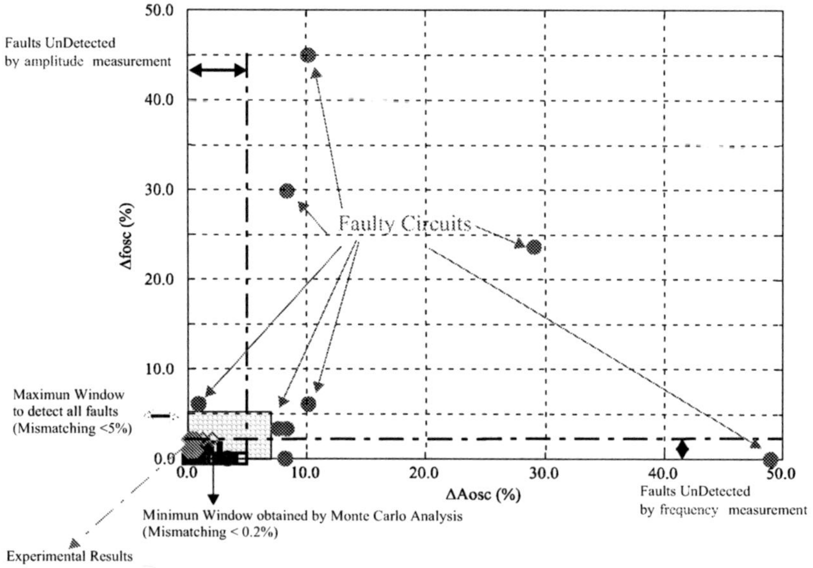

Figure 8-16. Typical frequency vs amplitude plot

Figure 8-17. DMTF Microphotograph

6. SYSTEMS WHERE THE BASIC HYPOTHESIS IS VIOLATED: THE CASE OF ΣΔ MODULATORS

The oscillation-based test concept (OBT) is specially tailored to active filters that can be decomposed into second-order sections [7]-[11], [12], [16]-[19]. Band-pass transfer functions are easily converted into oscillators just adding some feedback. This filter action eliminates higher frequency components and makes easier the oscillator design. When this is not the case, we have discussed above that it is still feasible to force oscillations in most cases. Additionally, the describing-function method can be used giving quite nice results for predicting the oscillator behavior.

Unfortunately, there exist circuits for which happens exactly the opposite. The filter action is not enough to sustain a first-order linear model. ΣΔ modulators represent a typical example of these circuits [34], [42]-[44]. These modulators are usually formed by a linear part (exercising some filtering) and a comparator. What is interesting in this case is that the nonlinear block is already available and connected in the SUT feedback loop, suggesting us to use the modulator structure without any extra nonlinear component [13]-[14]. However, the linear part is not capable of filtering higher-order tones out, giving rise to a more complex oscillation pattern. In what follows, we try to clarify this problem and propose a method for applying OBT to ΣΔ modulators. Although the method can be used even for first-order modulators, it is particularly intended for second-order and higher.

6.1 Basic approach in discrete-time ΣΔ modulators

A typical low-pass discrete-time second-order modulator is shown in solid lines in Fig. *8-18*. Parameters δ_0 and δ_1 are chosen to optimize the modulator operation without endangering it because of unwanted oscillations. To force oscillations, we add another feedback loop to modify the action of δ_0 and δ_1. This loop (in practice, a modification of the existing one) is intended for converting the linear block in a band-pass filter with a relatively large value for Q. The technique we first proposed in [13] is based in adding another feedback loop to force system oscillations in an almost sinusoidal regime.

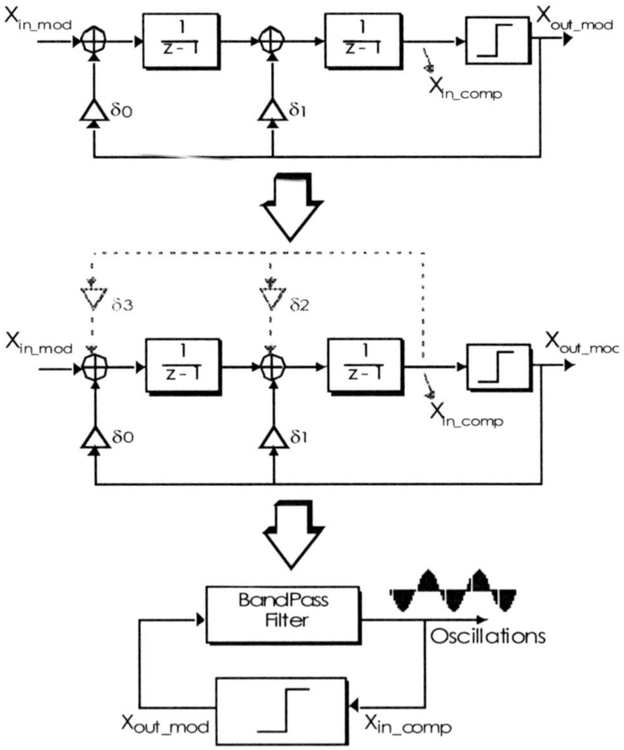

Figure 8-18. Closed-loop structure to force oscillations in a low-pass ΣΔ modulator

However, in the case of a modulator, the OBT method has tighter constraints than those in a filter. In essence, the difficulty relies in that higher-order harmonics can not be disregarded so simply, giving rise to nonlinear effects that must be taken into account. As discussed in [1]-[19], a key issue for applying OBT is the ability to accurately predict the oscillation parameters as well as the impact of any change in a given component in terms of these parameters. Unfortunately, there can be significant disparities between what is predicted by the linearized model and what is actually observed. To illustrate this point, let us modify the modulator in the top of Fig. *8-18* by adding some extra feedback as represented in the bottom of this Figure by dotted lines. Fig. *8-19* depicts both the oscillation frequency predicted by the DF and an accurate (nonlinear) simulation of the oscillatory behavior of this modulator for a particular feedback condition. As can be seen from this figure, there are discrete frequency variations instead of a continuous evolution as predicted by the describing function [13]. In addition, for values of $\delta_2 > -2$, chaotic behaviors can be observed that are not predicted by the linearized model. Although this effect can be minimized by changing δ_2, its existence is unavoidable since there is an intrinsic modeling inaccuracy due to a violation of the basic assumptions for the validity of the DF method. This is due to an incomplete filtering action that does not eliminate higher-order tones.

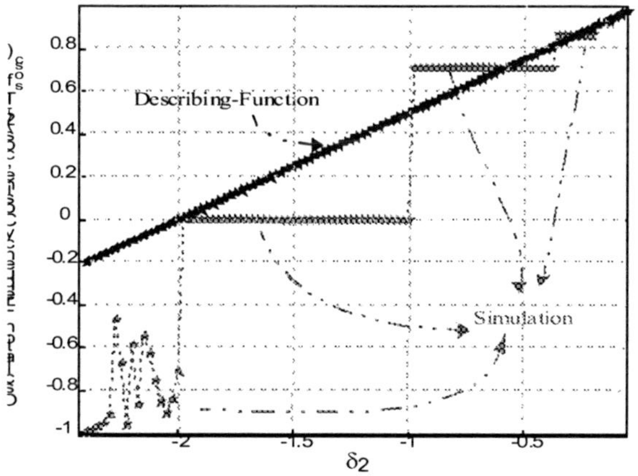

Figure 8-19. Model comparison

In practice, assuming negligible all other harmonics than the fundamental is an important error source since in conventional ΣΔ modulators the open-loop transfer function of its linear part does not filter out most harmonics. Instead, nonlinear oscillation modes can be easily excited as has been extensively proven in the literature [44], [63]-[64]. It is true that these nonlinear modes are usually more robust than almost-linear ones, but their robustness lead to an insensitivity against some faults that may not be simple to investigate.

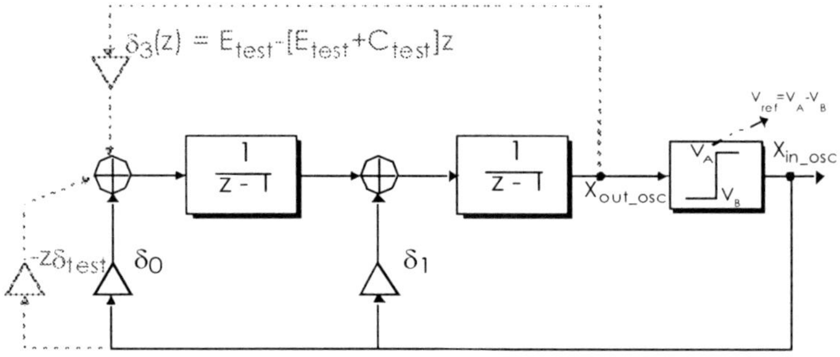

Figure 8-20. A convenient scheme for forcing almost sinusoidal oscillations in a low-pass ΣΔ modulator

To cope with this problem we can change the feedback path in order to provide the resulting linear transfer function with adequate properties to perform a filter action upon the undesired tones. Let us represent the oscillator as drawn in Fig. *8-20*, where the solid lines give an equivalent representation for a conventional low-pass modulator. Then, dashed lines correspond to extra feedback to force oscillations without spurious harmonics. The feedback path has been split in two and the result is a modified functionality for the linear component of this system, which allows increasing the filtering actions and thus prevent nonlinear oscillation modes as much as possible. Notice, moreover, that the nonlinear components required for the modulator normal operation have been re-used and only some linear extra elements have been added but no circuitry has been removed.

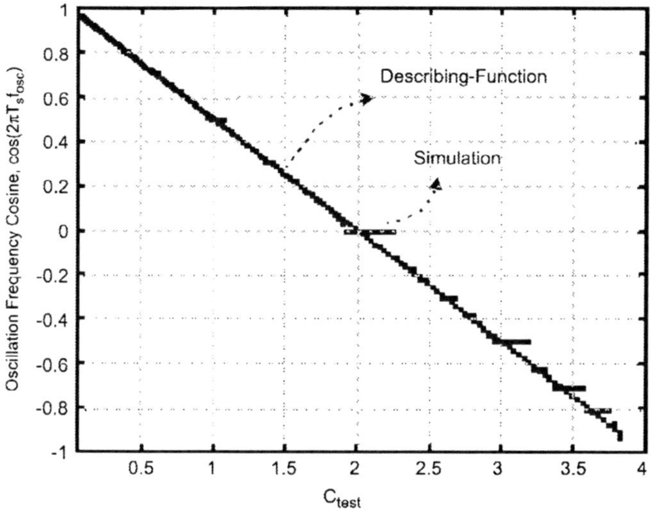

Figure 8-21. Model comparison for the circuit modifications in Fig. *8-20*

Comparing the results predicted in this case for the linearized model versus those obtained from a nonlinear simulation, much more consistent data are obtained, as drawn in Fig. *8-21*, where the oscillation frequency is represented as a function of parameter C_{test}, for values fulfilling the conditions in Table*8-6*. As has been shown elsewhere [14], this method can be extended to other modulators, no matter if they are low-pass or band-pass. The oscillator scheme would be as shown in Fig. *8-22* where R is 0 if we are representing a low-pass modulator and R is not equal to 0 if we are representing a band-pass modulator. Then, the oscillations are characterized by Table*8-6*.

Table 8-6. Oscillator features

Oscillation Parameters	Oscillation Conditions
$A_{osc} = \frac{2V_{ref}}{\Pi} \cdot \left(\frac{\delta_1 - \delta_0}{E_{test}}\right)$	$\delta_1 - \delta_0 \neq 0$ $sign(V_{ref}) = sign(\delta_1 - \delta_0)$
$f_{osc} = \frac{1}{2\pi T_s} \cdot \mathrm{acos}\left[\frac{2 + R - C_{test}}{2}\right]$	$-1 < \frac{2 + R - C_{test}}{2} < 1$

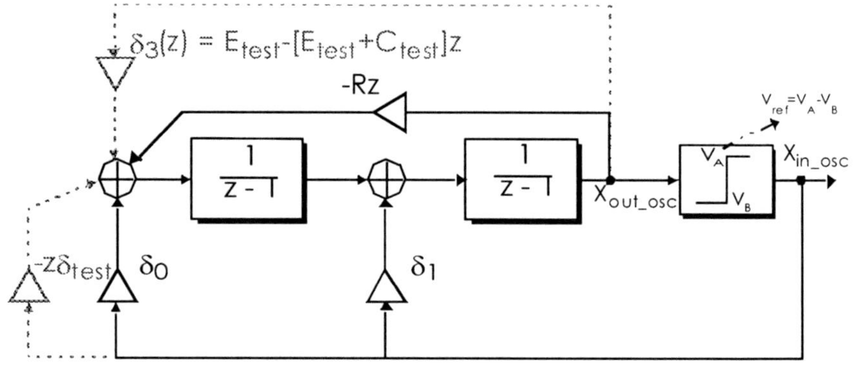

Figure 8-22. Oscillator scheme for a generic modulator

6.2 Fault Analysis

Two kinds of experiments have to be done in order to validate these results on modulators. First of all, the accuracy of the describing-function model has to be proven. Second, the quality of the OBT technique has to be demonstrated through fault simulation experiments.

To address the first issue, let us go back to Fig. *8-21*. From this Figure we can see that a judicious election of C_{test} may lead to operate in an oscillation region where the linearized model is quite accurate and the OBT methodology can be used as explained above [13].

Regarding the second point, the problem is whether or not the effect of most faults translates into a change in either the frequency or the amplitude of the oscillator likewise it happens in filters. Additional work remains to be done yet to qualify the use of OBT in the case of modulators, however it can be confirmed that an adequate election of the extra parameters (i.e., those not present in the original modulator structure) provides enough freedom to force oscillations which can be worthwhile for testing purposes.

To illustrate how faults are modifying the oscillation parameters and how such a modification depends on the selected coefficients, let us consider the case of a variation in the integrator gains. If we fix the value of those parameters imposed by the modulator design, we have to study how to choose the value of the remaining parameters (E_{test} and C_{test}) in order to obtain a high fault coverage. It can be proved that a 10% change in the first or second integrator gain will affect both the oscillation amplitude and the oscillation frequency as well. Fig. *8-23* depicts the movement of the normalized amplitude curves for this variation, and Fig. *8-24* shows the corresponding curves for the normalized frequency. In both figures, the central curves correspond to the nominal case and the predicted changes depend on the value selected for E_{test} and C_{test}.

Depending on the parameter, either amplitude or frequency can be preferential for test. For example, if one of the modulator coefficients, δ_0 or δ_1 is varied producing a ±10 % change in its nominal value, a variation is noticed in the amplitude but never in the frequency (see Table*8-6*). Examples where only the frequency is changed can be given; both situations are just limit cases for the OBT technique [13]-[14].

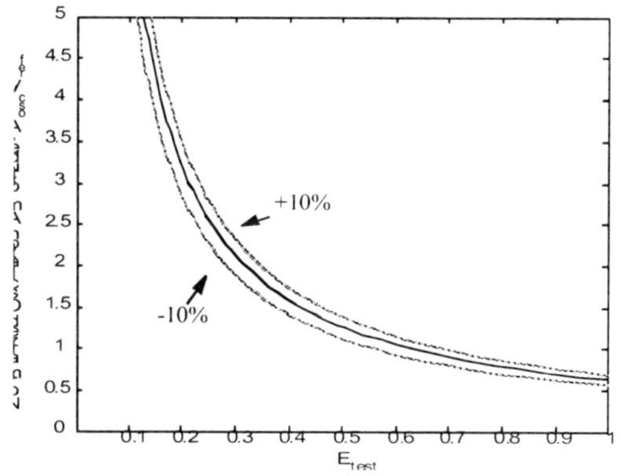

Figure 8-23. Changes in the amplitude due to a 10% change in an integrator gain

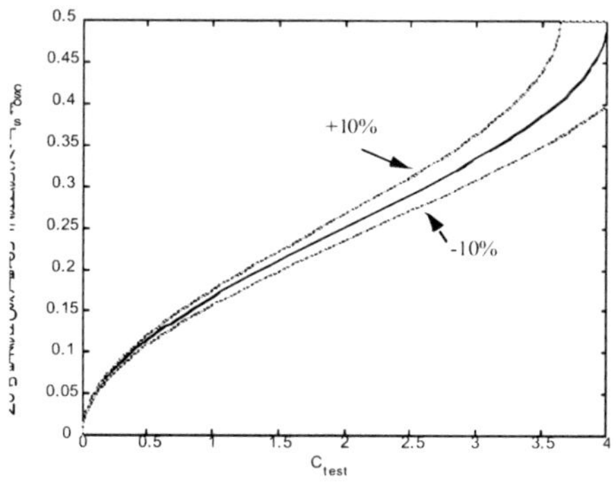

Figure 8-24. Changes in the frequency due to a 10% change in an integrator gain

7. CONCLUSIONS

In this chapter we have reviewed the so-called OBT approach. This is a test technique which falls into a structural test category, and relies in the ability to convert the SUT in an (almost) sinusoidal oscillator whose amplitude and frequency can be accurately predicted by quasi-linear methods. The steps to apply OBT have been presented as the bases for a pragmatical approach to apply Oscillation-Based Test to Mixed-signal integrated circuits. A rather general, yet simple to implement, mechanism has been proposed to guarantee oscillations in the SUT. Such a mechanism has to be nonlinear in order to force robust oscillations. Furthermore, because the resulting oscillator is quite predictable in terms of both frequency and amplitude, this makes feasible to use these two parameters to obtain very high fault coverage when typical faults are considered. This approach has been illustrated by practical examples, including a complete integrated DTMF. In addition, considerations have been made to cope with circuits for which the constraints required by the DF method have to be relaxed.

ACKNOWLEDGEMENT: This work was partially supported by the European ESPRIT Project ASTERIS: *Advanced Solutions in Test Engineering Research for next generation Integrated Systems*. ESPRIT-CEE, N° 26354-1293/AA.

8. REFERENCES

[1] K. Arabi and B. Kaminska. *"Oscillation-Based Test Strategy (OBTS) for Analog and Mixed-Signal Circuits"*. US Patent Application, 1995.

[2] K. Arabi and B. Kaminska."Oscillation-Test Strategy for Analog and Mixed-Signal Integrated Circuits". *Proc. 14th VLSI Test Symposium*, 1996, pp. 476-482.

[3] K. Arabi and B. Kaminska."Design for Testability of Integrated Operational Amplifiers Using Oscillation-Test Strategy". *Proc. International Conference on Computer Design*, Austin, October 7-9, 1996, pp. 40-45.

[4] K. Arabi, B. Kaminska and S. Sunter, "Testing integrated operational amplifier based on oscillation-test method". *Proc. International Mixed-signal Test Workshop*, Canada, May 15-18, 1996, pp. 221-226.

[5] K. Arabi and B. Kaminska. "Testing Analog and Mixed-Signal Integrated Circuits Using Oscillation-Test Method". *IEEE Transactions on Computer-Aided Design of Integrated Circuits and Systems*, vol. 16, no. 7, July 1997.

[6] K. Arabi and B. Kaminska. "Oscillation Built-In Self Test (OBIST) Scheme for Functional and Structural Testing of Analog and Mixed-Signal Integrated Circuits".*Proc. IEEE International Test Conference*, Nov. 1-6, 1997, pp. 578-586 .

[7] G. Huertas, D. Vazquez, A. Rueda and J.L. Huertas, "Oscillation-based Test Experiments in Filters: A DTMF Example", *Proc. International Mixed-signal Test Workshop*, Whistler, June 1999, pp. 249-253.
[8] G. Huertas, D. Vazquez, A. Rueda and J.L. Huertas. "Effective Oscillation-Based Test for application to a DTMF Filter Bank". *Proc. IEEE International Test Conference*, Atlantic City, USA, September 28-30, 1999, pp. 549-555.
[9] G. Huertas, D. Vazquez, E. Peralias, A. Rueda and J.L. Huertas. "Testing Mixed-Signal Cores". *Proc. 13th Symposium on Integrated Circuits and Systems Design*, Sept. 18-24, 2000, pp. 307-312.
[10] G. Huertas, D. Vazquez, A. Rueda, J.L. Huertas."Practical Oscillation-Based Test of Integrated Filters". *IEEE Design and Test of Computers,* November-December 2002, pp. 64-72.
[11] G. Huertas, D. Vazquez, A. Rueda, J.L. Huertas."Testing Mixed-Signal Cores: Practical Oscillation-Based Test in an Analog Macrocell". *IEEE Design and Test of Computers,* November-December 2002, pp. 73-82.
[12] G. Huertas, D. Vazquez, E. Peralias, A. Rueda and J.L. Huertas."Practical Oscillation-Based Test in Analog Filters: Experimental Results". *Proc. International Workshop on Electronic Design, Test, and Applications (DELTA 2002)*, Christchurch, New Zealand, January 29-31, 2002, pp.18-24.
[13] G. Huertas, D. Vazquez, E. Peralias, A. Rueda, J.L. Huertas. "Oscillation-based Test in Oversampling A/D Converters", *Microelectronics Journal, Elsevier,* 2002, vol. 33, no. 10, pp. 799-806.
[14] G. Huertas, D. Vazquez, A. Rueda, J.L. Huertas."Oscillation-based Test in Oversampling A/D Converters". *Microelectronics Journal, Elsevier,* 2003, vol. 34, no. 10, pp. 927-936.
[15] Marina Santo Zarnik, et al., *"Design of oscillation-based test structures"*, IJS Tech. Report 7479, 1996.
[16] Marina Santo Zarnik, Franc Novak and Srecko Macek. "Design of oscillation-based test structures for active RC filters". *Proc. European Design & Test Conference,* Paris, France, March 17-20, 1997, pp. 618.
[17] Marina Santo Zarnik, Franc Novak and Srecko Macek. "Oscillation test structures of actives RC filters". *Proc. IEE Colloquium on Testing mixed signal circuits and systems,* London, October 23, 1997, pp.4/1-4/6.
[18] M.S. Zarnik, F. Novak and S. Macek. "Design of oscillation-based test structures for active RC filters". *IEE Proc-Circuits Devices Syst.* , October 2000, vol. 147, no. 5, pp. 297-302.
[19] M.S. Zarnik and F. Novak. "On oscillation-based test structures of active RC filters". *International Journal of Numerical Modelling, Electronic networks, Devices and Fields*, 2001, vol. 14, no. 3, pp. 283-288.
[20] Mike W.T. Wong. "On the Issues of Oscillation Test Methodology". *IEEE Transaction on Instrumentation and Measurement*, April 2000, vol. 49, no. 2, pp. 240-245.
[21] John E. Gibson. *"Nonlinear Automatic Control"*. International Student Edition, McGraw-Hill Book Company, 1963.
[22] Katsuhiko Ogata. *"Ingenieria de Control Moderna"*. Editorial Prentice/Hall International, 1974.
[23] Aram Budak, *"Passive and Active Network Analysis and Synthesis"*. Houghton Mifflin Company, 1974.

[24] Paul E. Fleischer, A. Ganesan, and Kenneth R. Laker. "A switched Capacitor Oscillator with Precision Amplitude Control and Guaranteed Start-up". *IEEE J. of Solid -State Circuits*, April 1985, vol. SC-20, no. 2, pp. 641-647.

[25] P. E. Fleischer, K. R. Laker, "A Family of Active Switched Capacitor Biquad Building Blocks". *Bell Syst. Tech. J*, Dec. 1979, vol. 58, pp. 2235-2269.

[26] B. Linares-Barranco, A. Rodriguez-Vazquez, E. Sanchez-Sinencio, and J.L Huertas. "CMOS OTA-C high -frequency sinusoidal oscillators". *IEEE Journal of Solid-State Circuits*, February 1991, vol. 26, no. 2, pp. 160-165.

[27] M. Vidyasagar, "*Nonlinear Systems Analysis*". Prentice-Hall, New Jersey, 1978.

[28] Stephan J. G. Gift. "Multiphase Sinusoidal Oscillator Using Inverting-Mode Operational Amplifiers". *IEEE Transactions on Instrumentation and Measurement*, August 1998, vol. 47, no. 4, pp. 968-991.

[29] J. Huang, C.Ong, K. Cheng, "A BIST scheme for on-chip ADC and DAC testing". *Proc. Design, Automation and Test in Europe Conference (DATE)*, Paris, France, March 2000, pp. 216-220.

[30] J.L. Huertas, A. Rueda and D. Vazquez. "Improving the testability of switched-capacitor filters". *J. Electron. Test., Theory Appl.*, November 1993, vol. 4, No. 4, pp. 299-313.

[31] J.L. Huertas. "Test and design for testability of analog and mixed-signal integrated circuits: Theoretical basis and pragmatical approaches". *Selected Topics in Circuits and Systems, H. Dedieu, Ed. Amsterdam: Elsevier*, 1993.

[32] J.L. Huertas, A. Rueda and D. Vazquez. "Testable switched-capacitor filters". *IEEE J. Solid-State Circuits*, July 93, vol. 28, pp. 719-724.

[33] J.L. Huertas, A. Rueda and D. Vazquez."A new strategy for testing analog filters". *Proc. IEEE VLSI Test Symp.*, April 1994, pp. 36-41.

[34] S. A. Jantzi, W. M. Snelgrove, and P. F. Ferguson. "A Fourth-Order Bandpass Sigma-Delta Modulator". *IEEE Journal of Solid-State Circuits*, March 1993, vol. 28, no. 3, pp. 282-291.

[35] S.H.Lewis, R.Ramachadran, W.M.Snelgrove: "Indirect Testing of Digital-Correction Circuits in A/D Converters with Redundancy", *IEEE Trans. on Circ. and Syst.-II*, July 1995, vol. 42, Nº 7, pp. 437-445.

[36] *"Using SIMULINK: Dynamic System Simulation for MATLAB"*. The MathWorks Inc, December 1996.

[37] Milor, L. and V. Visanathan, "Detection of Catastrophic Faults in Analog Integrated Circuits". *IEEE Trans. on Computer Aided Design*, 1989, vol. 8, no. 2, pp. 114-130.

[38] Milor, L. et al. "Optimal Test Set Design for Analog Circuits". *IEEE ICCAD*, 1990, pp. 294-297.

[39] S. Mir, D. Vazquez, A.Rueda, J.L.Huertas. "Switch-Level Fault Coverage Analysis for Switched-Capacitor Systems". *Instituto de Microelectronica de Sevilla, Centro Nacional de Microelectronica, Universidad de Sevilla.*

[40] S. Mir, A.Rueda, T.Olbrich, E.Peralias, J.L.Huertas. "SWITTEST: Automatic Switch-level Fault Simulation and Test Evaluation of Switched-Capacitor Systems". *Proc. 34th Design Automation Conference*, June 1997, pp. 281-286.

[41] M. Renovell, F. Azais, and Y. Bertrand, "A DFT technique to fully access embedded modules in analog circuits under test". *Proc. Int. Mixed Signal Testing Workshop*, Grenoble, France, June 1995, pp. 172-177.

[42] Jose Manuel de la Rosa "Modelado y Diseño de Moduladores ΣΔ Paso de Banda para Comunicaciones Digitales usando Circuitos de Corrientes en Conmutacion" *P.H. Thesis*, Sevilla, Spain, December 2002.

[43] Shanthi Pavan and Yannis P. Tsividis. "An Analytical Solution for a Class of Oscillators, and Its Application to Filter Tunning". *IEEE Transactions on Circuits and Systems-I: Fundamental Theory and Applications,* May 1998, vol. 45, no. 5, pp.547-556.

[44] Steven R. Norsworthy, Richard Schreier, Gabor C. Temes. "Delta- Sigma Data Converters. Theory, Design and Simulation". *IEEE PRESS,* 1996.

[45] M. Soma. "A design-for-test methodology for active analog filters". *Proc. IEEE Int. Test Conf.*, 1990, pp. 183-192.

[46] M. Soma and V. Kolarik. "A design-for-test technique for switched-capacitor filters". *Proc. IEEE VLSI Test Symp.*, April 1994, pp. 42-47.

[47] *"Users' Manual for SWITCAP2 Version 1.1"*. Department of Electrical Engineering and Center for Telecommunications Research. Columbia University, February 1992.

[48] D.Vazquez , A. Rueda, and J.L Huertas. "A new Strategy for Testing Analog Filters". *Proc. IEEE VLSI Test Symp.,* April 1994, pp. 36-41.

[49] D.Vazquez, A.Rueda, and J.L. Huertas, "Reducing the impact of DfT on the performance of analog integrated circuits: Improved sw-opamp design". *Proc. VLSI Test Symp.,* 1996, pp. 42-47.

[50] D.Vazquez, A.Rueda, and J.L. Huertas, "Fully-differential SW-opamp for testing analog embeded modules". *Proc. Int. Mixed-Signal Testing Workshop*, Quebec, Canada, 1996, pp. 204-209.

[51] D.Vazquez, A.Rueda, J.L Huertas and E. Peralias. "A High-Q Bandpass Fully Differential SC Filter with Enhanced Testability". *IEEE Journal of Solid-State Circuits*, July 1998, vol. 33, no. 7, pp. 976-986.

[52] D. Vazquez, G. Huertas, J.L. Huertas: *"Circuitry for providing safe initial conditions in systems based on operational amplifiers"*. Patent application P200100996.

[53] Diego Vazquez, Maria J. Avedillo, Gloria Huertas, José M. Quintana, Manfred Pauritsch, Adoracion Rueda and Jose L. Huertas."A Low-Voltage Low-Power High Performance Fully Integrated DTMF Receiver". *Proc. 27th European Solid-State Circuits Conference ESSCIRC 2001*, Villach, Austria, Septiembre 18-20, 2001, pp. 368-371.

[54] Diego Vazquez, Gloria Huertas, Gildas Leger, Adoracion Rueda, Jose L. Huertas. "On-Chip Evaluation of Oscillation-Based-Test Output Signals". *Proc. Design of Circuits and Integrated Systems Conference*, Porto, November 2001, pp. 179-184.

[55] D. Vazquez, G. Huertas, A.Rueda, and J.L. Huertas, "A simple and Secure Start-up Circuitry for Oscillation-Based-Test Application", *Analog Integrated Circuits ans Signal Processing, Kluwer Academic Publishers*, 2002, vol. 32, pp. 187-190.

[56] D.Vazquez, G. Huertas, A.Rueda and J.L. Huertas, *"DTMF Silicon Demonstrator Report"*. Internal Report of the European Sprit Project ASTERIS.

[57] D. Vazquez, G. Huertas, G. Leger, A. Rueda and J.L. Huertas. "Low-Cost On-Chip Measurements for Oscillation-Based-Test in Analog Integrated Filters". *Proc. 3rd IEEE Latin American Test Workshop (LATW02)*, Montevideo, Uruguay, February. 2002, pp. 89-93.

[58] D. Vazquez, G. Huertas, G. Leger, A. Rueda and J.L. Huertas. "Practical Solutions for the Apliccation of the Oscillation-Based-Test: Start-Up and On_Chip Evaluation". In Proceedings of *the* VTS 2002 Symposium, Monterey, California, pp. 433-438, 28 April-2 May 2002.

[59] D. Vazquez, G. Huertas, G. Leger, A. Rueda and J.L. Huertas. "Practical Solutions for the Application of the Oscillation-Based-Test in Analog Integrated Circuits". *Proc. IEEE*

International Symposium on Circuits and Systems (ISCAS 2002), Scottsdale, Arizona, May 2002, pp. 589-592.

[60] D. Vazquez, G. Huertas, G. Leger, A. Rueda and J.L. Huertas. "On-chip Evaluation of Oscillation-Based-Test Output Signals for Switched-Capacitor Circuits". *International Journal of Analog Integrated Circuits and Signal Processing, Edit. Kluwer Academic Publishers*, Nov. 2002, vol. 33, no. 2, pp. 201-211.

[61] K. Y. Ko, N. S. Gorla, Mike Wong and Y. S. Lee. "Improving fault diagnostic resolution of an oscillation-based test methodology scheme for the threshold detector circuit". *Int. J. Electronics,* 2001, vol. 88, no. 2, pp. 175-187.

[62] J. Roh and J. Abraham. "A Comprehensive TDM Comparator Scheme for Effective Analysis of Oscillation-based Test". *Proc. IEEE VTS*, April 2000, pp. 143-148.

[63] V. Friedman. "The Structure of the Limit Cycles in Sigma Delta Modulation". *IEEE Transaction on Communications*, August 1988, vol. 36, no. 8, pp. 972-979.

[64] O. Feely and L. Chua. "The effect of Integration Leak in ΣΔ Modulation". *IEEE Transaction on Circuits and Systems,* November 1991, vol. 38, no. 11, pp. 1293-1305.

[65] S. Haykin. *"Communication Systems"*. John Wiley & Sons, Inc., 1993.

/2/A

9 781402 077241